Polymeric and Self Assembled Hydrogels

From Fundamental Understanding to Applications

Monographs in Supramolecular Chemistry

Series Editors:
Philip Gale, *University of Southampton, UK*
Jonathan Steed, *Durham University, UK*

How to obtain future titles on publication:
A standing order plan is available for this series. A standing order will bring
delivery of each new volume immediately on publication.

For further information please contact:
Book Sales Department, Royal Society of Chemistry, Thomas Graham House,
Science Park, Milton Road, Cambridge, CB4 0WF, UK
Telephone: +44 (0)1223 420066, Fax: +44 (0)1223 420247
Email: booksales@rsc.org
Visit our website at http://www.rsc.org/Shop/Books/

Polymeric and Self Assembled Hydrogels
From Fundamental Understanding to Applications

Edited by

Xian Jun Loh
Melville Laboratory for Polymer Synthesis,
Department of Chemistry, University of Cambridge,
Cambridge, UK
Email: XianJun_Loh@scholars.a-star.edu.sg, xjl21@cam.ac.uk

Oren A. Scherman
Melville Laboratory for Polymer Synthesis,
Department of Chemistry, University of Cambridge,
Cambridge, UK
Email: oas23@cam.ac.uk

RSC Publishing

Monographs in Supramolecular Chemistry No. 11

ISBN: 978-1-84973-561-2
ISSN: 1368-8642

A catalogue record for this book is available from the British Library

Published by The Royal Society of Chemistry,
Thomas Graham House, Science Park, Milton Road,
Cambridge CB4 0WF, UK

Registered Charity Number 207890

Visit our website at www.rsc.org/books

Printed in the United Kingdom by Henry Ling Limited, Dorchester, DT1 1HD, UK

Preface

Polymeric and self-assembled hydrogel research is a rapidly changing field and this book is intended to serve as a guide through the latest work, offering summaries of the current state-of-the-art cutting-edge research as well as extensive references to the latest breakthroughs. Each chapter is an authoritative treatise on its specific topic and can be read on its own. All the authors invited to write the chapters in this book have many years of experience in hydrogel research and their knowledge is succinctly condensed into this volume. Readers can be expected to be taken through the entire spectrum of hydrogel research as the book is structured into three major sections covering basic fundamental research, applied and platform technologies, and commercially viable applications.

Looking ahead to the topics covered in the book, we begin by giving a broad overview of the field of polymeric hydrogels and covering some of the newest and most exciting research in this field. Chapter 2 covers basic fundamental research on the hydrogel properties of tetra-PEG hydrogels and on using techniques such as small-angle neutron scattering to understand the internal hydrogel structure.

Laboratory work in the synthesis and characterization of novel hydrogels is discussed in Chapters 3–5 and exemplifies cutting-edge development in the field of hydrogels. Chapter 3 describes the recent developments in supramolecular approaches to the formation of hydrogels using host–guest chemistries, while Chapter 4 highlights a unique "slide-ring" hydrogel based on cyclodextrin inclusion complexes that has exceptional stretch abilities. Chapter 5 describes the growing area of peptide hydrogels, their preparation and their potential applications.

Practical applications are addressed in Chapters 6–8, with particular emphasis on the fast-moving chemo-sensing and biomedical fields. Chapter 6 presents the use of hydrogels in chemo-sensing applications and Chapter 7 is a

Monographs in Supramolecular Chemistry No. 11
Polymeric and Self Assembled Hydrogels: From Fundamental Understanding to Applications
Edited by Xian Jun Loh and Oren A. Scherman
© The Royal Society of Chemistry 2013
Published by the Royal Society of Chemistry, www.rsc.org

complete overview of a class of physical hydrogels known as thermogelling polymers and their applicative aspects. Chapter 8 represents an outline of the use of hydrogels in biomedical fields.

Finally, in order to showcase commercial applications, Chapters 9 and 10 present current commercial successes of temperature-sensitive hydrogels and hydrogel microspheres for drug delivery applications. To obtain a sense of the vast scope of hydrogel applications, a quick snapshot is also presented in the introduction to this book.

We would like to gratefully acknowledge the contributions of the well-regarded academics and industrialists who have taken time out of their busy schedules to embark on this project together with us. The contributions of Setu Kasera and Sandra Leytheauser in the production of the book cover are gratefully acknowledged. We would also like to acknowledge the help of the RSC, especially Leanne Marle, Alice Toby-Brant and Katrina Harding for their patience with us at various stages of the project. We thank Joo Gek Lim and Moi Joo Loh who have been actively involved in all stages of the layout, language, style and graphic editing of this book. This book would not have been possible without their collective inputs and, indeed, the book is now much better because of their contributions. Finally, we hope that this volume will serve as an indispensable reference for students, researchers, academics and industrialists in the field of hydrogel research.

Xian Jun Loh
Oren A. Scherman
Cambridge, UK

Contents

Monographs in Supramolecular Chemistry No. 11
Polymeric and Self Assembled Hydrogels: From Fundamental Understanding to Applications
Edited by Xian Jun Loh and Oren A. Scherman
© The Royal Society of Chemistry 2013
Published by the Royal Society of Chemistry, www.rsc.org

CHAPTER 1
Introduction

XIAN JUN LOH* AND OREN A. SCHERMAN

Melville Laboratory for Polymer Synthesis, Department of Chemistry, University of Cambridge, Lensfield Road, Cambridge CB2 1EW, UK
*Email: xianjun_loh@scholars.a-star.edu.sg

1.1 Origins

Contact lenses, shoe sole cushions, vitamin capsules, baby diapers and wound dressings, all these objects have a material in common: *hydrogels*. The use of hydrogels has experienced phenomenal growth and they are now widely used by almost everyone and in almost every imaginable application. Yet, in spite of its wide applicability, there is still intense research work being carried out on hydrogels. Every year, for the past five years, more than 2000 publications are reported in the scientific literature, with the number growing year on year (Figure 1.1). Hydrogels can generally be classified by their origin and the type of cross-linking.

Hydrogels can originate from either natural or synthetic sources. In general, natural hydrogels can be protein or polysaccharide derived. Amino acids and saccharide monomer units are generally water soluble. Thus, along with the high molecular weight of these components and their high degree of inter-molecular interactions, they have a tendency to form self-assembled 3D structures that can swell in water. For example, the basis of human tissue scaffold is the extracellular matrix (ECM). The ECM provides structural and mechanical support to the cells and is the main component of connective tissue in animals. Polysaccharide gels and fibrous proteins fill up the interstitial matrix

Monographs in Supramolecular Chemistry No. 11
Polymeric and Self Assembled Hydrogels: From Fundamental Understanding to Applications
Edited by Xian Jun Loh and Oren A. Scherman
© The Royal Society of Chemistry 2013
Published by the Royal Society of Chemistry, www.rsc.org

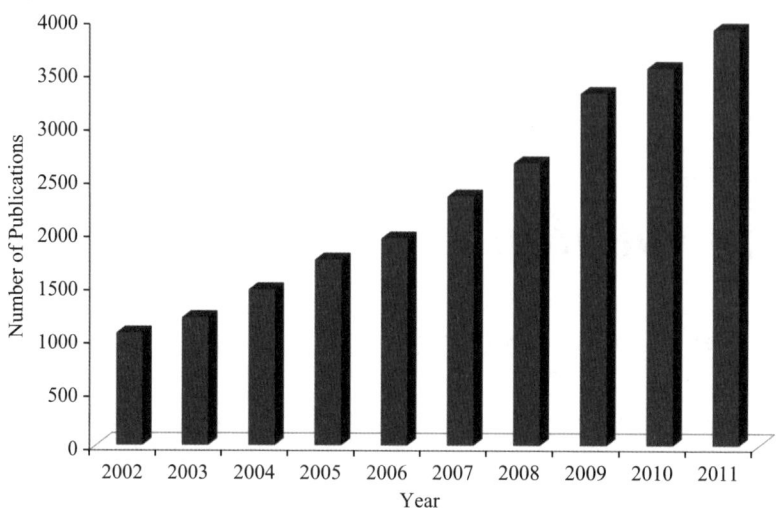

Figure 1.1 Number of publications on hydrogels (including journal articles, book chapters, conference proceedings and reviews).

and act as a compression buffer against the stress placed on the ECM. Some examples of naturally derived hydrogels include collagen, gelatin, fibrin, hyaluronic acid and chitosan.

Common synthetic hydrogels can be classified into stable, biodegradable, peptide and stimuli-responsive hydrogels. Examples of stable gels are poly(vinyl alcohol) gels and acrylamide/acrylate-based gels, which are typically very stable and do not readily degrade. On the other hand, biodegradable gels are designed to disintegrate into their monomeric components after some time. The main constituents of these gels are polyesters such as poly(lactic acid), poly(ε-caprolactone) or poly[(*R*)-3-hydroxybutyrate]. Peptide hydrogels sit at the interface between natural and synthetic hydrogels. They can be prepared by standard peptide synthetic methods, which have been gaining research interest in the past few years. These peptide hydrogels are generally cyto-compatible and do not show significant immune response in biological systems. As a result of their folding capabilities to form α-helices, β-sheets and random coils, they offer unique insights into the molecular construction of a hydrogel, which can be exploited to form a variety of soft materials. Stimuli-responsive hydrogels such Pluronics/Poloxamer or poly-(*N*-isopropylacrylamide) hydrogels offer a degree of control over the physical behaviour of the hydrogels upon application of an external stimulus, in this case, temperature. External control is useful for a variety of reasons, for example, the regulation of the behaviour of the hydrogels can allow users to fine-tune the rate of the release of a loaded cargo in the gel matrix. Potentially, dual-phased release can be realised with a system incorporating multiple stimuli-responsive moieties.

1.2 Type of Cross-Linking

Conventional hydrogels are 3D water-swollen structures that are cross-linked by non-reversible chemical bonds. However, these hydrogels can be brittle in nature, causing the swollen gels to shatter upon application of pressure. A non-covalent approach, based on physical interactions, offers an attractive alternative to producing hydrogels. This creates a self-correcting system which adjusts itself to external pressure, thus resolving the brittleness problem. Recently, a hydrogel system based on clay, water and a dendritic binder has been reported. The simple mixture of these three components leads to the formation of free-standing hydrogels with excellent mechanical properties.[1] Supramolecular hydrogels or gel assemblies based on host–guest interactions have also been reported.[2–7] Temperature-induced gelation based on hydrophobic/hydrophilic switching has also been investigated in great detail.[8–23] Triggered hydrogel systems based on a change in pH have also been investigated but these are less common.[24,25] Supramolecular hydrogels have been reported with low molecular weight hydrogelators and the interested reader is referred to several notable papers and reviews.[26–32]

1.3 Cutting-edge Research

Current cutting-edge research utilising hydrogels focuses on applications such as photonic gel films for mechanochromic sensing,[33] soft circuits,[34] 3D cell culture[35] and as a sensor for a cell's mechanical behaviour.[36] These new and exciting applications are all being developed to tackle real and very urgent world problems and some of these applications require hydrogels with unique physical properties. For example, there is a worldwide shortage of organs for transplant due to cultural and societal boundaries on organ donation and transplant acceptance. Tissue engineering has been proposed as a potential strategy to provide a partial solution to this problem. Instead of seeking organ donors, entire organs may be built in the laboratory (Figure 1.2). "Printable"

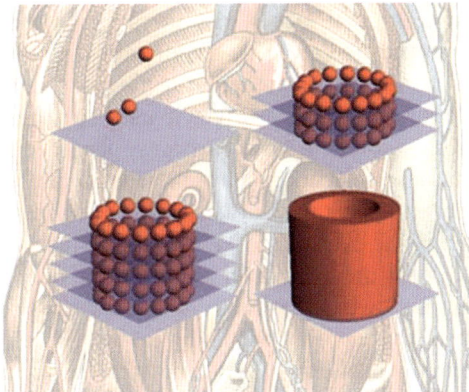

Figure 1.2 The printing of 3D hydrogel networks for the building of artificial organs. [Reference: *J. Mater. Chem.*, 2007, **17**, 2054–2060].

organs have been suggested as a very realistic and attractive idea.[37] The key components required to print an organ are a printer, ink, paper and a bioreactor. Recently, Alblas and her co-workers have developed a way to "print" stable cell-containing scaffolds.[38] Thermogelling polymers were used as the polymer hydrogel "ink". These printed gel structures were heated to form gels and further cross-linked to form permanent gel structures. Stem cells were incorporated into the hydrogels and can be used to initiate tissue generation within the scaffolds. A cell-laden hydrogel can be used to build much more complex structures such as tissue grafts that have blood vessels built into them. With this technology, even the fabrication of an entire organ may soon be possible *via* a bottom-up approach. However, there are concerns as to whether the material is entirely useful for clinical applications as these gels are not natural gels, an aspect that requires further development.

In another new breakthrough in the field of hydrogels, Varghese and co-workers have developed the hydrogel version of Velcro™, a self-healing hydrogel that binds in seconds and is able to be stretched repeatedly.[39] This material utilises the ability of the carboxylic acid group to form hydrogen bonds with each other to create the self-healing effect. By varying pH levels, the pieces are able to weld and separate very easily. The process was successfully repeated many times without any reduction in material strength. Numerous potential applications of this material include medical sutures, targeted drug delivery, industrial sealants and self-healing plastics.

Hydrogels have also been making inroads into the field of soft electronics. Supercapacitors are complementary devices to batteries in energy storage and delivery schemes, providing quick bursts of power when required. A poly(aniline) based hydrogel was prepared and used as electrodes in a supercapacitor.[40] The new material has a capacitance about three times greater than a typical carbon supercapacitor. Supercapacitors are fabricated from two closely spaced, porous carbon electrodes that charge and discharge rapidly. Poly(aniline) was cross-linked with phytic acid to form the conductive porous hydrogel. The polymer can be synthesised using inkjet printing or spray coating each solution on a surface. The fabrication of the hydrogel is feasible for large-scale energy storage applications or miniaturised for microelectronic applications. This hydrogel supercapacitor is more cost-effective than the carbon supercapacitor, as the electrolyte it uses is water-based and cheaper than the organic ionic liquids used in carbon supercapacitors.

Recently, Thomas and co-workers have unveiled a photonic-crystal hydrogel material that reflects a wide range of wavelengths in response to a variety of stimuli.[41] By changing the external gel environment, the wavelength of reflected light can be shifted from the ultraviolet region to the near infrared region. The polymer gel comprises two constituent polymers that self assemble into a structure that can interact with light and change the colour of the gel. This hydrogel can potentially be used as a colourimetric chemical sensor on clothing.

As a class of substances, hydrogels have been around for some time and still their applications continue to grow as the versatility of hydrogels is recognised and as new research and development increases their impact across ever-widening fields in science and technology: from aerospace to agriculture, from electricity to the environment, from food to footwear and from medical to military applications. This vibrant field is more than mere laboratory curiosity as in this increasingly dynamic and interdependent world, close interactions are needed between laboratory researchers and industrial partners to bring technology to fruition.

References

1. Q. Wang, J. L. Mynar, M. Yoshida, E. Lee, M. Lee, K. Okuro, K. Kinbara and T. Aida, *Nature*, 2010, **463**, 339–343.
2. M. Nakahata, Y. Takashima, H. Yamaguchi and A. Harada, *Nat. Commun.*, 2011, **2**, 511–516.
3. I. Tomatsu, A. Hashidzume and A. Harada, *Macromolecules*, 2005, **38**, 5223–5227.
4. I. Tomatsu, A. Hashidzume and A. Harada, *Macromol. Rapid Commun.*, 2005, **26**, 825–829.
5. H. Yamaguchi, Y. Kobayashi, R. Kobayashi, Y. Takashima, A. Hashidzume and A. Harada, *Nat. Commun.*, 2012, **3**, 603–607.
6. E. A. Appel, F. Biedermann, U. Rauwald, S. T. Jones, J. M. Zayed and O. A. Scherman, *J. Am. Chem. Soc.*, 2010, **132**, 14251–14260.
7. E. A. Appel, X. J. Loh, S. T. Jones, C. A. Dreiss and O. A. Scherman, *Biomaterials*, 2012, **33**, 4646–4652.
8. M. H. Cha, J. Choi, B. G. Choi, K. Park, I. H. Kim, B. Jeong and D. K. Han, *J. Colloid Interface Sci.*, 2011, **360**, 78–85.
9. E. Y. Kang, B. Yeon, H. J. Moon and B. Jeong, *Macromolecules*, 2012, **45**, 2007–2013.
10. H. Lee, B. G. Choi, H. J. Moon, J. Choi, K. Park, B. Jeong and D. K. Han, *Macromol. Res.*, 2012, **20**, 106–111.
11. H. J. Moon, B. G. Choi, M. H. Park, M. K. Joo and B. Jeong, *Biomacromolecules*, 2011, **12**, 1234–1242.
12. M. H. Park, M. K. Joo, B. G. Choi and B. Jeong, *Acc. Chem. Res.*, 2012, **45**, 424–433.
13. U. P. Shinde, M. K. Joo, H. J. Moon and B. Jeong, *J. Mater. Chem.*, 2012, **22**, 6072–6079.
14. E. J. Yun, B. Yon, M. K. Joo and B. Jeong, *Biomacromolecules*, 2012, **13**, 1106–1111.
15. X. J. Loh, S. H. Goh and J. Li, *Biomaterials*, 2007, **28**, 4113–4123.
16. X. J. Loh, S. H. Goh and J. Li, *Biomacromolecules*, 2007, **8**, 585–593.
17. X. J. Loh, S. H. Goh and J. Li, *J. Phys. Chem. B*, 2009, **113**, 11822–11830.
18. X. J. Loh and J. Li, *Expert Opin. Ther. Pat.*, 2007, **17**, 965–977.
19. X. J. Loh, P. Peh, S. Liao, C. Sng and J. Li, *J. Controlled Release*, 2010, **143**, 175–182.
20. X. J. Loh, K. B. C. Sng and J. Li, *Biomaterials*, 2008, **29**, 3185–3194.

21. X. J. Loh, Y. X. Tan, Z. Y. Li, L. S. Teo, S. H. Goh and J. Li, *Biomaterials*, 2008, **29**, 2164–2172.

22. X. J. Loh, P. N. N. Vu, N. Y. Kuo and J. Li, *J. Mater. Chem.*, 2011, **21**, 2246–2254.

23. V. P. N. Nguyen, N. Y. Kuo and X. J. Loh, *Soft Matter*, 2011, **7**, 2150–2159.

24. O. Borisova, L. Billon, M. Zaremski, B. Grassl, Z. Bakaeva, A. Lapp, P. Stepanek and O. Borisov, *Soft Matter*, 2011, **7**, 10824–10833.

25. D. C. Wu, X. J. Loh, Y. L. Wu, C. L. Lay and Y. Liu, *J. Am. Chem. Soc.*, 2010, **132**, 15140–15143.

26. G. L. Liang, Z. M. Yang, R. J. Zhang, L. H. Li, Y. J. Fan, Y. Kuang, Y. Gao, T. Wang, W. W. Lu and B. Xu, *Langmuir*, 2009, **25**, 8419–8422.

27. Q. G. Wang, Z. M. Yang, X. Q. Zhang, X. D. Xiao, C. K. Chang and B. Xu, *Angew. Chem. Int. Ed.*, 2007, **46**, 4285–4289.

28. Z. Yang and B. Xu, *J. Mater. Chem.*, 2007, **17**, 2385–2393.

29. Z. M. Yang, P. L. Ho, G. L. Liang, K. H. Chow, Q. G. Wang, Y. Cao, Z. H. Guo and B. Xu, *J. Am. Chem. Soc.*, 2007, **129**, 266–267.

30. Z. M. Yang, G. L. Liang, M. L. Ma, A. S. Abbah, W. W. Lu and B. Xu, *Chem. Commun.*, 2007, 843–845.

31. Y. Zhang, Y. Kuang, Y. A. Gao and B. Xu, *Langmuir*, 2011, **27**, 529–537.

32. L. A. Estroff and A. D. Hamilton, *Chem. Rev.*, 2004, **104**, 1201–1217.

33. E. P. Chan, J. J. Walish, E. L. Thomas and C. M. Stafford, *Adv. Mater.*, 2011, **23**, 4702–4706.

34. H. J. Koo, J. H. So, M. D. Dickey and O. D. Velev, *Adv. Mater.*, 2011, **23**, 3559–3564.

35. J. J. Zhang, T. Tokatlian, J. Zhong, Q. K. T. Ng, M. Patterson, W. E. Lowry, S. T. Carmichael and T. Segura, *Adv. Mater.*, 2011, **23**, 5098–5103.

36. H. Y. Yoshikawa, F. F. Rossetti, S. Kaufmann, T. Kaindl, J. Madsen, U. Engel, A. L. Lewis, S. P. Armes and M. Tanaka, *J. Am. Chem. Soc.*, 2011, **133**, 1367–1374.

37. V. Mironov, G. Prestwich and G. Forgacs, *J. Mater. Chem.*, 2007, **17**, 2054–2060.

38. N. E. Fedorovich, I. Swennen, J. Girones, L. Moroni, C. A. van Blitterswijk, E. Schacht, J. Alblas and W. J. A. Dhert, *Biomacromolecules*, 2009, **10**, 1689–1696.

39. A. Phadke, C. Zhang, B. Arman, C. C. Hsu, R. A. Mashelkar, A. K. Lele, M. J. Tauber, G. Arya and S. Varghese, *Proc. Natl. Acad. Sci. U. S. A.*, 2012, **109**, 4383–4388.

40. L. Pan, G. Yu, D. Zhai, H. R. Lee, W. Zhao, N. Liu, H. Wang, B. C. K. Tee, Y. Shi, Y. Cui and Z. Bao, *Proc. Natl. Acad. Sci. U. S. A.*, 2012, **109**, 9287–9292.

41. Y. Kang, J. J. Walish, T. Gorishnyy and E. L. Thomas, *Nat. Mater.*, 2007, **6**, 957–960.

CHAPTER 2

Fabrication, Structure, Mechanical Properties, and Applications of Tetra-PEG Hydrogels

MITSUHIRO SHIBAYAMA*[a] AND TAKAMASA SAKAI[b]

[a] Institute for Solid State Physics, The University of Tokyo,
5-1-5 Kashiwanoha, Kashiwa, Chiba 277-8581, Japan; [b] Department of
Bioengineering, School of Engineering, The University of Tokyo,
7-3-1 Hongo, Bunkyo-ku, Tokyo 113-8656, Japan
*Email: sibayama@issp.u-tokyo.ac.jp

2.1 Introduction

Hydrogels have been used in daily life for many years such as in diapers, contact lenses, cosmetics, foods, drug reservoirs, *etc.*[1,2] Despite the unique characteristics of high water absorbency and water-retention capability, practical applications of hydrogels, especially as structural materials, are restricted because of their low mechanical strength and brittleness. There are several reasons why hydrogels are mechanically weak. In addition to their low content of polymer, *e.g.* 10 wt% of polymer in aqueous media, inhomogeneities in the network structure of gels lower their mechanical properties far below what they should be. The inhomogeneities are categorized into spatial, topological, connectivity, and motility inhomogeneities.[3] Figure 2.1 (top) indicates

Monographs in Supramolecular Chemistry No. 11
Polymeric and Self Assembled Hydrogels: From Fundamental Understanding to Applications
Edited by Xian Jun Loh and Oren A. Scherman
© The Royal Society of Chemistry 2013
Published by the Royal Society of Chemistry, www.rsc.org

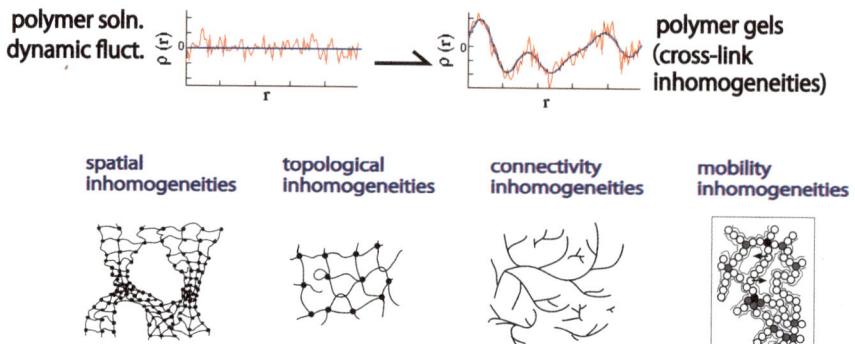

Figure 2.1 Schematic representation of concentration fluctuations in polymer gels. *Top*: concentration fluctuations in polymer solution (*left*) and in polymer gels (*right*). *Bottom*: various types of inhomogeneities; spatial, topological, connectivity, and mobility inhomogeneities.
(Reproduced from Shibayama and Norisuye[4] with permission from the Chemical Society of Japan.)

concentration fluctuations in polymer solutions (left) and in gels (right).[4] In polymer solutions, only thermal concentration fluctuations exist, of which the average is zero. On the other hand, gels contain both frozen concentration fluctuations (the low-frequency component in this figure; blue) introduced by cross-linking and thermal concentration fluctuations (high-frequency component; red). The introduction of cross-links brings about various types of inhomogeneities as shown in Figure 2.1 (bottom), *i.e.* spatial inhomogeneities (nonrandom spatial variations of cross-link density in a gel), topological inhomogeneities (defects of network, such as dangling chains, loops, chain entrapment, *etc.*), connectivity inhomogeneities, and motility inhomogeneities (local degree of mobility). In general, it is difficult to control and/or reduce these inhomogeneities, resulting in complexity and inferior physical properties of polymer gels. For example, because of the presence of dangling chains (topological inhomogeneities), they begin to break from the weakest link, thus reducing the whole mechanical strength.

Many attempts were made to obtain an ideal gel network consisting of unimodal strands, free from defects or entanglements. Some of these attempts included gelation by physical cross-linking, gelation from prepolymers,[5] and gelation from polymers by γ-ray irradiation.[6] Homogeneity could be relatively easily obtained in thermoreversible physical gels. However, the homogeneity cannot be controlled strictly even in physical gels because cross-links are introduced randomly. In addition, because physical cross-linking is generally weaker than chemical cross-linking, the resulting physical gels are soft and weak. Gelation from prepolymers seemed more promising than the others because the detailed network structure was controllable by designing constitutional units, and the gelation process was well predicted by Flory's tree approximation.[7,8] He *et al.* reported the synthesis of jungle-gym-type polyimide organgels using tri-functional cross-linkers and telechelic rigid aromatic oligomers as backbones, which had a high compression modulus.[9] As for hydrogels, there were numerous studies to obtain the homogeneous biocompatible hydrogels.[10–13] As far as can be determined, no hydrogels formed from macromers have compressive strength reaching a MPa range, which is one of the most important criteria for practical use. The fragility likely comes from inhomogeneity of the network structure as discussed above.

It should be noted that most gels made from prepolymers are obtained by coupling asymmetrical components such as multi-functional cross-linkers and telechelic polymers (Figure 2.2a). These asymmetrical combinations should give the network a high degree of freedom, allowing various micro-structures including loops and defects. These micro-structures deprive a gel of cooperativeness, weakening the gel. A new strategy for forming a homogeneous network by decreasing the degree of freedom of the micro-network structure has been employed.[14] A gel was formed by combining two well-defined symmetrical tetra-functional precursors (*i.e.* modules) of the same size (Figure 2.2b). As each prepolymers has four end-linking groups reacting with each other, these two prepolymers must connect alternately to avoid the self-biting reaction. This gelation process is a simple A–B type reaction in accordance with Flory's classical theory.[7] This gel is named "Tetra-PEG gel" after its structure [PEG = poly(ethylene glycol)] and the name of the components. The constitutional prepolymers and the reaction were biocompatible, and the compressive strength of resulting gel was in the MPa range, which was much superior to

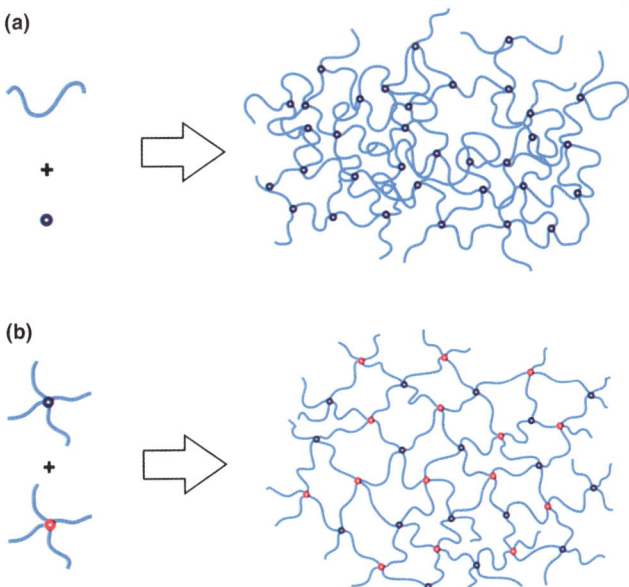

Figure 2.2 (a) Cross-linking of telechelic polymers with multi-functional cross-linkers *vs.* (b) cross-end-coupling of multi-functional modules.

those of agarose gels or acrylamide gels having the same network concentrations. It is believed that this methodology, *i.e.* module assembling by cross-end-coupling, opens a new paradigm of gel preparation.[14,15]

Section 2.2 will describe the synthesis and characterization of Tetra-PEG macromers and gels. In Section 2.3, the structure of Tetra-PEG gels is discussed on the basis of small-angle neutron scattering (SANS) results. Section 2.4 is devoted to the discussion on the relationship between the mechanical properties and structure of Tetra-PEG gels. In Section 2.5, a variety of medical applications of Tetra-PEG gels are demonstrated. Finally, future directions of the study and application of Tetra-PEG gels are given.

2.2 Fabrication

2.2.1 Synthesis of Precursors

Tetrahydroxyl-terminated PEG (THPEG; Figure 2.3a) was synthesized by successive anionic polymerization reactions of ethylene oxide from the sodium alkoxide of pentaerythritol. Tetraamine-terminated PEG (TAPEG; Figure 2.3b) and tetra-NHS-glutarate-terminated PEG (TNPEG; Figure 2.3c) were successfully synthesized by changing the end-group of THPEG to amine and *N*-hydroxysuccinimidyl (NHS) ester, respectively. Previous studies reported by Sakai *et al.* provide the detailed synthesis method that was carried out.[14]

2.2.2 Characterization

The molecular weight (M_w) and the functionality of a Tetra-PEG module were estimated by ^{1}H NMR measurements. The polydispersity was determined using a gel permeation chromatography system. The molecular weights were calibrated with poly(ethylene glycol) standards.

2.2.3 Synthesis of Tetra-PEG Gels: Cross-End-Coupling

Constant amounts of TAPEG and TNPEG were dissolved in phosphate buffer (pH 7.4) and phosphate/citric acid buffer (pH 5.8), respectively.[14] The two solutions were mixed and the resulting solution was poured into the mold with

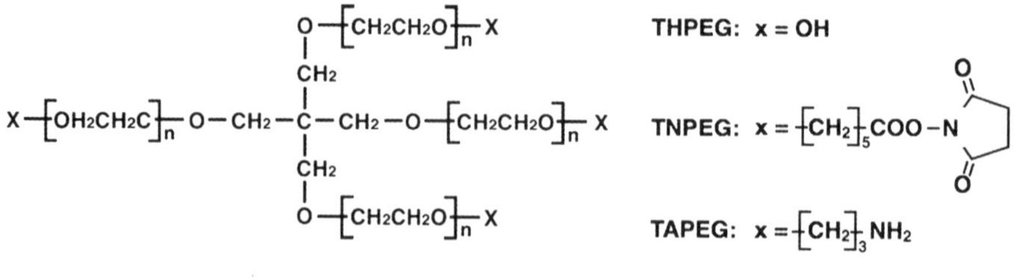

Figure 2.3 Molecular structures of THPEG, TAPEG, and TNPEG.

the desired shape. At least 12 h was allowed for the completion of the reaction before the following experiment was performed. Typical monomer concentrations were from 10 to 160 g L^{-1}.

2.2.4 Gelation Kinetics

The main reaction of the Tetra-PEG gel system is an aminolysis reaction between the amine group within TAPEG and the activated ester group within TNPEG (Figure 2.4).[16] Here two points should be noted. First, the activated ester group within TNPEG gradually dissociates in aqueous solution due to a hydrolysis reaction. In addition, the terminal NH$_2$ group within TAPEG coexists with the protonated NH$_3^+$ in equilibrium. Only the un-ionized amine group reacts with the activated ester group because the ionized amine group does not have an unshared electron pair. Therefore, the rate equation for the gelation is described as follows:

$$-d[-NH_2]_{total}/dt = k_{gel}[-NH_2][-NHS] \qquad (2.1)$$

$$-d[-NHS]/dt = k_{gel}[-NH_2][-NHS] + k_{deg}[-NHS] \qquad (2.2)$$

where k_{gel} is the rate constant for the gelation. It should be noted here that the pH of the solution is very important to control the rate of the coupling reaction.[17]

2.2.4.1 Degradation Kinetics of TNPEG

Prior to estimating the rate constant for gelation in the Tetra-PEG system, the rate for the degradation of TNPEG was evaluated by infrared attenuated total

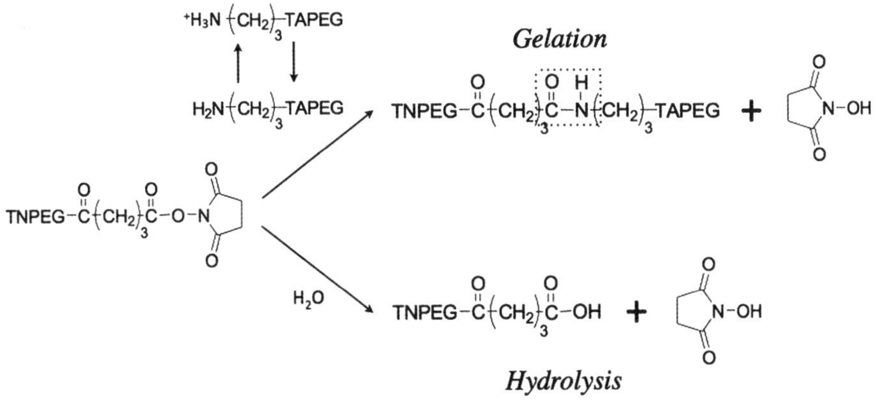

Figure 2.4 Reaction scheme for Tetra-PEG gel: gelation *vs.* hydrolysis.

reflection (IR-ATR) measurements.[16] On the basis of theoretical IR bands, the observed IR spectrum can be satisfactorily deconvoluted into seven bands. Three bands at 1728, 1778, and 1812 cm^{-1} are assigned to the terminal NHS group within TNPEG and three bands at 1646, 1670, and 1704 cm^{-1} to the dissociated NHS ion. The weak and broad band at 1555 cm^{-1} is ascribed to the terminal ionized carboxyl group. The peak intensities of the terminal NHS group decreased and those of dissociated NHS increased with increasing t, suggesting the hydrolysis of an activated ester. The kinetic trace of [NHS$^-$] was measured. The apparent rate constant, k_{deg}, was estimated by least-squares-fit analysis on the basis of the following rate equation: $-d[-NHS]/dt = d[NHS^-]/dt = k_{deg}[-NHS]$. The fit works well and the linear relationship between $\ln([-NHS]/[-NHS]_0)$ and time is observed, suggesting that the TNPEG in solution dissociates under a pseudo-first-order kinetics reaction to give NHS and TNPEG-COO ions.

2.2.4.2   Polycondensation Kinetics

Figure 2.5(a) shows the time dependence of IR spectra for the mixture of Tetra-PEG precursors (TAPEG and TNPEG) at 20 °C, obtained by subtracting those for buffer solutions.[16] According to theoretical IR bands, the IR spectrum can be deconvoluted into eight bands, *i.e.* three bands at 1728, 1778, and 1812 cm^{-1} originated from the NHS ester, three at 1646, 1670, and 1704 cm^{-1} from the dissociated NHS ion, and a 1624 cm^{-1} band from a newly formed amide bond. It is found that there is no peak at 1555 cm^{-1} assigned to the ionized carbonyl group, which is clearly seen in Figure 2.5(a). This indicates that the formation of an amide bond (–CONH–) on polycondensation successfully proceeds, while hydration of terminal NHS groups does not proceed practically.

Figure 2.5(b) shows the kinetic traces for the dissociated NHS ion, the second-order plots of [NHS$^-$] at various temperatures.[16] It was found that the gelation rate increases with increasing temperature. The fit results using

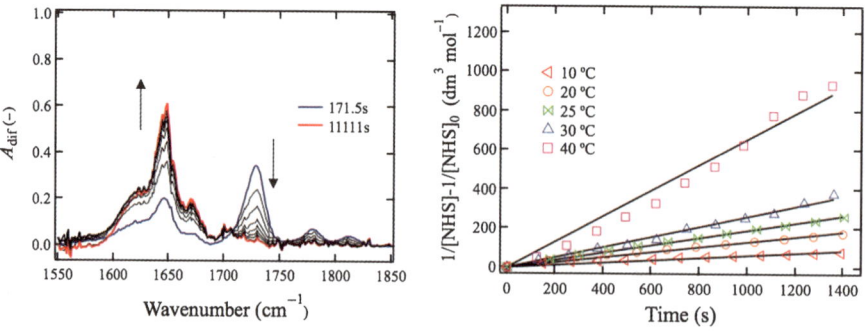

Figure 2.5   Time evolution of (a) IR spectra and (b) [NHS] during the gelation of Tetra-PEG gel.
(Reproduced from Nishi *et al.*[16] with permission from the American Chemical Society.)

eqns (2.1) and (2.2) are also shown in Figure 2.5(b). Here, k_{deg} is set as a fixed parameter obtained by degradation analysis. The fit worked well from the initiation to near completion of the reaction. Moreover, the second-order plot gives a straight line passing through the origin in Figure 2.5(b). It should be noted that the gelation occurred at 1187, 496, 194, and 100 s after initiation at 283, 293, 303, and 313 K, respectively. These results suggest that this polycondensation reaction obeys simple second-order kinetics from initiation to the end regardless of the gelation threshold. The reason is suspected to be that the polycondensation kinetics of Tetra-PEG gel are not affected by the substitution effect or steric hindrance.

2.2.5 Final Conversion

Final conversion was also estimated by IR spectroscopy.[18] The reaction conversion, p, was estimated from the peak intensity of the ionized carbonyl group (I_{CO}, 1555 cm^{-1}) and that of the amide bond (I_{amide}, 1624 cm^{-1}) as follows:

$$p = \frac{I_{amide}}{I_{CO} + I_{amide}} \tag{2.3}$$

The ratio of the molar absorbance coefficient of ionized carbonyl group to amide bond was estimated to be 1.0:1.0 in advance. The value of p was almost constant against the polymer volume fraction, ϕ_0, for both Tetra-PEG-10k and −20k, and was practically constant and close to 0.9,[18] suggesting the near absence of dangling chains ($\sim 10\%$) in Tetra-PEG gel.

2.3 Structure Characterization by SANS

Structural characterizations of polymer gels have been carried out mainly by scattering methods, such as small-angle X-ray scattering (SAXS), light scattering (LS), small-angle neutron scattering (SANS), NMR, *etc.*[19] Microscopy is, in principle, a powerful technique. However, it is rather difficult to apply microscopy for investigations of wet systems like hydrogels except for some examples.[20] SANS has been a complementary technique of SAXS[21,22] and LS.[23,24] Polymer gels are suitable to be investigated by SANS because of the following reasons.[25,26] (1) Scattering contrast can be introduced to the sample simply by making a gel in a deuterated solvent or by immersing in a deuterated solvent. (2) Special sample environments, such as shear deformation measurements and stretching measurements under temperature and humidity control, can be easily realized with quartz and/or metal windows because of the strong penetration power of neutron beams. Hence, SANS has been intensively utilized for structural characterization of polymer gels.[19,25–27] Figure 2.6 shows a schematic representation of a SANS measurement for a polymer solution. A monochoromated neutron beam with a wavelength of λ is incident on a polymer solution with a scattering length density of $\rho_{solvent}$ in which neutrons

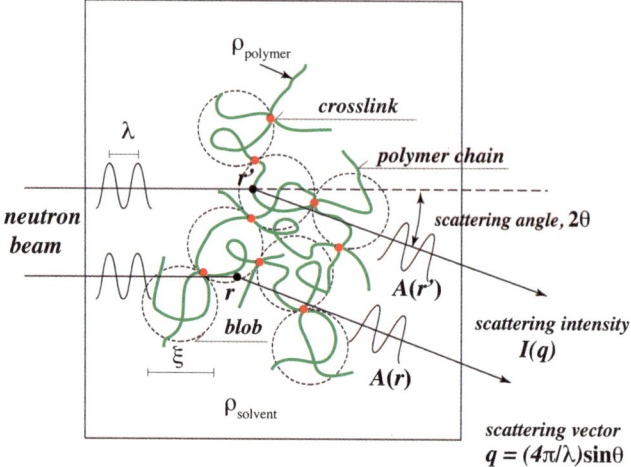

Figure 2.6 Principle of neutron scattering from polymer gels. The scattering intensity is obtained as a function of the scattering vector, q.

are scattered in a direction given by a scattering angle of 2θ. Because of the wave nature of neutrons, the scattered neutron beams at arbitrary positions r and r' characterized by the scattering contrast, $A(r)$ and $A(r')$, interfere with each other depending on their phase difference. By summing up the product $A(r)A(r')$ over the irradiated volume of the sample by taking account of the phase difference, $e^{iq\cdot(r-r')}$, the scattering intensity is obtained as a function of the scattering vector, q $[=(4\pi/\lambda)\sin\theta]$. The scattering intensity is proportional to the scattering contrast, *i.e.* the square of the scattering length density difference, $(\Delta\rho)^2 = (\rho_{\text{solvent}} - \rho_{\text{polymer}})^2$, which will be described below. In order to obtain useful information, it is necessary to design the combination of the solute polymer and the solvent to obtain a high contrast with low background noise that arises from the incoherent scattering of hydrogen atoms. Higgins and Benoit[28] provide additional information on neutron scattering from polymers and Imae *et al.*[29] deals with soft matter.

2.3.1 Theoretical Background

2.3.1.1 Scattering Function for a Single Polymer Chain

The basic scattering function for polymer chains is a "Debye function". The form factor of a single Gaussian chain is given by:

$$P_{\text{D}}(q) = \frac{2Z}{u^2}\{\exp(-u) - 1 + u\} \qquad (2.4)$$

Here, Z is the degree of polymerization of the polymer chains and $u \equiv Za^2q^2/6$, with a being the segment length. In the case of f-armed star polymer chains, eqn (2.4) needs to be modified to the form factor of multi-arm star polymer chains, where $P_{star}(q)$ is given by:[30,31]

$$P_{star}(q) = \frac{2Z}{fu^2}\left\{u - [1 - \exp(-u)] + \frac{f-1}{2}[1 - \exp(-u)]^2\right\} \tag{2.5}$$

where f is the number of arms of the star polymer. Eqn (2.5) is a "Debye function" for f-arm polymer chains. The inter-polymer interaction can be taken into consideration in terms of the Flory–Huggins interaction parameter, χ, and the scattering intensity is given by:[32]

$$I(q) = \frac{(\Delta\rho)^2}{N_A}\frac{V_2\phi P_{star}(q)}{1 + (1 - 2\chi)\left(\frac{V_2}{V_1}\right)\phi P_{star}(q)} \tag{2.6}$$

where N_A is Avogadro's number and ϕ is the volume fraction of the solute; V_1 and V_2 are the molar volumes of the solvent and the monomer unit of the solute, respectively; $(\Delta\rho)^2$ is given by:

$$(\Delta\rho)^2 = (\rho_2 - \rho_1)^2 = \left[\left(\frac{b_2}{\tilde{V}_2}\right) - \left(\frac{b_1}{\tilde{V}_1}\right)\right]^2 \tag{2.7}$$

where b_i and $\tilde{V}_i (\equiv V_i/N_A)$ are the scattering length and the volume of the solvent ($i=1$) and the monomeric unit of the component ($i=2$).

2.3.1.2 Scattering Function for Semi-Dilute Polymer Solutions

The real-space concentration correlation function of polymer solutions is given by:

$$\langle\phi(r)\phi(0)\rangle \approx \frac{1}{r}\exp\left(-\frac{\xi}{r}\right). \tag{2.8}$$

where ξ is the correlation length and r is the distance between monomers; angular bracket means ensemble averaging. The space characterized by the length scale ξ is often called "blob" in which the polymer chain does not interact with other chains.[33] By taking a Fourier transform of eqn (2.8), the scattering function for polymer solutions is obtained as the Ornstein–Zernike (OZ) function, *i.e.*:

$$I(q) = \frac{(\Delta\rho)^2 RT\phi^2}{N_A K_{os}}\frac{1}{1 + \xi^2 q^2} \tag{2.9}$$

where R is the gas constant and T is the absolute temperature; K_{os} is the osmotic modulus of the polymer solution.

2.3.1.3   *Scattering Functions of Polymer Gels*

According to de Gennes,[33] the scattering intensity function of polymer gels in swelling equilibrium is the same as that of polymer solutions at the same concentration and is given by:

$$I(q) = \frac{(\Delta\rho)^2 RT\phi^2}{N_A M_{os}} \frac{1}{1 + \xi^2 q^2} \tag{2.10}$$

where M_{os} is the osmotic longitudinal modulus of the gel, given by:[34]

$$M_{os} = \frac{RT\phi^2}{V_1}\left(\frac{1}{1-\phi} - 2\chi\right) + v_e RT\left[\frac{1}{2}\left(\frac{\phi}{\phi_0}\right) + \left(\frac{\phi}{\phi_0}\right)^{1/3}\right] + \frac{4}{3}G \tag{2.11}$$

where v_e is the number density of the effective elastic chains in the network and G is the shear modulus of the gel. In principle, the scattering function can be represented by the so-called Ornstein–Zernike function,[35] *i.e.* eqn (2.10). However, it is known that scattering intensity functions of gels have significant forward scattering at low-q regions due to cross-linking inhomogeneities. As a result, by adding an extra term representing the inhomogeneities, eqn (2.10) is modified to:[36]

$$I(q) = \frac{(\Delta\rho)^2 RT\phi^2}{N_A M_{os}}\left[\frac{1}{1 + \xi^2 q^2} + \frac{A_{inhom}}{\left(1 + \Xi^2 q^2\right)^2}\right] \tag{2.12}$$

The extra term A_{inhom} is dependent on the chemistry of the gel preparation since it represents how the component polymer chains are topologically frozen by cross-linking.[3]

2.3.2   Macromer

2.3.2.1   *Macromer Solutions*

SANS experiments were carried out for (a) TAPEG-5k, (b) TAPEG-10k, (c) TAPEG-20k, and (d) TAPEG-40k macromer solutions in as-prepared states with various concentrations, ϕ ($= \phi_0$).[32,36] All observed SANS functions were well fitted by the scattering function for tetra-arm polymer chains (solid lines), *i.e.* eqn (2.6). The molecular weight dependence of the interaction parameter χ was a decreasing function of M_w [from 0.495 (TAPEG-5k) to 0.465 (TAPEG-40k)]. The reasons for the M_w dependence may be the presence of arm-ends,[14] and the chain-end effect becomes insignificant by increasing M_w. As a matter of fact, the value of χ approaches the value of linear PEG chains[11] ($\chi = 0.43$ at $T = 20\,^{\circ}$C) by increasing M_w. It was found that R_g seems to be a decreasing function of the initial polymer fraction, ϕ_0 ($R_g \approx \phi_0^{-1/3}$), and is an increasing function of M_w ($R_g \approx M_w^{3/5}$), indicating that a simple contraction occurs in Tetra-PEG chains as a function of ϕ_0. This exponent indicates that the Tetra-PEG chains behave as compressive elastic modules dependent on the concentration, and no inter-penetration occurs even at $\phi_0 > \phi_0{}^*$, where $\phi_0{}^*$ is the overlap concentration. The absence of inter-penetration may be due to the presence of a large number of end groups, as discussed by Matsunaga *et al.*[32,36]

2.3.2.2 As-Prepared Gels

Figure 2.7 shows SANS curves of as-prepared gels for (a) Tetra-PEG-5k gels, (b) Tetra-PEG-10k gels, (c) Tetra-PEG-20k gels, and (d) Tetra-PEG-40k gels. The solid lines are the fits with eqn (2.10), *i.e.* the OZ function. In the case of as-prepared Tetra-PEG-5k gel, a strong upturn in $I(q)$ for $q \leq 0.01 \, \text{Å}^{-1}$ was observed for $\phi_0 \geq 0.0531$, and a fitting with eqn (2.10) was poor. However, for larger q and for higher ϕ_0 values, each $I(q)$ can be fitted with an OZ function. Unlike the case of Tetra-PEG-5k gels, no significant upturn in $I(q)$ is found at

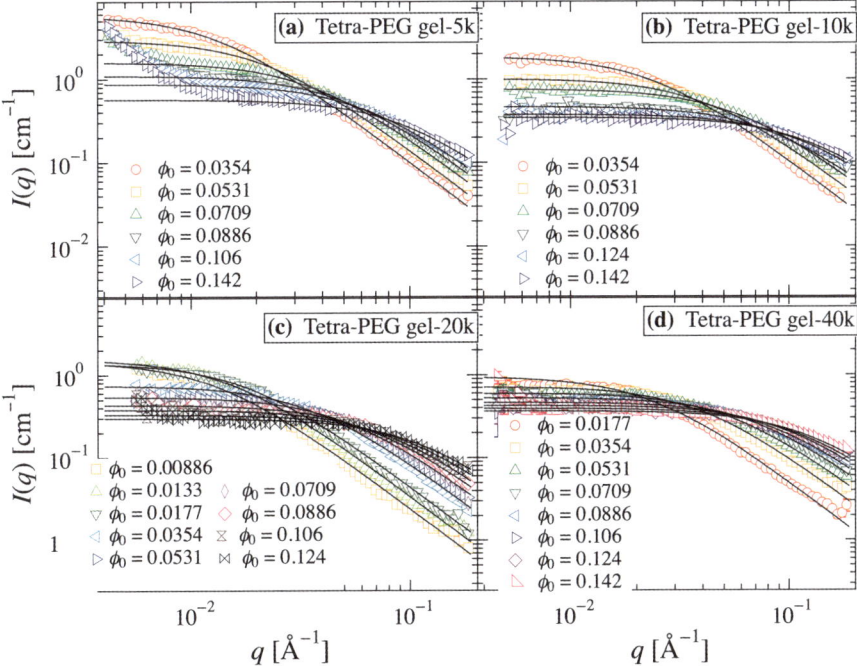

Figure 2.7   SANS intensity functions of Tetra-PEG gels at various ϕ_0 values: (a) 5k, (b) 10k, (c) 20k, and (d) 40k. The solid lines are the fit with the OZ function.
(Reproduced from Matsunaga *et al.*[36] with permission from the American Chemical Society.)

low-*q* regions in Tetra-PEG-10k, −20k, or −40k in this *q* region. It should be noted that polymer gels usually exhibit a strong upturn in low-*q* regions due to the presence of inhomogeneities.[3,37] Even in the case of "ideal polymer networks" made by end-linking of telechelic polymer chains, significant inhomogeneities are reported.[38,39]

2.3.2.3 Swollen Gels

According to de Gennes' so-called *c** theorem, a gel automatically maintains a concentration *c* proportional to *c**.[33] In order to confirm this hypothesis, SANS experiments were carried out for swollen gels prepared at different initial concentrations, ϕ_0. Figure 2.8 shows a series of SANS intensity functions for Tetra-PEG gels in swelling equilibrium for (a) Tetra-PEG gel-5k, (b) −10k, (c) −20k, and (d) −40k, where ϕ is the volume fraction of polymer network in the solvent at observation. As shown in these figures, *I*(*q*) values for the swollen gels are satisfactorily fitted with OZ functions irrespective of ϕ. As also shown in these figures, cross-linking inhomogeneities (Figure 2.7a) seem to disappear by swelling (Figure 2.8a), which is opposite to the case of randomly

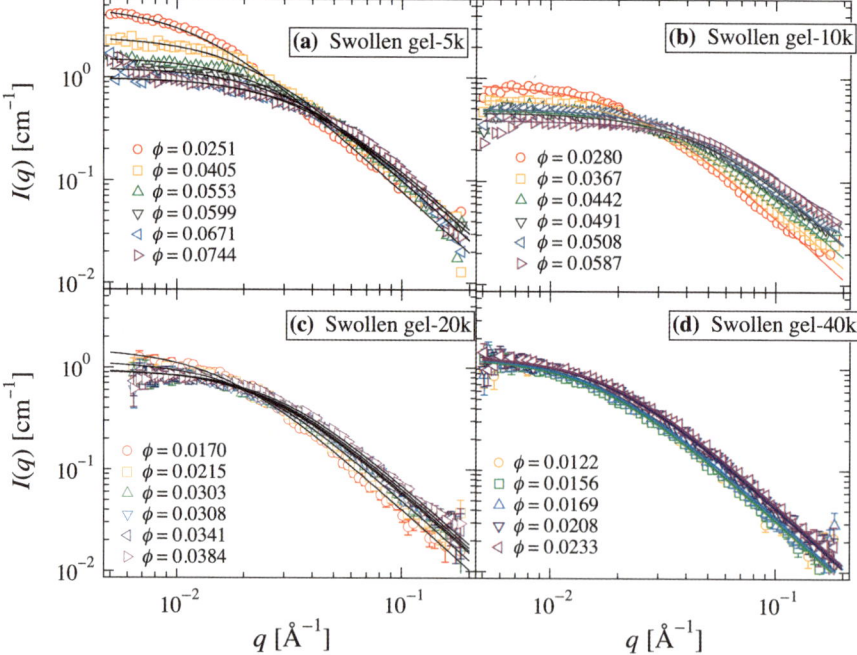

Figure 2.8 SANS intensity functions of Tetra-PEG gels at swelling equilibrium: (a) 5k, (b) 10k, (c) 20k, and (d) 40k; ϕ is the polymer volume fraction and the solid lines are the fit with the OZ function.
(Reproduced from Matsunaga *et al.*[36] with permission from the American Chemical Society.)

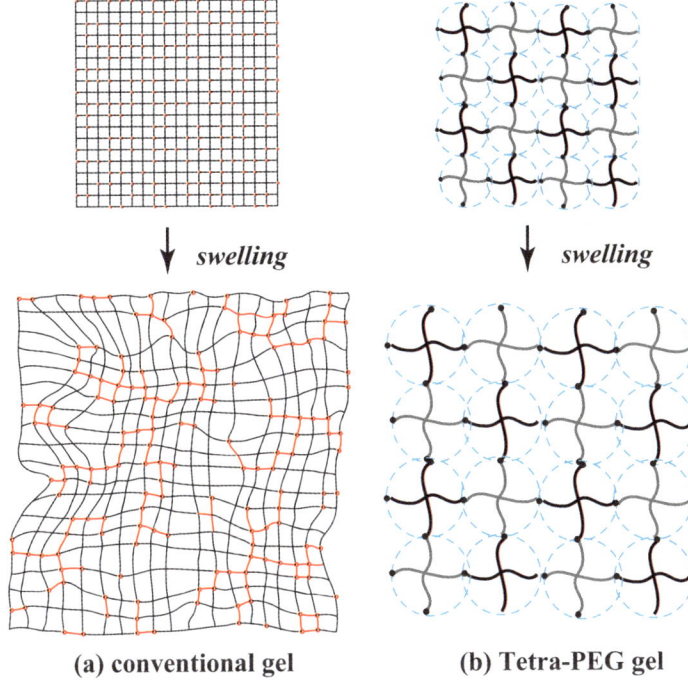

(a) conventional gel   (b) Tetra-PEG gel

Figure 2.9   Schematic illustration of the comparison of concentration fluctuations in (a) conventional gels and (b) Tetra-PEG gels. In the case of conventional gels, inhomogeneities increases by swelling, while Tetra-PEG gels do not.

cross-linked conventional gels (statistical gels).[40–45] According to Bastide and Leibler,[40] cross-linking inhomogeneities are expected to increase by swelling due to inhomogeneous swelling according to the difference in the local cross-link density (Figure 2.9a). Hence, the absence of an upturn in the SANS intensity functions observed in Tetra-PEG gels is quite unusual and indicates a very uniform network structure, as illustrated by Figure 2.9(b).

Another interesting feature of Tetra-PEG gels is that the SANS intensity functions become ϕ-independent by increasing M_n (see Figure 2.8d). Figure 2.10 schematically shows as-prepared Tetra-PEG-40k gels as a function of ϕ_0, and those in swelling equilibrium at the polymer volume fraction ϕ.[36] In the case of as-prepared gels, the characteristic size of the network scales with ϕ^α, where the exponent α is close to $-3/4$. When the gels are swollen in water, they reach a unique swollen state depending on their molecular weights, as shown schematically in the bottom figure of Figure 2.10. The presence of a unique swollen state independent of ϕ_0 suggests that cross-linking occurs exclusively at the surface of Tetra-PEG macromers and no entanglement formation takes place. Hence, it is concluded that the cross-end-coupling employed in this system is an effective method to prepare an ideal network without defects or trapped entanglements.

tetra-PEG gel-40k

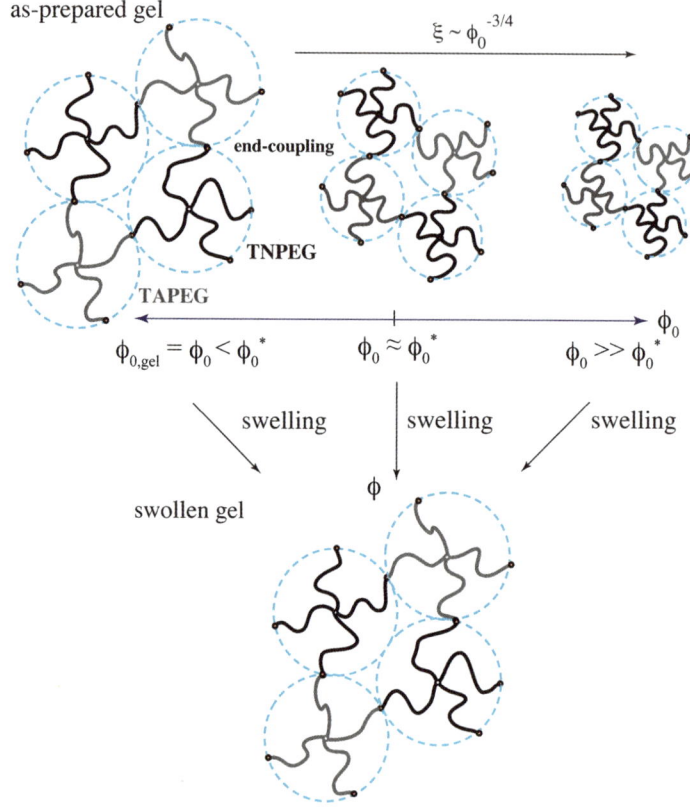

Figure 2.10   Scheme showing the formation of unique gel structure by swelling from
gels prepared at different initial concentrations. ϕ_0 and ϕ are the polymer
volume fractions at preparation and at swelling equilibrium, respectively.
Note that no trapped entanglements exist so as to attain a unique
structure.
(Reproduced from Matsunaga *et al.*[36] with permission from the
American Chemical Society.)

2.3.2.4   Master Relationship

Figure 2.11 shows the master curves obtained for Tetra-PEG gel-5k (open
squares), −10k (open triangles), −20k (open inverted triangles), and −40k
(open rhombus). Note that LS data are added to each SANS curve to cover
low-q regions, and the curves are shifted vertically so as to be connected with
the corresponding SANS data. Interestingly, by scaling $I(q)/\phi_0\xi^2$ and ξq, all
scattering intensity curves fall onto a single maser curve, as shown in the figure.
The curves denoted by crosses indicate the scattering intensity function of
linear PEG chains ($M_w = 102 \times 10^3$ at 10 °C) reported by Hammouda.[46] Surprisingly enough, the SANS function for the linear PEG chains is also nicely

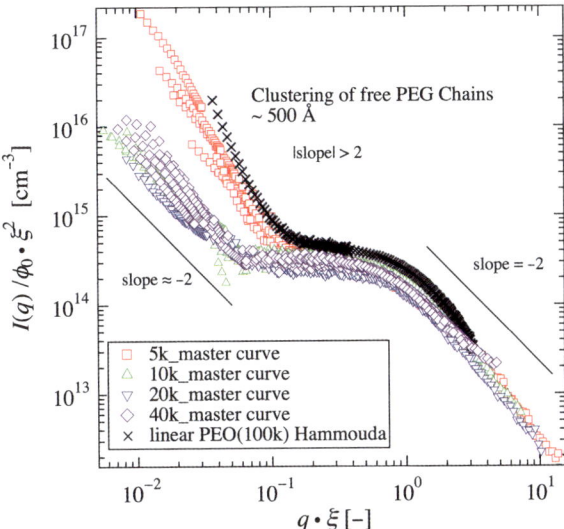

Figure 2.11   Master curve of SANS intensity functions of Tetra-PEG gels made from
different molecular weights and initial concentrations. Crosses are the
data for linear PEG [PEO; poly(ethylene oxide)] solutions.
(Reproduced from Matsunaga *et al.*[36] with permission from the
American Chemical Society.)

superimposed onto the master curve by employing the reduced variables as
shown in Figure 2.11. It is reported that this upturn corresponds to clustering
of PEG chains in water with a typical size of 500 Å, and the major reason of
clustering is the presence of hydrophobic end groups, such as $-OCH_3$. It is of
particular interest that the behavior of the linear PEG is quite similar to Tetra-
PEG-5k gel, suggesting that Tetra-PEG gel-5k has a clustered structure similar
to linear PEG chains in water. This is probably due to the clustering of arm
PEG chains with unreacted end groups. It is noted that the presence of such
clustered (or aggregated) structures is typical in PEG aqueous solutions.[47–49]

A more rigorous discussion on the vertical shift factor can be done with the
argument on the scattering intensity function for semi-dilute polymer solutions.
The scattering intensity for polymer solutions in a semi-dilute regime is
inversely proportional to the osmotic modulus, M_{os}, as given by eqn (2.11).
Since M_{os} is given by $M_{os} = \phi(\partial\Pi/\partial\phi)_T + (4G/3)$, and Π and ξ are scaled with
ϕ by $\Pi \approx \phi^{-3}$ and $\xi \approx \phi^{-3/4}$, one obtains:

$$I(q) = \frac{(\Delta\rho)^2 \phi V_2 \xi^{5/3}}{N_A} \frac{1}{1 + \xi^2 q^2} \qquad (2.13)$$

The value of $(\Delta\rho)$ for the system of Tetra-PEG and deuterated water,
$\Delta\rho = \rho_{D2O} - \rho_{PEG} = 5.72 \times 10^{10}$ cm^{-2}, was carefully evaluated. Figure 2.11 can

be obtained without any adjustable parameter. Hence, the scattering intensity function for polymer gels can be scaled with $I(q)/\phi_0\xi^{5/3}$ and ξq.[50,51]

2.3.2.5 Deformed Gels

SANS is recognized to be one of the most suitable means to elucidate the relationship between the microscopic and macroscopic behaviors of deformation, because polymer chains in a polymer network can be selectively labeled and the change of polymer conformation can be studied as a function of macroscopic deformation.[52,53] Benoit *et al.* performed a pioneering work on deformation of polymer network with SANS.[54] They carried out SANS experiments on branch-labeled and labeled-chain networks of polystyrene. The former gave the information on the inter-crosslink distance and the latter on the radius of gyration of labeled chains. They reported that both the change of the inter-crosslink distance and that of the radius of gyration were far below those predicted by affine or junction-affine models. In 1980s, a number of SANS investigations were carried out on deformed polymer networks. Ullman proposed scattering functions from labeled chains in polymeric networks for swollen and stretched polymer networks and compared them with those obtained by SANS experiments.[55–57] One of the important findings was that the chain deformation was less than that calculated from the phantom network model.

Tetra-PEG gels are suitable systems to study (1) the relationship of spatial inhomogeneities and the gel structure and (2) deformation mechanism of Tetra-PEG gels. In order to answer these questions, SANS experiments were carried out on uniaxially stretched Tetra-PEG gels as a function of the stretching ratio. Surprisingly, the results showed no noticeable increase in scattering intensity by deformation.[58] This is because Tetra-PEG gels inherently do not have inhomogeneities, resulting in a uniform deformation by stretching.

2.4 Mechanical Properties

According to the theory of rubber elasticity,[34] structural parameters such as the concentration of elastically effective chains and the molecular weight of network strands determine the elastic modulus and ultimate elongation, respectively. However, in general, theories with structural parameters cannot predict the mechanical properties of real polymer gels. This is because of the inhomogeneities inherently introduced into polymer gels during the cross-linking process, as discussed in Sections 2.2 and 2.3. These inhomogeneities introduce a variety of substructures and complicate the structure of polymer networks, rendering it difficult to understand the relationship between mechanical properties and structural parameters. Thus, development of a proper model describing the mechanical properties of polymer gels is one of the ultimate goals of polymer science. Furthermore, owing to the lack of effective methods for controlling these inhomogeneities, there has been no universal design guide for fabricating the polymeric materials with the required mechanical properties. Therefore, an ideal polymer network free from inhomogeneities has been desired from both the academic and industrial points of view. Tetra-PEG gel is a promising candidate for an ideal polymer network along this line. This section discusses the relationship between mechanical properties and those predicted by structural parameters.

2.4.1 Theoretical Background

In the original theory of rubber elasticity proposed by Flory [neo-Hookean (NH) model], the relationship between nominal stress (σ) and principal stretching ratio (λ) of rubbery materials is given by:[59]

$$\sigma = G\left(\lambda - \lambda^{-2}\right) \tag{2.14}$$

where G is the elastic modulus. The material, the stress–elongation relationship of which is represented by eqn (2.14), is called the incompressible neo-Hookean material. The value of G is related to structural parameters such as the number density of elastically effective chains (v) and cross-links (μ), where elastically effective chains are the network strands connecting to cross-links.

2.4.1.1 Elastic Modulus

There are three conventional models describing the contribution of chemical cross-links to rubber elasticity. From the elementary theory of Kuhn,[60] G ($= G_{Af}$) is found to be:

$$G_{Af} = vRT \tag{2.15}$$

This model is called the "affine network" model. James and Guth proposed a theory of elastic properties of rubbers which allow thermal motions of cross-links in the rubber, the so-called the "phantom network" model.[61] In their theory, G ($= G_{Ph}$) is found to be:

$$G_{Ph} = \left(1 - \frac{2}{f}\right)vRT = (v - \mu)RT \tag{2.16}$$

A third model is the "junction affine network" model, which predicts the intermediate value of G ($= G_{JA}$) as follows:[8]

$$G_{JA} = (v - h\mu)RT \tag{2.17}$$

where h is a constant indicating the thermal fluctuation of chemical cross-links.

2.4.1.2 Miller–Macosko Model

The Miller–Macosko (MM) model is the simplest model that can provide an estimate of v and μ from known structural parameters.[62] The MM model was developed based on the recursive nature of the branching process of the Bethe lattice. It assumes that no small loops are formed and all functional groups are equally reactive throughout the reaction process. Θ was defined as a variable with the probability that an arbitrary site is not connected to infinity through one fixed bond originating from this site. For the f-functional polycondensation, p is related to Θ by the following self-consistent equation:

$$\Theta = 1 - p + p\Theta^{f-1} \tag{2.18}$$

The MM theory assumes that an elastically effective chain identifies with a chain that is connected to infinity. So the elastically effective cross-link is defined as the site originating from which more than three bonds are connected to infinity and the elastically effective chain is defined as a bond which links two elastically effective cross-links. Using Θ calculated from eqn (2.18), the concentration of elastically effective chains and cross-links are given by:

$$v = \psi \times \sum_{i=3}^{f} \frac{i}{2} \times {}_f C_i \times (1 - \Theta)^i \times \Theta^{f-i} \tag{2.19}$$

$$\mu = \psi \times \sum_{i=3}^{f} {}_f C_i \times (1 - \Theta)^i \times \Theta^{f-i} \tag{2.20}$$

where ψ denotes the number density of the f-functional precursor (Figure 2.12).[63]

2.4.2 Small Deformation

The elastic moduli of Tetra-PEG gel-10k and -20k obtained by stress–strain measurements are shown in Figure 2.13 against ϕ_0.[18] The value of G roughly increases linearly with an increase in ϕ_0, and the slope of Tetra-PEG gel-10k is approximately double that of Tetra-PEG gel-20k, confirming the relationship $G \approx \phi_0/M_n$. As far as the authors are aware, this relationship has never been observed for any polymer network because of the existence of inhomogeneities affecting G in unknown ways. The elastically ineffective loops increase with a decrease in ϕ_0, especially near ϕ^*, because the probability of intramolecular reactions increases with a decrease in ϕ_0.[64] The number density of entanglements (ε) increases with an increase in ϕ_0 in the semi-dilute region as $\varepsilon \approx \phi_0^{9/4}$.[65] Therefore, this agreement strongly suggests that the mechanical response of Tetra-PEG gel is governed by the chemical cross-links, and so inhomogeneities are strongly suppressed.

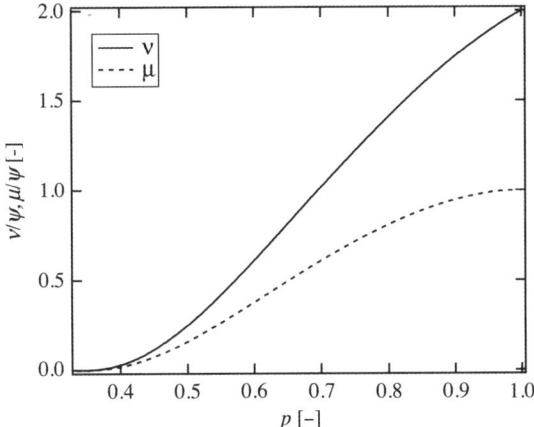

Figure 2.12 The variation of v and μ predicted by the MM model. (Reproduced from Akagi *et al.*[63] with permission from the American Chemical Society.)

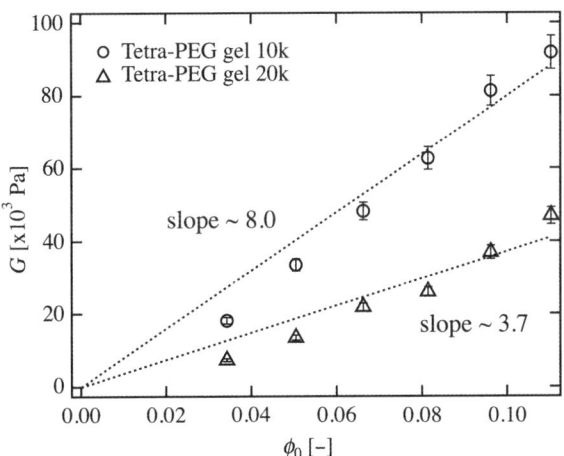

Figure 2.13 Elastic modulus, G, as a function of ϕ_0 for Tetra-PEG gel-10k and -20k.

Figure 2.14 shows the variation of G, G_{Af}, and G_{Ph} against ϕ_0, where G_{Af} and G_{Ph} are calculated from eqns (2.16)–(2.20) using p. G_{Af} and G_{Ph} also increased linearly with an increase in ϕ_0, reflecting the constant p. G and G_{Ph} corresponded well with each other in the region above ϕ^* (also displayed in Figure 2.14), suggesting that the elasticity of Tetra-PEG gel is well predicted by the phantom network model. The downward deviation of G from G_{Ph} around and below ϕ^* (~ 0.04) may suggest the formation of elastically ineffective loops as it corresponds well to the nature of these loops. Again, these data strongly

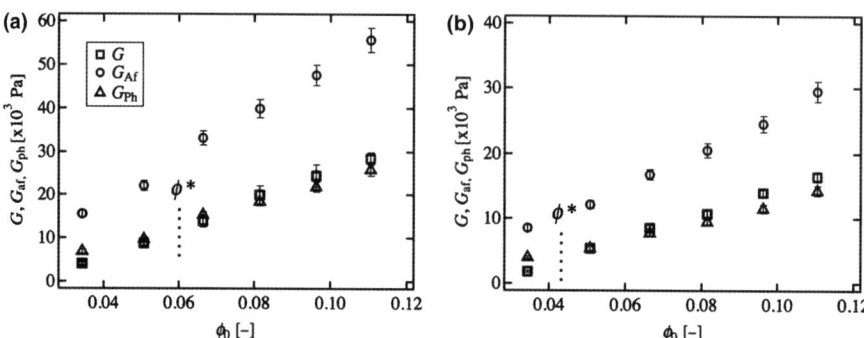

Figure 2.14 G, G_{Af}, and G_{Ph} as a function of ϕ_0 for (a) Tetra-PEG gel-10k and (b) −20k.

suggest that Tetra-PEG gel formed in the semi-dilute condition ($0.05 < \phi_0 < 0.11$) is practically free from any defects, and the phantom network model is the true model predicting the elastic modulus of an ideally homogeneous polymer network.

2.4.3   *p*-Controlled Gels

In the preceding section, the agreement between the elastic modulus and that estimated from the MM model and phantom network model was confirmed. However, the validity of these models was only confirmed by near-ideal polymer networks. In order to investigate the validity of the MM model and the phantom network model, an experiment was conducted on connectivity (*p*)-tuned Tetra-PEG gel. The value of *p* for the Tetra-PEG gel can be easily tuned by hydrolyzing the activated ester prior to the initiation of the gelation reaction. The experimental procedure is schematically depicted in Figure 2.15. By using partly hydrolyzed TNPEG as a precursor, the final reaction probability can be controlled and this corresponds to the ratio of the concentration of amide bond to the initial concentration of NHS ester. In the near future, these experiments will be conducted to examine the validity of the MM model and the phantom network model.

2.4.4   **Large Deformation**

An example of a stress–strain curve for the Tetra-PEG gel-20k ($\phi_0 = 0.14$) for a large deformation is shown in Figure 2.16. The picture of stretching measurement is also shown in the inset of Figure 2.16. Initially, the specimen was stretched to $\lambda = 6.5$ and released; this stretch and release was performed three times, and then the network was stretched to the limit. The

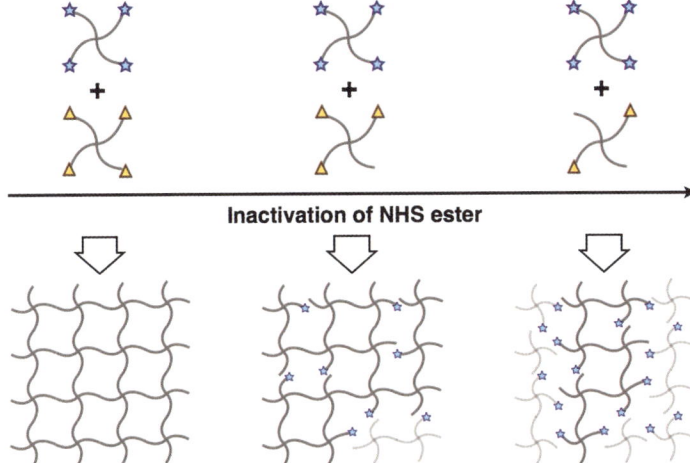

Figure 2.15 The schematic picture of sample preparation. The stars and triangles represent the amine and NHS ester terminal groups, respectively. The NHS ester on TNPEG was hydrolyzed in pre-gel solution for a given time in order to control the reaction conversion.

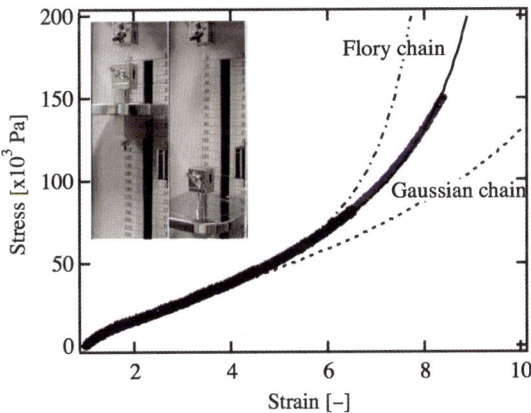

Figure 2.16 Stretching stress–strain curve of Tetra-PEG gel-20k ($\phi_0 = 0.14$). The solid line, dashed line, and chain lines represent the fitting result by eqn (2.2) using R_0 as a free parameter, the fitting result assuming the Gaussian chain, and the fitting result assuming the Flory chain, respectively. The *inset* shows the pictures of the stretching measurement.

average value of λ_{max} was 8.1 and the maximum value was 9.9. Practically no hysteresis was observed even after the elongation just before the fracture. In order to predict the full scale of the stress–elongation relationship, the effect of finite extensibility of network strands should be taken account.

The Langevin chain statistics is one of the conventional ideas along this line. In this model, the force applied to the single chain (F) is represented as:

$$F = \frac{kT}{b} L^{-1} \left(\frac{R_0 \lambda}{Nb} \right) \tag{2.21}$$

where R_0 is the initial end-to-end distance of a polymer, N is the degree of the polymerization of a network strand, and b is the monomer length. The use of the Langevin chain statistics in polymer networks was considered in the three-chain model by James and Guth.[61] In this model, the relationship between σ and λ of a polymer network was constructed by v inverse-Langevin elastic blobs and is represented by:

$$\sigma = \frac{vkT}{3} \frac{R_0}{b} \left\{ L^{-1} \left(\frac{R_0 \lambda}{Nb} \right) - \lambda^{-3/2} L^{-1} \left(\frac{R_0 \lambda^{-1/2}}{Nb} \right) \right\} \tag{2.22}$$

According to the literature, b was set to 3.5 Å.[66] The solid line represents a curve fit with eqn (2.22). Surprisingly, the stress–elongation relationship of the Tetra-PEG gel is nicely fitted with eqn (2.22) using the intrinsic fitting parameters. The resultant R_0 was 7.4 nm; it was in the range between the Gaussian (5.3 nm) and the Flory (9.0 nm) limits, suggesting that elastic blobs take the intermediate conformation between the two models.[15] This result is also in good agreement with the previous SANS results, showing that the Tetra-arm module takes a more contracted configuration than the Flory chain.[32] λ_{max} ($= Nb/R_0$) obtained from the fitting result was 10.8, corresponding well to the experimental results. It should be noted that the stress–elongation relationships for polymer networks in general cannot be represented by eqn (2.22) using the intrinsic fitting parameters.[67,68] This is probably because the exact value of N cannot be defined due to a variety of inhomogeneities, and/or because the polymers are likely to be crystallized in the larger λ region. Thus, when a large strain is applied to conventional networks, weak regions are broken at a lower elongation ratio, and the crystallized regions are unlikely to relax to the initial state, so that hysteresis is observed.[69–71] In contrast, no hysteresis was observed in the Tetra-PEG gel. These experimental results strongly suggest that the Tetra-PEG gel has an extraordinarily uniform network structure, and that its elasticity can be represented by uniformly packed elastic blobs.

2.4.5 Viscoelastic Properties

The dynamic mechanical response ($\omega = 1.0$ rad s^{-1}) of the Tetra-PEG gel-10k with various ϕ_0, where ω is the angular frequency,[15] was measured. The storage modulus (G'), loss modulus (G''), and loss tangent (tan δ) are shown in Figure 2.17 as a function of the feed polymer fraction (ϕ_0). The inset shows a series of stroboscopic photographs of a Tetra-PEG ball [Tetra-PEG gel-10k

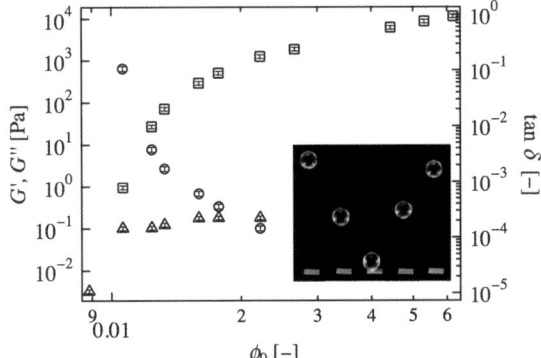

Figure 2.17 *G′*, *G″*, and tan δ of Tetra-PEG gel-10k as a function of ϕ_0. The *squares*, *triangles*, and *circles* represent *G′*, *G″* (*left axis*), and tan δ (*right axis*), respectively. The *inset* shows the picture of the sphere-shaped Tetra-PEG gel dropped onto a hard floor.

($\phi_0 = 0.12$)]. Theoretically, *G′* is the same with *G* obtained by the stretching measurement and *G″* is known to appear when a polymer network has inhomogeneities.[33,72–74] The tan δ represents the *G″* normalized by *G′*, *i.e.* *G″/G′*. Thus, the tan δ of the ideal network is expected to be zero. As shown in Figure 2.17, the elasticity appeared for the first time at $\phi_0 = 0.011$; namely, the gelation threshold was reached. As the ϕ_0 increased above the gelation threshold, *G′* increased, while tan δ decreased, suggesting the proceeding of the network formation and the elimination of inhomogeneities. At a concentration approximately above $\phi_0 = 0.022$, the *G″* and tan δ became extremely low, below the lower measurable limit of the rheometer (tan δ $= 10^{-4}$). The value of tan δ was by one to three orders smaller than that of the other model networks and ringing gels;[75–77] that is, practically no external energy applied to the network was dissipated. Indeed, when a Tetra-PEG gel ball was dropped onto a hard floor, it sprang back to near the initial height (the reflection coefficient ≈ 0.84) in a similar fashion to a power ball made of vulcanized polybutadiene (inset of Figure 2.17). Furthermore, when the Tetra-PEG gel-10k ($\phi_0 = 0.12$) was tapped with a finger, a ringing sound was emitted in the same manner as for ringing gels,[78] suggesting that thousands of oscillations were permitted before the deformation energy was dissipated into frictional energy.[78] These results also strongly suggest that the Tetra-PEG gel is free from inhomogeneities.

2.5 Medical Applications

Hydrogels are promising candidate materials for use in drug delivery systems and regenerative medicine, because of their high biocompatibility that stems from their similarities to human tissue.[79–81] The largely water-filled structure enables the containment and release of drugs, and allows biological substances to permeate throughout its body. For practical applications, however, there are major problems to resolve, including mechanical fragility, poor control of elastic modulus, poor control of drug diffusibility, and poor control of degradation time. These problems mainly arise from the inhomogeneities in the structures of hydrogels.

As discussed in the previous sections, Tetra-PEG gels are suitable candidates for medical use. A Tetra-PEG gel with precisely controlled and well predicted degradation properties has been introduced.[82] With precisely controlled degradation behavior, this Tetra-PEG gel is an excellent material for use in drug delivery systems and regenerative medicine. In order to control the degradability of Tetra-PEG gel, a third unit that contained a degradation site upstream of the terminal activated ester group (TGPEG; Figure 2.18) was designed.[82] TGPEG is expected to react with TAPEG to form an amide bond in the same manner as TNPEG; the only difference is the coincidental installation of one cleavable site per amide bond. This design enabled the degradability to be controlled by simply changing the ratio of TCPEG to TGPEG without changing the original reaction system.

2.5.1 Elastic Modulus

The Tetra-PEG gel was formed by mixing three kinds of tetra-PEG units, *i.e.* TAPEG, TGPEG, and TNPEG, at a 1 to r_{deg} to $(1 - r_{deg})$ ratio (see Figure 2.18).[82] In order to investigate the effect of r_{deg} on the physical structure of the

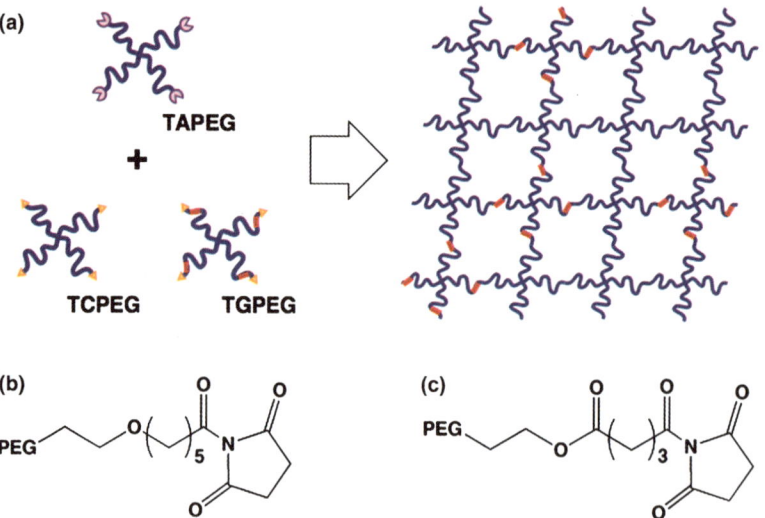

Figure 2.18   (a) Design of degradability tunable Tetra-PEG gel, together with the end group structures of (b) TNPEG and (c) TGPEG. TGPEG introduced the cleavable carbonyl site (shown as *red square*) into the network structure. (Reproduced from Li *et al.*[82] with permission from the American Chemical Society.)

polymer network in the as-prepared state, the G values of Tetra-PEG gels with different r_{deg} were measured. According to the phantom network model, the number density of cycle rank (Ω), which is the maximum number of independent loops contributing to the elasticity of polymer network, is related to G as follows:

$$G = (v - \mu)k_B T = \Omega k_B T \qquad (2.23)$$

The values of Ω were almost constant, suggesting that r_{deg} does not affect the physical structure of the polymer network. In other words, TGPEG and TNPEG have practically the same reactivity to TAPEG, and so these hydrogels also have the same elastic moduli, mesh size, and water content.

2.5.2 Degradation Behavior

The gel samples were immersed in phosphate buffer (pH 7.4) at 37 °C and the degradation behavior was measured. The variation of the swelling ratio (Q), which is the ratio of the gel volume in the equilibrium-swollen state to that in the as-prepared state, of the Tetra-PEG gels with various r_{deg} is shown in Figure 2.19.[82] The gel samples ($r_{deg} \geq 0.69$) swelled with time and finally disintegrated, showing bulk degradation behavior. The degradation time (t_{deg}) was determined as the point at which no gel could be observed.

According to the Flory–Rehner equation with the phantom model,[59] Ω is related to Q as:

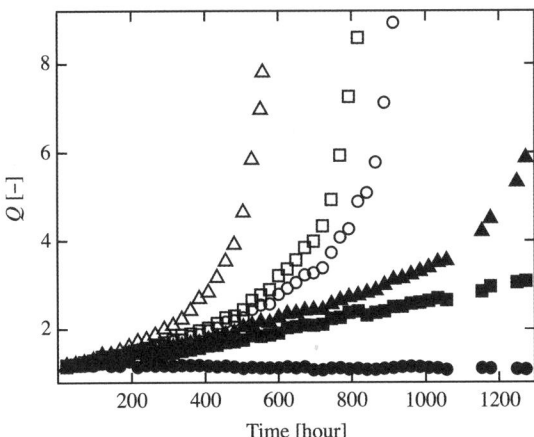

Figure 2.19 The time course of the swelling ratio (Q) of Tetra-PEG gel with different r_{deg} values. *Filled circles*: $r_{deg} = 0$; *filled squares*: $r_{deg} = 0.63$; *filled triangles*: $r_{deg} = 0.69$; *open circles*: $r_{deg} = 0.81$; *open squares*: $r_{deg} = 0.88$; *open triangles*: $r_{deg} = 1.0$.
(Reproduced from Li *et al.*[82] with permission from the American Chemical Society.)

$$\Omega = \frac{\ln\left(1 - \frac{\phi_0}{Q}\right) + \frac{\phi_0}{Q} + \chi\left(\frac{\phi_0}{Q}\right)^2}{-V_1 Q^{-\frac{1}{3}}}   \qquad (2.24)$$

The value of χ was estimated to be 0.45 using the values of Ω and Q for non-degradable Tetra-PEG gel ($r_{deg} = 0$). The value of χ corresponded well with the value obtained by other authors.[83,84] The time courses of Ω estimated by eqn (2.24) are shown in Figure 2.20.[82]

Following this, a new model that predicts the degradation behavior of Tetra-PEG gel was set up. According to the theory of tree-like structures for tetra-functional networks, Ω is represented as:

$$\Omega = \psi\left(\frac{1}{2} + \left(\frac{1}{p} - \frac{3}{4}\right)^{\frac{1}{2}}\right)\left(\frac{3}{2} - \left(\frac{1}{p} - \frac{3}{4}\right)^{\frac{1}{2}}\right)^3   \qquad (2.25)$$

Here, it is assumed that the reactivities of TCPEG and TGPEG to the TAPEG are the same, and that the degradable units dispersed homogeneously and were hydrolyzed over time according to pseudo-first-order kinetics. Considering the case where Tetra-PEG gel has the constant fraction of degradable unit (r_{deg}), p is represented as:

$$p = p_0\left((1 - r_{deg}) + r_{deg}\exp(-k_{dis}t)\right)   \qquad (2.26)$$

where p_0 and k_{dis} are the initial fraction of connected bonds and the degradation rate constant, respectively. By substituting eqn (2.26) into eqn (2.25), the time variation of Ω is represented as:

$$\xi = U\left(\frac{1}{2} + \left(\frac{1}{p_0\left((1 - r_{deg}) + r_{deg}\exp(-k_{deg}t)\right)} - \frac{3}{4}\right)^{\frac{1}{2}}\right)$$
$$\times \left(\frac{3}{2} - \left(\frac{1}{p_0\left((1 - r_{deg}) + r_{deg}\exp(-k_{deg}t)\right)} - \frac{3}{4}\right)^{\frac{1}{2}}\right)^3   \qquad (2.27)$$

Experimental data ($\Omega > 2$) using eqn (2.27) were used to evaluate this model (solid lines in Figure 2.20). The curve fit worked well for all samples; the slight upward deviation of points was attributed to the dissociation of the sol fraction. The critical fraction of connection at disintegration (p_c) was estimated to be 0.46.

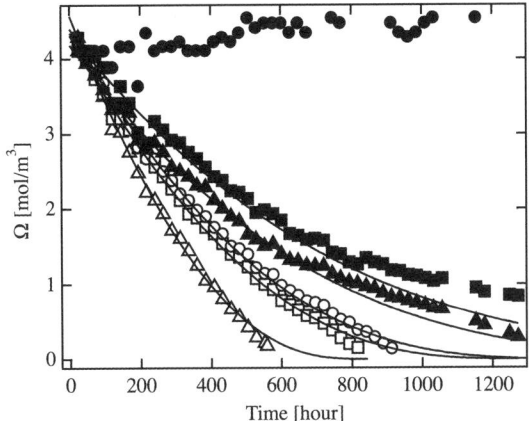

Figure 2.20   The variation of the cycle rank (Ω) of Tetra-PEG gel having different degradable unit fractions (r_{deg}). The symbols show the same samples as in Figure 2.19.
(Reproduced from Li *et al.*[82] with permission from the American Chemical Society.)

In the above discussion, the assumptions and the estimated degradation criterion, p_c, was validated and confirmed. By substituting $p = p_c$ into eqn (2.26), the equation describing the relationship between r_{deg} and t_{deg} is:

$$t_{\text{deg}} = \frac{1}{k_{\text{deg}}} \ln \frac{r_{\text{deg}}}{\dfrac{p_c}{p_0} - \left(1 - r_{\text{deg}}\right)} \tag{2.28}$$

The theoretical curve and experimental data of t_{deg} as a function of r_{deg} are shown in Figure 2.21. Here, values of $k_{\text{dis}} = 9.76 \times 10^{-4}$, $p_0 = 0.86$, and $p_c = 0.46$ were used. As clearly shown in Figure 2.21,[82] this model precisely describes the degradation behavior of Tetra-PEG gel and is able to precisely predict and control t_{deg} by tuning r_{deg}.

The ability to control the degradability of hydrogels has been sought after for many years. In previous attempts, the methods to control the degradation rate affected other important properties that need to be controlled independently. In addition, the precise prediction of degradation behavior over the full timescale using an equation with intrinsic parameters has never been achieved; there were only scaling arguments and kinetic-co-statistical models predicting initial degradation.[84,85] In contrast, a novel system has been invented in which the degradation behavior can be tuned without affecting other properties.[82] Using this system, the degradation behavior over the full timescale of a hydrogel has been, for the first time, precisely predicted by a universal equation with one intrinsic parameter, k_{dis}. Because the precise prediction and control of degradation behavior is an essential property for materials used in drug delivery

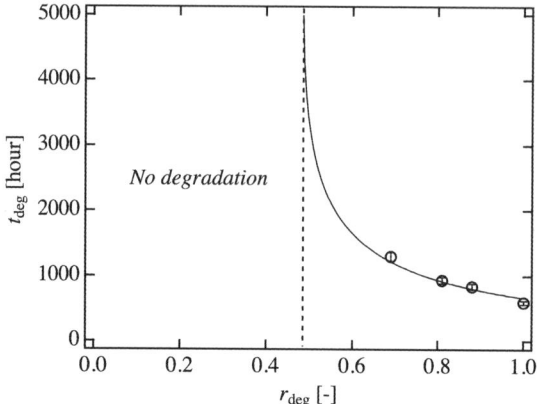

Figure 2.21   The degradation time (t_{deg}) as a function of the degradable unit (r_{deg}).
(Reproduced from Li *et al.*[82] with permission from the American
Chemical Society.)

systems and regenerative medicine, Tetra-PEG gels will make a promising
candidate material for these applications.

2.6   Conclusions

"Tetra-PEG gels", which have an extremely homogeneous and high performance polymer network, have been successfully fabricated. The uniqueness of
Tetra-PEG gels is the concept of cross-linking, *i.e.* module assembling by cross-
end-coupling (cross-end-coupling of symmetric macromers carrying complementary functional groups). The rate of gelation can be controlled by pH.
The uniformity and homogeneity are well characterized by SANS. The SANS
functions of Tetra-PEG gels are similar to those of polymer solutions in a semi-
dilute regime, *i.e.* the Ornstein–Zernike function. Hence, the SANS functions
can be reduced to a master curve irrespective of the molecular weight and
polymer volume fraction. This indicates that the network structure of Tetra-
PEG gels is scalable with respect to the mesh size, ξ.

The mechanical properties are more superior to those of conventional chemical gels. The gel is stretchable up to eight times when the polymer concentration is about 10 wt%. The gel is very elastic with a negligible loss tangent
as low as 10^{-4}. The bond probability can reach 0.95 or more and is tunable.
Since the network structure can be tailored, Tetra-PEG gels provide a model
for studying the theory of rubber elasticity for polymer networks as a function
of the bond probability, a phantom percolation model. There are a variety of
practical applications of Tetra-PEG gels, especially in the area of medical science. This strongly suggests that Tetra-PEG gels provide a new paradigm in
polymer science.

Acknowledgements

This work was partially supported by the Ministry of Education, Science, Sports and Culture, Japan [Grant-in-Aid for Scientific Research (A), 2010–2012, No. 22245018, and for Scientific Research on Priority Areas, 2006–2010, No. 18068004)]. The SANS experiment was performed at SANS-U with the approval of Institute for Solid State Physics, The University of Tokyo (Proposal No. 10621 and 10629), at the Japan Atomic Energy Agency, Tokai, Japan. This work was carried out by the support of the Atomic Energy Initiative, MEXT. T.M. acknowledges the support from the Research Fellowship for Young Scientists of the Japan Society for the Promotion of Science.

References

1. *Polymer Gels*, ed. D. Rossi, K. Kajiwara, Y. Osada and A. Yamauchi, Plenum, New York, 1991.
2. *Gel Handbook*, ed. Y. Osada and K. Kajiwara, Academic Press, New York, 2001.
3. M. Shibayama, *Macromol. Chem. Phys.*, 1998, **199**, 1–30.
4. M. Shibayama and T. Norisuye, *Bull. Chem. Soc. Jpn.*, 2002, **75**, 641–659.
5. M. Shibayama, H. Takahashi and S. Nomura, *Macromolecules*, 1995, **28**, 6860–6864.
6. T. Norisuye, N. Masui, Y. Kida, M. Shibayama, D. Ikuta, E. Kokufuta, S. Ito and S. Panyukov, *Polymer*, 2002, **43**, 5289–5297.
7. P. J. Flory, *J. Am. Chem. Soc.*, 1941, **63**, 3038.
8. P. J. Flory, *J. Chem. Phys.*, 1977, **66**, 5720–5729.
9. J. He, S. Machida, H. Kishi, K. Horie, H. Furukawa and R. Yokota, *J. Polym. Sci., Part A: Polym. Chem.*, 2002, **40**, 2501–2512.
10. P. Martens and K. S. Anseth, *Polymer*, 2000, **41**, 7715–7722.
11. M. P. Lutolf and J. A. Hubbell, *Biomacromolecules*, 2003, **4**, 713–722.
12. A. K. Azab, B. Orkin, V. Doviner, A. Nissan, M. Klein, M. Srebnik and A. Rubinstein, *J. Controlled Release*, 2006, **11**, 281–289.
13. M. Malkoch, R. Vestberg, N. Gupta, L. Mespouille, P. Dubois, A. F. Mason, J. L. Hedrick, Q. Liao, C. W. Frank, K. Kingsbury and C. Hawker, *Chem. Commun.*, 2006, 2774–2776.
14. T. Sakai, T. Matsunaga, Y. Yamamoto, C. Ito, R. Yoshida, S. Suzuki, N. Sasaki, M. Shibayama and U. I. Chung, *Macromolecules*, 2008, **41**, 5379–5384.
15. T. Sakai, Y. Akagi, T. Matsunaga, M. Kurakazu, U. Chung and M. Shibayama, *Macromol. Rapid Commun.*, 2010, **31**, 1954–1959.
16. K. Nishi, K. Fujii, M. Chijiishi, Y. Katsumoto, U. Chung, T. Sakai and M. Shibayama, *Macromolecules*, 2012, **45**, 1031–1036.
17. M. Kurakazu, T. Katashima, M. Chijiishi, K. Nishi, Y. Akagi, T. Matsunaga, M. Shibayama, U. Chung and T. Sakai, *Macromolecules*, 2010, **43**, 3935–3940.

18. Y. Akagi, T. Katashima, K. Fujii, T. Matsunaga, U. Chung, M. Shibayama and T. Sakai, *Macromolecules*, 2011, **44**, 5817–5821.
19. *Physical Properties of Polymer Gels*, ed. J. P. Cohen Addad, Wiley, New York, 1996.
20. Y. Hirokawa, H. Jinnai, Y. Nishikawa, T. Okamoto and T. Hashimoto, *Macromolecules*, 1999, **32**, 7093.
21. A. Guinier and G. Fournet, *Small-Angle Scattering of X-rays*, Wiley, New York, 1955.
22. T. A. Ezquerra, M. C. Garcia-Gutierrez, A. Nogales and M. Gomez, *Applications of Synchrotron Light to Scattering and Diffraction in Materials and Life Sciences*, Springer, Berlin, 2009.
23. B. Chu, *Annu. Rev. Phys. Chem.*, 1970, **21**, 145–174.
24. R. Borsali, in *Light Scattering: Principles and Development*, ed. W. Brown, Oxford University Press, Oxford, 1996, pp. 255–289.
25. M. Shibayama, in *Soft Matter Characterization*, ed. P. Pecora and R. Borsali, Springer, Berlin, 2008, vol. 2, pp. 783–832.
26. M. Shibayama, *Polym. J.*, 2011, **43**, 18–34.
27. *Molecular Basis of Polymer Networks*, ed. A. Baumgartner and C. E. Picot, Springer, Berlin, 1989.
28. J. S. Higgins and H. C. Benoit, *Polymers and Neutron Scattering*, Clarendon Press, Oxford, 1994.
29. *Neutrons in Soft Matter*, ed. T. Imae, T. Kanaya, M. Furusaka and N. Torikai, Wiley, Hoboken, NJ, 2011.
30. H. Benoit, *J. Polym. Sci.*, 1953, **11**, 507–510.
31. D. Richter, B. Farago, J. S. Huang, L. J. Fetters and B. Ewen, *Macromoleules*, 1989, **22**, 468–472.
32. T. Matsunaga, T. Sakai, Y. Akagi, U. Chung and M. Shibayama, *Macromolecules*, 2009, **42**, 1344–1351.
33. P. G. de Gennes, *Scaling Concepts in Polymer Physics*, Cornell University Press, Ithaca, NY, 1979.
34. L. R. G. Treloar, *The Physics of Rubber Elasticity*, Clarendon Press, Oxford, 1975.
35. L. S. Ornstein and F. Zernike, *Proc. Acad. Sci., Amsterdam*, 1914, **17** , 793.
36. T. Matsunaga, T. Sakai, Y. Akagi, U. I. Chung and M. Shibayama, *Macromolecules*, 2009, **42**, 6245–6252.
37. J. Bastide and S. J. Candau, in *The Physical Properties of Polymer Gels*, ed. J. P. Cohen Addad, Wiley, New York, 1996, p. 143.
38. E. Mendes, A. Hakiki, J. Herz, F. Boué and J. Bastide, *Macromolecules*, 2004, **37**, 2643–2649.
39. S. K. Sukumaran, G. Beaucage, J. E. Mark and B. Viers, *Eur. Phys. J. E*, 2005, **18**, 29–36.
40. J. Bastide and L. Leibler, *Macromolecules*, 1988, **21**, 2647.
41. E. J. Mendes, P. Lindner, M. Buzier, F. Boué and J. Bastide, *Phys. Rev. Lett.*, 1991, **66**, 1595.
42. E. Mendes, B. Girard, C. Picot, M. Buzier, F. Boué and J. Bastide, *Macromolecules*, 1993, **26**, 6873–6877.

43. E. Mendes, R. Oeser, C. Hayes, F. Boué and J. Bastide, *Macromolecules*, 1996, **29**, 5574.
44. M. Shibayama, F. Ikkai, Y. Shiwa and Y. Rabin, *J. Chem. Phys.*, 1997, **107**, 5227–5235.
45. M. Shibayama, Y. Shirotani and Y. Shiwa, *J. Chem. Phys.*, 2000, **112**, 442–449.
46. B. Hammouda, D. Ho and S. Kline, *Macromolecules*, 2002, **35**, 8578–8585.
47. P. Zhou and W. Brown, *Macromolecules*, 1990, **23**, 1131–1139.
48. B. Hammouda, D. L. Ho and S. Kline, *Macromolecules*, 2004, **37**, 6932–6937.
49. W. F. Polik and W. Burchard, *Macromolecules*, 1983, **16**, 978–982.
50. H. Asai, K. Fujii, T. Ueki, T. Sakai, U. Chung, M. Watanabe, Y. S. Han, T. H. Kim and M. Shibayama, *Macromolecules*, 2012, **45**, 3902–3909.
51. M. Shibayama, *Soft Matter*, 2012, **8**, 8030–8038.
52. M. Beltzung, C. Picot and J. Herz, *Macromolecules*, 1984, **17**, 663.
53. F. Boué, J. Bastide and M. Buzier, in *Molecular Basis of Polymer Networks*, ed. A. Baumgartner and C. E. Picot, Springer, Berlin, 1989, pp. 65–81.
54. H. Benoit, D. Decker, C. Duplessix, C. Picot and P. Rempp, *J. Polym. Sci., Polym. Phys. Ed.*, 1976, **14**, 2199–2128.
55. R. Ullman, *Macromolecules*, 1982, **15**, 582–588.
56. R. Ullman, *Macromolecules*, 1982, **15**, 1395–1402.
57. H. M. Tsay and R. Ullman, *Macromolecules*, 1988, **21**, 2963–2972.
58. T. Matsunaga, H. Asai, Y. Akagi, T. Sakai, U. Chung and M. Shibayama, *Macromolecules*, 2011, **44**, 1203–1210.
59. P. J. Flory, *Principles of Polymer Chemistry*, Cornell University Press, Ithaca, NY, 1953.
60. W. Kuhn, *Naturwissenschaften*, 1936, **24**, 346.
61. H. M. James and E. Guth, *J. Chem. Phys.*, 1943, **10**, 455–481.
62. D. R. Miller and C. W. Macosko, *Macromolecules*, 1976, **9**, 206–211.
63. Y. Akagi, T. Matsunaga, M. Shibayama, U. Chung and T. Sakai, *Macromolecules*, 2010, **43**, 488–493.
64. V. G. Vasiliev, L. Z. Rogovina and G. L. Slonimsky, *Polymer*, 1985, **26**, 1667–1676.
65. J. D. Ferry, *Viscoelastic Properties of Polymers*, Wiley, New York, 3rd. edn., 1980.
66. A. Harada, J. Li and M. Kamachi, *Macromolecules*, 1993, **26**, 5698–5703.
67. B. Vorselaars, A. V. Lyulin and M. A. J. Michels, *Macromolecules*, 2009, **42**, 5829–5842.
68. R. D. Groot, A. Bot and W. G. M. Agterof, *J. Chem. Phys.*, 1996, **104**, 9202–9219.
69. J. S. Bergstrom and M. C. Boyce, *J. Mech. Phys. Solids*, 1998, **46**, 931–954.
70. C. Miehe and J. Keck, *J. Mech. Phys. Solids*, 2000, **48**, 323–365.
71. C. O. Horgan, R. W. Ogden and G. Saccomandi, *Science*, 2004, **460**, 1737–1754.

72. S. F. Edwards, H. Takano and E. M. Terentjev, *J. Chem. Phys.*, 2000, **113**, 5531–5538.
73. M. A. Villar and E. M. Valles, *Macromolecules*, 1996, **29**, 4081–4089.
74. L. E. Roth, D. A. Vega, E. A. Valles and M. A. Villar, *Polymer*, 2004, **45**, 5923–5931.
75. M. Gradzielski, M. Muller, M. Bergmeier, H. Hoffmann and E. Hoinkis, *J. Phys. Chem. B*, 1999, **103**, 1416–1424.
76. K. Urayama, R. Yokoyama and S. Kohjiya, *Macromolecules*, 2001, **34**, 4513–4518.
77. H. Takahashi, Y. Ishimuro and H. Watanabe, *Nihon Reoroji Gakkaishi*, 2006, **34**, 135–145.
78. M. Gradzielski, H. Hoffmann and G. Oetter, *Colloid Polym. Sci.*, 1990, **268**, 167–178.
79. C. C. Lin and A. T. Metters, *Adv. Drug Delivery Rev.*, 2006, **58**, 1379–1408.
80. B. V. Slaughter, S. S. Khurshid, O. Z. Fisher, A. Khademhosseini and N. A. Peppas, *Adv. Mater.*, 2009, **21**, 3307–3329.
81. J. Jagur-Grodzinski, *Polym. Adv. Technol.*, 2010, **21**, 27–47.
82. X. Li, Y. Tsutsui, T. Matsunaga, M. Shibayama, U. Chung and T. Sakai, *Macromolecules*, 2011, **44**, 3567–3571.
83. A. G. Mikos and N. A. Peppas, *Biomaterials*, 1988, **9**, 419–423.
84. A. Metters and J. Hubbell, *Biomacromolecules*, 2005, **6**, 290–301.
85. A. T. Metters, K. S. Anseth and C. N. Bowman, *J. Phys. Chem. B*, 2001, **105**, 8069–8076.

CHAPTER 3
Supramolecular Hydrogels

JESÚS DEL BARRIO, ERIC A. APPEL, XIAN JUN LOH
AND OREN A. SCHERMAN*

Melville Laboratory for Polymer Synthesis, Department of Chemistry, University of Cambridge, Cambridge CB2 1EW, UK
*Email: oas23@cam.ac.uk

3.1 Introduction

Polymer gels consist of three-dimensional, percolated assemblies of cross-linked macromolecules which can absorb large amounts of solvent, typically through surface tension and capillary forces.[1] They have defined a growing research area for over a century, in part due to their wide range of applications, including matrices in analytical chemistry and biology, superabsorbers and media for storage and delivery of active substances in biomedicine.[2–7] Hydrogels are a particular class of polymer networks that can absorb substantial amounts of aqueous solutions and simultaneously maintain their structural integrity. Synthetic hydrogels were first introduced in the late 1950s by Wichterle and Lim, who proposed the use of hydrophilic polymer networks of poly(2-hydroxyethyl methacrylate) in contact lenses.[8] Although today's contact lens materials constitute a minute application within the larger domain of biomaterials, the pioneering investigations of Wichterle and Lim ignited a whole wave of investigations on hydrogels with improved capabilities, such as enhanced cell–material interactions, tunable mechanical properties and functionality. The high water content of hydrogels leads to soft and elastic materials with similar physical properties to those of human tissues and results in minimal irritation of surrounding tissue and good biocompatibility.[9,10]

Monographs in Supramolecular Chemistry No. 11
Polymeric and Self Assembled Hydrogels: From Fundamental Understanding to Applications
Edited by Xian Jun Loh and Oren A. Scherman
© The Royal Society of Chemistry 2013
Published by the Royal Society of Chemistry, www.rsc.org

Hence, hydrogels have recently shown the most promising applications in biomedical related research fields, and more specifically in the controlled delivery of drugs and tissue engineering.

Hydrogels can be classified according to the type of cross-linking within the gel, which can be either chemical or physical in nature.[11] Chemically cross-linked hydrogels consist of polymer chains interconnected by permanent non-reversible bonds. They have been fabricated using numerous covalent reactions, including Michael-type addition,[12,13] Schiff base formation,[14,15] Photopolymerisation of thiol and enes,[16] free radical chain growth photopolymerization,[17] enzyme-catalysed reactions[18,19] and 1,3-dipolar cycloaddition between azides and alkynes.[20] Chemical cross-linking can be easily tuned in order to adequately alter the mechanical properties of the final material and has been commonly utilized when tough and stable hydrogels are required.[21] However, the suitability of these systems is limited by the utilization of metal catalysts, photoinitiators, ultraviolet light to induce gelation, or the incomplete conversion of reactive functional groups which may create biocompatibility concerns.[22] Additionally, structural limitations caused by network defects and problematic large equilibrium volume swelling have also hampered *in vivo* material performance. In contrast, physical gelation relies on transient cross-links between polymer chains arising from either polymer entanglements or non-covalent non-specific interactions such as hydrophobic interactions, hydrogen bonding and electrostatic interactions.[23,24] Physical cross-linking often leads to weaker gels more susceptible to shearing by mechanical forces. Nevertheless, their dynamic nature can be regarded as an advantageous characteristic as it is the basis for both shear-thinning (viscous flow under shear stress) and self-healing (time-dependent recovery upon relaxation) properties, two desirable characteristics of injectable materials.[25–27] Additionally, physical hydrogels formed from aqueous media do not require the addition of organic solvents or cross-linking reagents and their transition from solution (sol) to gel occurs without a significant volume change.

Recently, polymeric hydrogels have emerged as highly promising scaffolds to reconstitute artificial 3D environments that mimic the physical properties of the extracellular matrix through appropriate design.[28–33] Such materials do not require surgical or implantation procedures as they can be preformed *ex vivo* and then delivered *in vivo* by applying shear stress when injected through a syringe. Apart from injectability and self-healing properties, these materials must meet several other criteria. Their precursors should be miscible with bioactive molecules, including drugs, proteins, DNA or cells prior to gelation or injection, so that the gels can act as drug carriers for localized delivery or as scaffolds for tissue regeneration. They should be biocompatible and biode-gradable, and their rate of degradation should be tuned towards specific tissue development. Additionally, the hydrogel must be customized in order to ade-quately interact with the specific type of cells to be encapsulated and/or delivered and to promote cell migration, differentiation and proliferation. However, most physical polymeric hydrogels, such as biopolymers or poly-electrolytes, are cross-linked in a non-specific manner, thus hindering the tweaking of their physical properties. In this sense, "dynamic chemistries",

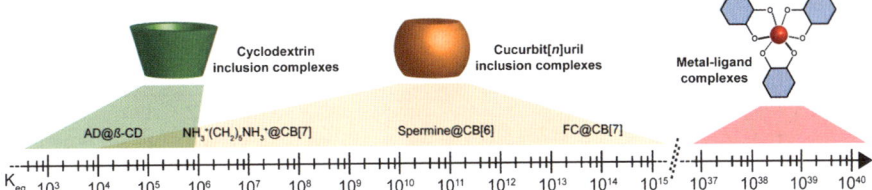

Figure 3.1 Some selected examples of supramolecular motifs, namely cyclodextrin (*left*) and cucurbituril (*middle*) inclusion complexes together with catechol–iron complexes (*right*), which have been utilized in the preparation of polymer hydrogels and their typical association constant ranges.

where a specific bond can be broken and reformed under a certain environmental stimulus, have emerged as an attractive and increasingly viable option for the preparation of a broad variety of materials, including supramolecular polymeric hydrogels. These networks rely on highly specific non-covalent binding motifs which tie the polymer chains together. The resultant hydrogels combine the characteristics of both chemical and physical networks and their mechanical properties can be readily tuned by modifying the cross-linking density and composition of the polymeric material. Typical supramolecular binding motifs capable of operating in aqueous media that have been employed in the preparation of non-covalent polymer hydrogels include host–guest, ionic, metal–ligand and ''biomimetic'' interactions, as well as hydrogen bonding and formation of stereo complexes (Figure 3.1). These specific-binding motifs have a wide range of binding strengths. This allows researchers to unlock a toolbox for the design and manipulation of the properties of the hydrogels. For example, the binding strength of cyclodextrins with different guests can be tuned over three orders of magnitude, whereas for cucurbiturils the specific host–guest interactions can be tuned over an impressive nine orders of magnitude! The dynamic and specific interaction between two communicating motifs can be exploited for the design of a wide variety of hydrogel materials.[34]

Within the topic of supramolecular gels, we summarise here the most important and recent results dealing with aqueous transient polymeric networks cross-linked through non-covalent interactions, with an emphasis on those based on host–guest inclusion complexes. The design, synthesis and physical properties of hydrogels formed by low molecular weight gelators have already been discussed in excellent reviews[35,36] and will not be treated in this chapter.

3.2 Supramolecular Polymeric Hydrogels Based On Host–Guest Inclusion Complexes

3.2.1 Hydrogels from Cyclodextrins

Cyclodextrins (CDs), also known as cycloamyloses, cyclomaltoses or Schardinger dextrins, are cyclic oligosaccharides composed of D-glucose

(a)   **(b)**

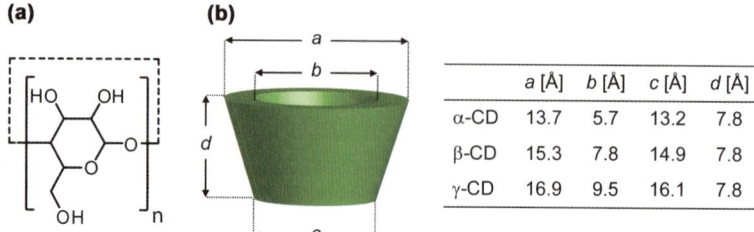

	a [Å]	b [Å]	c [Å]	d [Å]
α-CD	13.7	5.7	13.2	7.8
β-CD	15.3	7.8	14.9	7.8
γ-CD	16.9	9.5	16.1	7.8

Figure 3.2 (a) Schematic chemical structure of a cyclodextrin molecule and (b) dimensions for α-CD ($n = 6$), β-CD ($n = 7$) and γ-CD ($n = 8$).

repeating units coupled through α-1,4-glucosidic linkages (Figure 3.2a).[37,38] Commonly used CDs are α-, β- and γ-CDs, which are composed of six, seven and eight D-glucose repeating units, respectively, although larger CDs have been reported as well. Their 3D structure can be represented as a truncated cone with the primary hydroxyl groups on the smaller cone rim exposed to the solvent (Figure 3.2b). This particular arrangement makes the interior of the macrocycle less hydrophilic relative to the aqueous media and favours the hosting of hydrophobic molecules. Thus the main driving forces behind the formation of CD host–guest inclusion complexes are hydrophobic and van der Waals inter-actions, although other factors also play a role and include the release of CD ring strain, changes in solvent–surface tensions and hydrogen bonding with CD's hydroxyl groups.[39–41] The formation of the inclusion complex alters the physical and chemical properties of the guest molecule, which normally shows enhanced water solubility. Owing to their low price, good availability and the capability of forming inclusion complexes with high water solubility, CDs have proven to be very useful compounds in a wide range of areas, including analytical science, pharmacy, improved separation techniques and catalysis, as well as in the food, textile and cosmetic industries.[38,42–49]

The inclusion complex formation capability of CDs has only recently been utilized as a non-covalent binding motif for the development of a wide variety of dynamic polymeric networks and assemblies in aqueous media.[50,51] These polymeric systems have been frequently investigated in terms of pharmaceutical and biomedical applications, including sustained and targeted release of bioactive substances, biocompatible scaffolds for tissue engineering and med-ical diagnostics. The following section is an overview of the different polymeric systems that incorporate CDs in order to induce gelation. From a topological point of view it is possible to differentiate two families of non-covalent CD polymeric hydrogels: poly(pseudorotaxane) hydrogels containing CDs threaded onto one or two polymer chains and hydrogels in which the polymer chains are held together by host–guest inclusion between CDs and small organic molecule guests. Research in supramolecular CD polymeric hydrogels has been broadly developed since the 1990s, and since it has been reviewed on several occasions[52–55] the following two sections will only cover a few selected investigations on CD polymeric hydrogels.

3.2.1.1 Supramolecular Hydrogels Based On Cyclodextrin– Polymer Poly(pseudorotaxanes)

One group of supramolecular polymeric hydrogels is based on the self-assembly between CDs and linear polymers to form inclusion complexes in aqueous media (Figure 3.3a). Harada and Li reported in the 1990s the threading of CDs by linear polymers, including poly(ethylene glycol) (PEG), poly(propylene glycol) (PPG) and poly(methyl vinyl ether), to form poly(pseudorotaxane) inclusion complexes.[56–59] The enthalpy-favoured threading process of CDs onto polymer chains in polar solvents is driven by van der Waals and hydrophobic interactions, as well as hydrogen bonding between neighbouring threaded CD units.[60] This process is also dependent on the dimensions of the CDs and the cross-sectional areas of the polymer chains. For instance, the cross-sectional area of PPG is too large for it to form an inclusion complex with α-CD, but it easily penetrates the cavities of β- and γ-CDs.[56,61,62] Moreover, PEG, with a smaller cross-sectional area than PPG, readily threads α-CD but not the larger γ-CD.[63,64] In general, the use of relatively hydrophobic polymers results in complexes with higher thermodynamic stability. However, it is also associated with slow CD threading kinetics and hampers the water solubility of the final material. Therefore, most CD poly(pseudorotaxane) complexes for hydrogel formation are composed of block copolymers with hydrophilic and hydrophobic segments and show both moderate threading kinetics and enhanced water solubility.

Physical hydrogels from CD poly(pseudorotaxane)s rely on the formation of crystalline domains of tightly packed threaded CDs through hydrogen bonding interactions (Figure 3.4).[68] In general, these gels composed of PEG homopolymers or PEG-containing polymers and α-CD show the characteristic thixotropic behaviour of supramolecular hydrogels. Their preparation, properties and applications have been described in the past.[60,65–73] Therefore, we will only highlight here some representative examples and recent investigations

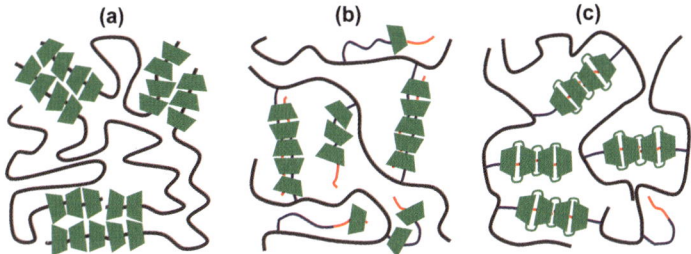

Figure 3.3   Schematic representations of some examples of cyclodextrin – polymer poly(pseudorotaxane) hydrogels: (a) Typical inclusion complexes from linear polymers and CDs; (b) dextran-g-(PEI-PEG) copolymers – gamma-CD system (dextran, PEI and PEG polymer blocks are represented by grey, blue and red lines, respectively) and (c) poly(ethylene oxide) monocetyl ether-g-dextran – CD nanotubes system (dextran, PEI and PEG polymer blocks are represented by grey, blue and red lines, respectively).

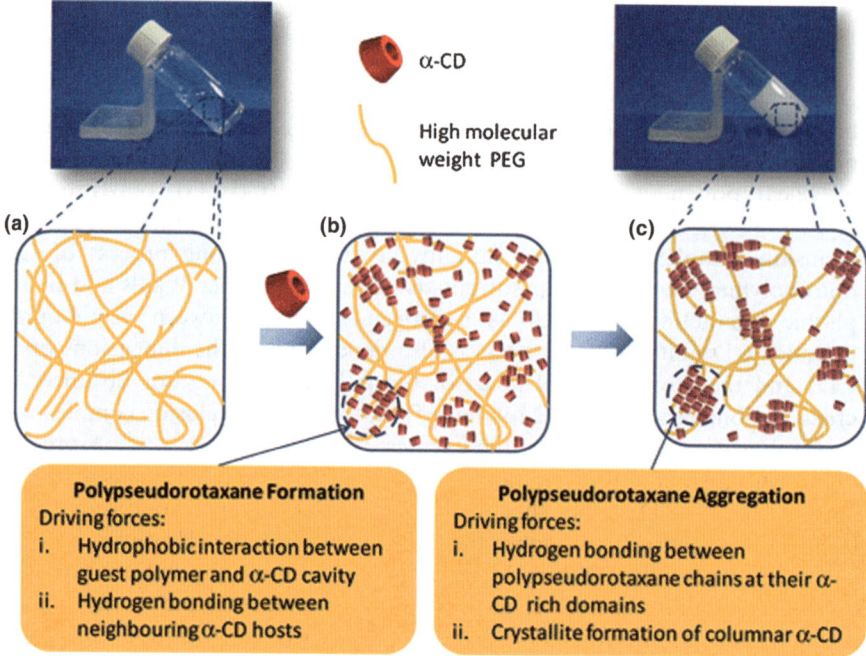

Figure 3.4 Formation of poly(pseudorotaxane) hydrogel from α-CD and high molecular weight PEG. (a) α-CD is added to a clear PEG aqueous solution, then (b) the host threads onto the polymer to form a poly(pseudorotaxane), which eventually (c) produces a transient hydrogel due to poly(pseudorotaxane) aggregation.
Reproduced from ref. 52 [K. L. Liu, Z. Zhang and J. Li, *Soft Matter*, 2011, 7, 11290–11297.]

on this fascinating type of hydrogel. Relatively unstable hydrogels have been prepared from α-CD and linear, hyperbranched PEG and PEG-grafted polysaccharides including dextran, hyaluronic acid and chitosan.[74–76] High molecular weight ($>10\,kDa$) PEG is required for hydrogel formation as lower molecular weights only lead to precipitation of α-CD. However, when considering biomedical applications, low molecular weight polymers are preferred as they can be readily excreted by renal filtration.[68,77] More stable hydrogels can be obtained by mixing α-CD and amphiphilic block copolymers composed of poly[(R)-3-hydroxybutyrate] (PHB), PPG or poly(ε-caprolactone) (PCL) as the hydrophobic block and PEG as the hydrophilic block. In these systems, hydrogel formation is supported by crystalline α-CD/PEG domains as well as aggregation of the hydrophobic polymer blocks.[78–82] Moreover, some of the polymers utilized as the hydrophobic segment are biodegradable, a highly desired quality when designing stable and biocompatible hydrogels.[67–69,83]

Most of the investigations dealing with CD poly(pseudorotaxane) hydrogels in drug delivery are limited to *in vitro* evaluation of release rates using model drugs such as fluorescein isothiocyanate labelled dextran (dextran-FITC) and FITC-labelled bovine serum albumin (BSA-FITC). Li and co-workers have investigated the dependence of the dextran-FITC release rate from α-CD/PEG hydrogels with PEG molecular weight. The authors showed that the rate increases with increasing molecular weight from 8 kDa up to 35 kDa and stays unchanged between 35 and 100 kDa.[84] Nevertheless, hydrogels composed of low molecular weight PEG showed very rapid release kinetics for drug delivery applications, whereas high molecular weight PEG may create biocompatibility concerns due to its difficulty to be excreted from the body. Supramolecular hydrogels were also made with α-CD/PEG-*b*-PPG-*b*-PEG triblock copolymer.[81] Over one month, sustained FITC-dextran release was observed for hydrogels of α-CD and a biodegradable PEG-*b*-PHB-*b*-PEG triblock copolymer.[79] These triblock-based hydrogels are promising materials for relatively long-term sustained controlled delivery of drugs as an injectable formulation. Additionally, their properties can be finely tuned by modifying the composition, molecular and chemical structures of the copolymer.

"Smart" polymers have shown the ability to penetrate the cavity of CDs, thus allowing for binding modulation and the development of stimuli-responsive hydrogels. For example, Wenz and co-workers demonstrated that α-CDs can be threaded onto linear poly(ethyleneimine) (PEI) chains at low pH with a binding affinity largely affected by the number of protonated amino groups along the polymeric backbone.[85,86] Yui and co-workers showed later that the ratio of complexation between PEI repeating units and α-CD at high pH is 2:1 but for γ-CD it is 4:1, *i.e.* only one linear PEI chain can penetrate the cavity of α-CD while γ-CD can be threaded onto two polymer chains simultaneously. Additionally, the amino groups trigger the CD dethreading when they become protonated.[87] This stimuli-responsive binding affinity was utilized in the preparation of pH-responsive γ-CD polymeric networks.[88] Dextran-*g*-(PEI-PEG) copolymers can be cross-linked by complexing PEG-PEI pendant blocks with γ-CD at both high and low pH values (Figure 3.3b). However, CDs only threaded PEG chains at low pH whereas both the PEI and the PEG blocks were complexed at high pH, which resulted in an increase in the viscoelastic properties of the polymeric network. Harada and co-workers have shown the formation of hollow CD nanotubes by cross-linking adjacent α-CD threaded onto PEG and subsequent unthreading of the guest polymer chain.[89] Yui and co-workers also showed that these CD tubes can form an inclusion complex with two amphiphilic linear chains which penetrate both ends simultaneously, and utilized this binding motif to supramolecularly cross-link poly(ethylene oxide) monocetyl ether-*g*-dextran (Figure 3.3c).[90]

3.2.1.2 Supramolecular Polymeric Hydrogels Driven By Cyclodextrin Inclusion Complexes

The formation of stable inclusion complexes between β-CD and adamantine (AD) or cholesterol (CH) derivatives ($K_a \approx 5 \times 10^4 \, M^{-1}$ and $2 \times 10^4 \, M^{-1}$ for adamantane carboxylate/β-CD and CH/2-hydroxypropyl-β-CD pairs,

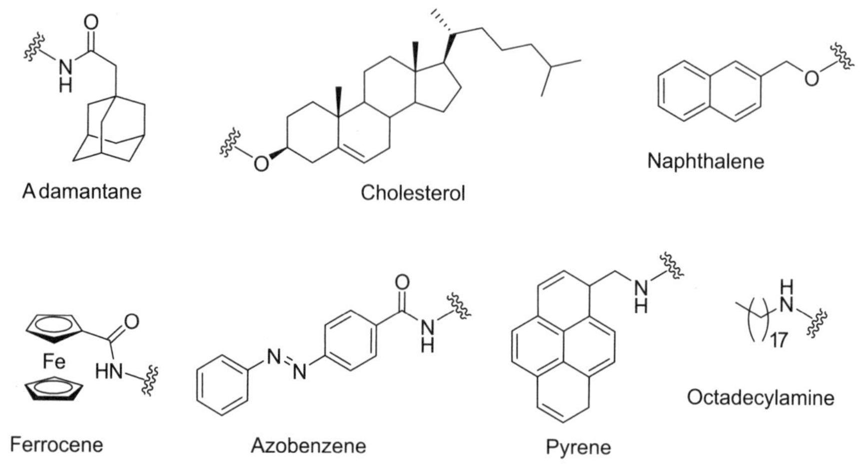

Adamantane   Cholesterol   Naphthalene

Ferrocene   Azobenzene   Pyrene   Octadecylamine

Figure 3.5   Chemical structures of some selected cyclodextrin guests utilized for hydrogel formation.

respectively)[39,41,91] has been exploited in the preparation of a wide variety supramolecular polymeric hydrogels. Besides AD and CH, other guest molecules for β-CD include linear alkyl chains, *N*-acylurea and naphthyl groups,[92–95] pH-sensitive 3-(trimethylsilyl)propionic acid derivatives,[96] light-responsive azobenzene units[97–100] and redox-responsive ferrocene (FC) units (Figure 3.5).[101]

One common approach to these types of hydrogel systems involves the mixing of linear polymers bearing complementary β-CD host and guest molecules, mainly AD derivatives (Figure 3.6a). Auzély-Velty and co-workers have functionalized PEG and naturally occurring polysaccharides, chitosan and hyaluronic acid with both β-CD and AD that form hydrogels after being mixed in aqueous solution.[102–106] In a similar fashion, Li and co-workers have prepared hydrogels from β-CD and AD-grafted poly(acrylic acid).[107] The strength and dynamic of these polymeric networks are dependent on the number of host–guest interactions per polymer chain, polymer concentration, temperature and the presence of free competitive host. The same authors have also investigated the viscosity of systems composed of β-CD containing chitosan cross-linked with bivalent AD-modified PEGs.[105] The authors found that the viscosity increased proportionally with the molecular weight of the bifunctional cross-linker and that the increase was not dependent on the inherent viscosity of the functional PEGs. This means that the effective connectivity of the system is enhanced by the use of longer linkers. Unusual viscoelastic properties were observed for a polymeric network in the semi-dilute regime which contained double-chain strands connected by "fourfold junction points" (Figure 3.6b).[106] Another more specialized system from the same group deals with networks of hyaluronic acid modified with bivalent complementary β-CD and AD binding motifs (Figure 3.6c).[106,108] In this case, the network

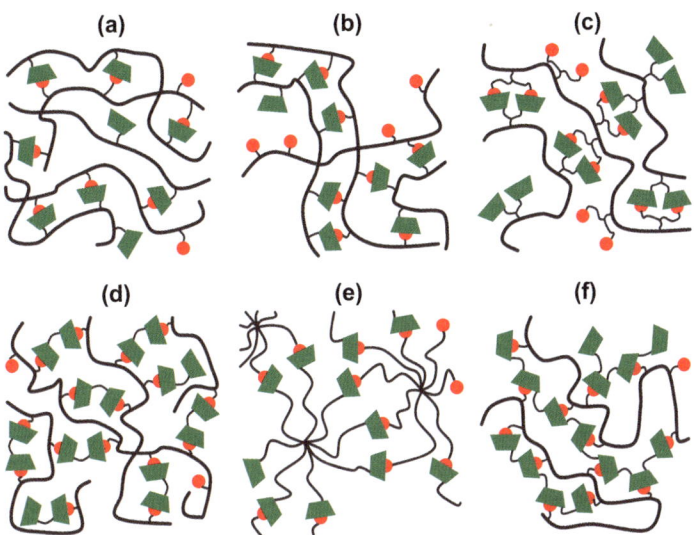

Figure 3.6 Schematic representations of some examples of polymeric hydrogels based on cyclodextrin inclusion complex cross-links. (a) Typical network based on CD- and guest-functionalised polymers, (b) double chain strands connected by fourfold junction points, (c) network based on CD and guest dimers attached to polymer chains (d) small molecule CD dimer and guest-functionalised polymer (e) complementary functional dendrimers and (f) main chain type CD polymer and guest-functionalised polymer.

consists of linear chains held together by duplex host–guest complexes, a feature that slows down the dynamics of the polymer system when compared with previous examples based on AD/β-CD single pairs.[109]

Gelation can also be efficiently induced by mixing polymer chains containing CD guest and CD dimers as cross-linkers (Figure 3.6d). Ritter and co-workers have obtained supramolecular hydrogels by the formation of inclusion complexes between AD-containing *N,N*-dimethylacrylamide or *N*-isopropylacrylamide copolymers and β-CD dimers based on a rigid spacer.[110] Sinaÿ, Sollogoub and co-workers have shown the synthesis and supramolecular properties of CD dimers in which the two primary rims are singly or doubly linked through flexible alkyl or oligo(ethylene oxide) spacers.[111,112] The binding properties of these dimers are greatly affected by their architecture and also their ability to act as effective cross-linkers.[112,113] The doubly bridged dimer with hydrophilic spacers showed more efficient cross-linking of AD-containing polymer chains in water than the singly bridge analogue. Additionally, the doubly bridged dimer with oligo(ethylene oxide) spacers showed higher solubility in water than its counterpart containing aliphatic spacers, which resulted in an enhancement of the association with AD groups.

Hennink and co-workers have developed a series of self-assembled hydrogels based on CH/β-CD inclusion complexes.[114,115] These gels were prepared by mixing PEG star polymers bearing both CH and β-CD moieties at the chain ends and subsequent hydration (Figure 3.6e). Similarly to other systems, the mechanical properties of the gels were dependent on the polymer concentration, CH/β-CD ratio, polymer molecular weight, temperature and the presence of competing guest. Storage moduli (G') as high as 30 kPa were achieved by some of these systems.[116] The authors also showed that hydrogel degradation was mainly due to surface erosion, which depends on the network swelling stresses and initial cross-link density. The degradation mechanism led to a quantitative and nearly zero-order release of entrapped proteins.[117] In a later report, the same authors showed that gelation could also be attained by mixing CH-modified eight-arm star-shaped PEG with β-CD (0.5 to 4 equivalents of β-CD per CH moiety).[118] Network formation was attributed to the formation of crystalline β-CD clusters, with some CD molecules bonded to the polymeric matrix *via* host–guest interactions. Li and co-workers have recently shown gelation from pyrene-terminated PEG star polymers through ternary γ-CD inclusion complexes: two pyrene molecules and one γ-CD.[119] This is an advantageous system when compared to β-CD hydrogels because the γ-CD ternary inclusion complexes effectively cross-link the polymer chains and β-CD dimers/β-CD polymers are not required. Additionally, gelation is fast as only the pyrene terminal chains are involved in the complexation processes.

A different approach involves the combination of polymers bearing CD guests and the so-called "mainly linear β-CD polymers" (Figure 3.6f). Although the generation of covalent networks of CDs in alkaline media by using epichlorohydrin has been known since the 1980s, the presence of toluene, which effectively binds β-CD, can lead to the formation of preferentially linear main-chain β-CD polymers.[120] While these host polymers show a relatively poorly defined structure, they are easy to synthesize and have been broadly utilized in the preparation of physical hydrogels through blending with a wide variety of water-soluble AD-containing polymers.[120–127]

Physical hydrogels, including CD hydrogels, are inherently sensitive to changes in temperature on account of the enthalpic nature of the host – guest inclusion complexes. Indeed, temperature-responsive behaviour has been demonstrated for many of the systems discussed previously; therefore only a few representative examples are discussed here. Auzély-Velty and co-workers have studied the influence of temperature on G' and G'' of aqueous mixtures of CD- and AD-containing polysaccharides.[103,106] Temperature affects the dynamics of the gel by changing not only the exchange rate and number of CD complexes but also chain mobility. Both the number of complexes as well as the longest interaction timescale decreased with increasing temperature. It was also found that the apparent activation energy, ΔH_r (ΔH_r can be considered as the potential barrier that a polymer chains must overcome to diffuse across the other chains), for some mixtures of complementary polymers were higher than that of a solution of the unmodified parent polymer (at the same concentration) and those corresponding to individual binding events (previously determined by calorimetric experiments). Similar trends were observed on hydrogels composed of multi-armed PEG polymers bearing complementary β-CD and CH units.[115]

The above-mentioned polymer networks are sensitive to external stimuli such as mechanical forces and temperature by virtue of the non-covalent nature of the cross-links. Nevertheless, when designing and fabricating new supramolecular constructs, the realization of reversibility is particularly important as it could enable these materials to be superior to more conventional ones. The host–guest supramolecular chemistry of CDs allows for controlled binding under stimuli other than heat, shearing or the presence of competing guest. Several investigations have exploited this point and have demonstrated pH, redox potential and light-responsive sol–gel transition.

3.2.1.2.1 Light-responsive CD Polymeric Hydrogels. The most common strategy uses the preparation of water-soluble polymers bearing selected guest molecules which can change their binding affinity for CD upon environmental change. Among the different investigated stimuli, light is particularly interesting as it is a remote stimulus that can be controlled spatially and temporally with great ease and convenience. More importantly, light irradiation does not necessarily have a harmful effect on the activity of most bioactive compounds. The reversible *trans–cis* photoisomerization of azobenzene derivatives is at the origin of a series of interesting phenomena with numerous potential applications, including holographic optical storage, the preparation of photomechanical actuators and molecular motors and the photocontrol of hierarchical phase-separated structures and chirality transfer processes, among others.[128–131] In general, the *trans* isomer is the most stable isomer and the *trans–cis* isomerization can be photo-induced by illumination with appropriate light in the absorption bands. *Cis–trans* isomerization can be both optically and thermally induced. Azobenzene derivatives can also bind CDs on account of their bulky and hydrophobic structure; however, the *trans*- and *cis*-azobenzene isomers exhibit different binding abilities. Both α-CD and β-CD show much higher affinity for the *trans* isomer than for the *cis* isomer. Harada and co-workers have exploited this unique feature to prepare polymeric hydrogels with a photo-triggered azobenzene-mediated sol–gel transition. Poly(acrylic acid) modified with linear dodecyl alkyl chains (PAAC12) self-assemble in water through solvophobic interactions between the aliphatic chains to afford physical gels (Figure 3.7a).[97] The addition of α-CD to the system induced a gel-to-sol transition as the CD is able to bind the aliphatic chains and completely disrupt the solvophobic interactions (Figure 3.7b). Subsequent addition of azobenzene-4,4'-dicarboxylic (AZ) acid reverts the solution to the original gel state (Figure 3.7c). The *trans*-AZ isomer can bind tightly to α-CD; it replaces the alkyl chains from the CD and triggers polymer aggregation. This ternary aqueous mixture of modified poly(acrylic acid), α-CD and AZ is therefore capable of photo-switching between the gel and the solution states on account of the photo-tunable binding affinity between the host and the guest (Figures 3.7c and 3.7d). Indeed, UV irradiation resulted in the conversion of the gel to a viscous fluid which could be reverted back by irradiation with visible light.

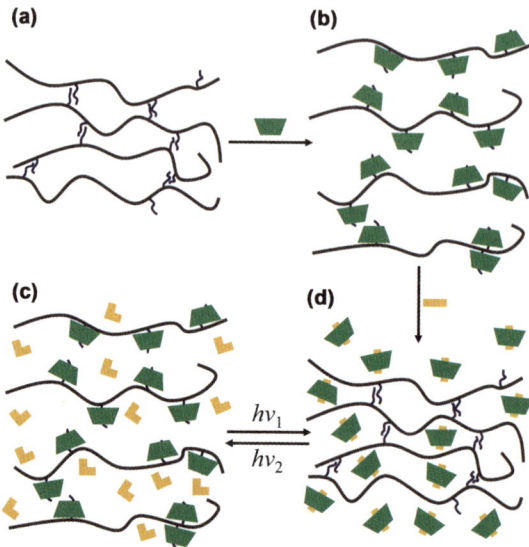

Figure 3.7   Schematic representation of a CD self-assembled hydrogel controlled by the photoresponsive complexation between an azobenzene derivative and α-CD (the polymeric backbone and the alkyl chains are represented by grey and blue lines).

The same authors have also shown drastic viscosity changes upon irradiation of mixtures of CD- and azobenzene-containing polymers in water.[98] In this study, two different α-CD-containing poly(acrylic acid)s were prepared for which the attachment of the host to the polymer backbone was done through the 3-position (p3CD) or 6-position (p6CD) of a glucose repeating unit. These polymers were able to form polymeric networks in water when mixed with photoresponsive complementary poly(acrylic acid)s containing azobenzene moieties that were linked to the main polymer chain through a dodecamethylene spacer (p12AZ). Steady-shear viscosity measurements of semi-dilute solutions of the polymer mixtures showed dramatic changes, despite the fact that both systems contain the same type of host and guest molecules. Viscosity values of the p6CD/p12AZ system were two orders of magnitude higher than those corresponding to the p3CD/p12AZ system, thus suggesting that the local binding affinities of each polymer mixture are different. NMR investigations of the two polymer networks revealed different binding interactions. The CD host molecule was predominantly interacting with the azobenzene derivative in the p3CD/p12AZ system, whereas the binding in the p6CD/p12AZ system involved mainly the α-CD/linker interaction and only partially the azobenzene moiety. Interestingly, these relatively small differences in binding significantly influenced the viscosity changes of upon irradiation. Whereas UV irradiation of the p3CD/p12AZ system induced a decrease in viscosity, that of the p6CD/p12AZ system slightly increased. The different response was related to the different interaction of the photoinduced *cis*-azobenzene moieties and the CD-polymers. The *cis* isomer does not fit into the cavity of α-CD, so the decrease in viscosity of the p3CD/p12AZ system upon UV irradiation can be directly attributed to the dissociation of the inclusion complex. However, interlocked inclusion complexes were formed in the p6CD/p12AZ system, with the azobenzene moiety completely excluded from the cavity of the host. Harada and co-workers have recently reported the fabrication of a physical polymeric hydrogel composed of both an azobenzene-containing poly(acrylic acid) and α-CD-containing bio-polymer.[99] In this study, curdlan (β-1,3 glucan) was modified with α-CD *via* click chemistry and, as the polymer possesses a rigid structure, a host moiety was introduced in every repeating unit, thus allowing for a high degree of cross-linking. The mixture of the two complementary polymers in appropriate ratios resulted in hydrogels with zero-shear viscosity values as high as 54 Pa s. UV irradiation of the hydrogel caused a decrease of its viscosity to give the sol state. The average viscosity value of the system after irradiation was 9 Pa s. In contrast, irradiation at $\lambda = 430$ nm of the sol state recovered the viscosity of the system and the hydrogel was regenerated within two minutes. The viscosity changes of the system could be repeatedly induced by using both UV and visible light. The association constant of α-CD with the *trans*-azobenzene unit ($K_a = 1100$ M^{-1}) is much larger than that with the *cis*-azobenzene ($K_a = 4.1$ M^{-1}). These results indicate that the control of the association and dissociation between the α-CD and the azo moieties by photoirradiation affected the phase transition. Jiang and co-workers have recently developed photoresponsive CD poly(pseudorotaxane) hydrogels based on azobenzene.[100] The authors demonstrated how the addition of a low molecular weight azobenzene derivative to an α-CD poly-(pseudorotaxane) hydrogel could effectively enable the control of the sol–gel transition *via* photoirradiation. In their system a 10 kDa PEG forms a hydrogel in the presence of α-CD. The hydrogel reverts into solution by adding the competitive *trans*-azobenzene derivative, which replaces PEG to form low molecular weight inclusion complexes. After UV irradiation, the poly-(pseudorotaxane) hydrogel regenerates as the *cis* isomer and has a small binding affinity for α-CD. The studies showed that the strength of the interactions follows the sequence: *trans*-azobenzene/α-CD < PEG/α-CD < *cis*-azobenzene/α-CD. Stoddart and co-workers have also obtained photoresponsive hydrogels from a two-component system of an azobenzene-containing polymer and a deoxycholic acid-modified β-CD derivative.[132] When the pendant azobenzene groups are in their *trans* state, CD can bind to them and the hydrophobic deoxycholate moieties aggregate and induce the gelation of the system. After *trans–cis* photoisomerization, the bent-shaped *cis*-azobenzene groups are expelled from the CDs, which form intramolecular inclusion complexes with the deoxy-cholate moieties and the hydrogel reverts to the sol state. Use of the erosion of the hydrogel network is a long-standing method to regulate compound release. Since the release of compounds from the hydrogel matrix can be controlled by the change of mesh size of the network, photoresponsive hydrogel systems are potentially useful for drug delivery systems with controlled release. Kros and co-workers used CD-modified dextran hydrogels for the *in vivo* release of hydro-phobic drugs.[133,134] In a recent example, a photoresponsive hydrogel composed of both a CD- and an azobenzene-modified dextran was studied as a controlled release system for proteins.[135]

3.2.1.2.2 pH-responsive CD Polymeric Hydrogels. Besides the photo-responsive systems mentioned above, other studies have reported physical hydrogels based on CDs that are responsive to pH.[96,125,136–142] Amiel and co-workers have demonstrated the formation of polymeric networks in water composed of an AD-containing PEG β-CD containing poly[(methyl vinyl ether)-*alt*-(maleic acid)] terpolymer which showed pH-sensitive behaviour due to the presence of the carboxylic acid groups.[142] The terpolymer behaved as a polyacid at pH 7 and it had neutral properties at pH 2. The mixture of the complementary polymers showed an associative phase separation sensitive to pH and the concentration of the medium. The associative phase separation at both the dilute and intermediate ranges was reversed by decreasing the pH from 7 to 2. The networks showed a much stronger dynamic modulus at pH 2 than at higher pH 7. This investigation showed that the coupling of host–guest inclusion complexation and hydrogen bonding (dimers of carboxylic acid groups and ethylene oxide repeating units bounded to carboxylic acid groups) may be an interesting concept to modulate the strength of supra-molecular polymer networks by changing the pH. The same group has showed that the viscosity of a aqueous solution containing poly(β-malic acid-*co*-β-ethyladamantyl malate) and a β-CD polymer was pH dependent.[125] The conformation of the AD-containing polymer was dependent on its degree of ionization. At low pH, the polymer adopts a compact coiled conformation with hydrophobic AD microdomains, whereas the repulsion of the ionized carboxylic acid groups induces the "stretching" of the polymer chains. Hence, the mobility and accessibility degree of the AD groups are lower at low pH, as is the probability of complexation. At pH values higher than 4 the degree of ionization of the polyester repeating units is increased, the polymer swells and the hydrophobic microdomains are progressively broken, thus enhancing the associative behaviour of the polymer chains.

Yui and co-workers have developed a system where α- or β-CD-containing poly(ε-lysine) are combined with 3-(trimethylsilyl)propionic acid (TPA) or TPA analogues.[96] The addition of TPA to an aqueous solution of the CD polymer leads to gel formation due to both the formation of host–guest complexes between α-CD and the TPA trimethylsilyl group and ionic interaction between the negatively charged carboxylic acid group of TPA and the positively charged amines of the poly(ε-lysine).[136] The same effect was observed when β-CD-containing poly(ε-lysine) was used instead of the α-CD poly(ε-lysine).[137–139] TPA can be bound to more than one α-CD simultaneously, whereas β-CD is large enough to form 1-to-1 inclusion complexes exclusively, which suggests that the gel formation is mainly the result of ionic interactions between the amino groups of the polymeric backbone and the carboxylic acid groups of the TPA. The combination of both ionic and host–guest interactions makes the system responsive to temperature as well as pH. The amino groups of the poly(ε-lysine) are positively charged at neutral pH and the ionic interaction with the CD-bound TPA holds together the polymeric network. At pH > 8, repulsive ionic interactions between the negatively charged carboxylic acid groups of the TPA and the polymer chains induces gel disassembly. Increasing the temperature also makes the polymeric network to fall apart due to the dissociation of the TPA/β-CD inclusion complexes. Additionally, the sensitivity of the system towards pH, temperature and ionic strength could be finely tuned by replacing TPA with other derivatives showing different affinity for β-CD, such as tri-methylacetic acid, *tert*-butylacetic acid or trimethylhexanoic acid.[140]

3.2.1.2.3 Redox-responsive CD Polymeric Hydrogels.

CDs can also form inclusion complexes with a variety of metallocenes and other types of redox-responsive derivatives, depending on the size of the CD cavity.[143–149] Ferrocene (FC) and its derivatives are among the most studied guest molecules in electrochemically driven molecular complexes.[149–151] In 1985, Osa, Evans and co-workers demonstrated that ferrocenecarboxylic acid is effectively bound by β-CD whereas its oxidized form is not.[152] Several groups have also investigated a number of FC derivatives and found similar results, *i.e.* one-electron oxidation greatly diminishes the stability of the inclusion complex between FC and β-CD.[143,153–155] FC/CD redox-responsive systems have been intensively studied for the past two decades; however, recently, redox-responsive polymer–CD systems have gained increasing attention. Harada and co-workers expanded their system composed of PAAC12 and CD to include a redox-responsive CD guest, ferrocenecarboxylic acid.[101] Similar to the photoresponsive system discussed previously, aqueous solutions of PAAC12 exhibited high viscosity and gel-like behaviour due to hydrophobic interactions between the alkyl chains. With the addition of β-CD, the viscosity of the system decreased due to the complexation of the alkyl side chains. When ferrocenecarboxylic acid was added to the aqueous solution, the system recovered its gel behaviour as CD preferentially binds to the FC derivative, thus releasing the alkyl chains. Finally, ferrocene-carboxylic acid was oxidized with sodium hypochlorite and the system underwent a gel-to-sol transition as the alkyl chains replaced the oxidized FC molecules from the cavity of β-CD. Although cyclic conversion between the sol and the gel states was not achieved due to the utilization of chemical redox stimulus, this type of responsive polymeric system could be interesting for the implementation of gel actuators controlled by electric potential.

Harada and co-workers have also shown recently that redox-responsive self-healing materials can be created based on FC/CD host–guest interactions.[156] In this investigation, a transparent supramolecular hydrogel was formed by mixing poly(acrylic acid) modified with β-CD as a host polymer with FC-containing poly(acrylic acid). Both chemical and electrochemical redox stimuli were found to induce a sol–gel phase transition in the supramolecular hydrogel (a ~35% decrease in the storage elastic modulus value was observed after electrochemical reduction, which was fully recovered after heating at 50 °C). One of the hydrogel systems (2% in weight) was subjected to small (0.1%) and large (200%) strains and the rheological properties were measured. Under a 0.1% strain, G' was larger than G'', indicating that the system formed a self-standing hydrogel. However, the G' and G'' values were inverted under 200% strain, indicating that the hydrogel was converted into the sol state, presumably on account of the breaking of the FC/CD host–guest interactions. When placed under 0.1% strain, G' and G'' returned to their original values. G' values were recovered to 90% of the initial state in 20 seconds, and hydrogel formation was observed in 500 seconds. The self-healing ability of the system was shown to work in the macroscopic re-adhesion of two cut surfaces of a hydrogel sample, which could be controlled by redox reactions. Recently, Harada and co-workers have demonstrated that well-defined molecular recognition events can be used to direct the reversible assembly of macroscopic gel blocks.[157] The authors demonstrated the synthesis of covalently cross-linked acrylamide-based gels bearing CDs and small guest moieties. Pieces of host and guest gels were shown to adhere to one another through the mutual molecular recognition between CDs and guest molecules at the gel surface. By changing the size and shape of the host and guest units, different gel blocks can be selectively assembled and sorted into distinct macroscopic structures that are of the order of millimetres to centimetres in size.

3.2.2   Hydrogels from Cucurbiturils

Cucurbit[*n*]urils (CB[*n*], $n = 5$–8, 10), named after the genus *Cucurbita* (genus in the gourd family *Cucurbitaceae*) because of its resemblance, are macrocyclic oligomers based on repeating monomer units of glycoluril (Figure 3.8a). They are prepared by an acid-catalysed condensation reaction of formaldehyde and glycoluril, a process first reported in 1905 by Behrend and co-workers.[158] The substance formed during this reaction (referred to as Behrend's polymer in the literature) was insoluble in all common solvents but could be recrystallized from sulfuric acid. Although Behrend and co-workers were not able to structurally characterize this substance, they did show its ability to form co-crystals with a variety of substances, including sodium permanganate, silver nitrate and Congo red. In 1981, Mock reported that the product of Behrend's reaction was

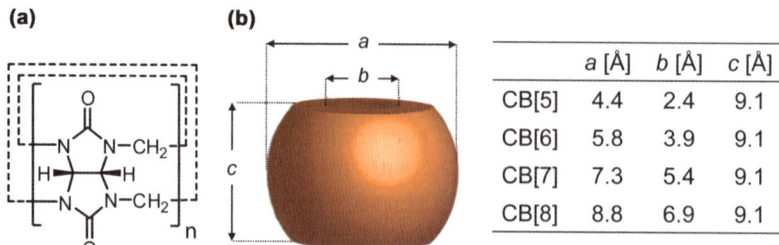

	a [Å]	b [Å]	c [Å]
CB[5]	4.4	2.4	9.1
CB[6]	5.8	3.9	9.1
CB[7]	7.3	5.4	9.1
CB[8]	8.8	6.9	9.1

Figure 3.8   Chemical structure of (a) a cucurbituril molecule and (b) dimensions of CB[5]–CB[8].

the cyclic hexameric CB[6], comprising six equivalents of glycoluril and 12 methylene linkages.[159] It was not until nearly 20 years later, when this reaction was conducted under milder conditions, by the research groups of Kim and Day, that CB[5]–CB[8] and CB[5]@CB[10] were isolated.[160–162]

The common characteristic structural features of CB[*n*]s are the hydrophobic cavity and the polar carbonyl groups surrounding the portals. They have a cavity width ranging from 4.4 to 8.8 Å (for CB[*n*], *n* = 5–8) and a portal diameter of 2.4–6.9 Å (Figure 3.8b). This corresponds to a cavity volume range of 82–479 Å^3, which results in discernible features in the host–guest chemistry of the different CB[*n*] homologues.[163] The carbonyl-laced portals are weakly basic, with CB[6]'s conjugate acid having a pK_a of 3.02. The water solubility of CB[n]s varies across the family in an odd-even fashion -CB[6] and CB[8] form stronger intermolecular interactions in the solid state than CB[5] and CB[7], and consequently are less soluble then the odd-membered homologues.[164] CB[5] and CB[7] show relatively high solubilities of 20–30 mM in neutral water, whereas CB[6] and CB[8] have much lower solubilities of 0.018 and <0.01 mM, respectively.[163] All CB[*n*]s are soluble in acidic water, as well as in aqueous solutions of alkali metals on account of protonation or coordination of the metal ions to the carbonyl oxygen atoms. The solubility of CB[*n*]s in common organic solvents is less than 10^{-5} M, and therefore the host–guest chemistry of CB[*n*]s has mainly been studied in aqueous media.

There have been few examples of singly or multiply substituted CBs being synthesized in order to modify their solubility to a range of solvents as well as to increase their aqueous solubility. An additional potential gain is to provide a route to more complex structures.[165–169] Substitution on a CB at this point in their development is defined as where the methine carbon of the glycoluril moiety (at the CB equator) bears a group, which can be either an alkyl group or an oxy group. CB homologues can be synthesized through the condensation of any suitably substituted glycoluril bearing a group(s) at either or both carbons of the fuse junction of the two imadazolone rings. This approach has met with limited success, as the predominant homologues are CB[5]s with only small amounts of CB[6]s. The alternative method for the introduction of substitution is through direct oxidation of the regular CB to introduce an oxy group. In 2003, Kim *et al.* reported an oxidative method to introduce hydroxyl moieties at the equatorial positions of CB[6],[168] which were used in a range of supramolecular systems from synthetic pores to surface-modified vesicles.[170] Although multivalent CB[6] has great potential, devising a straightforward procedure to introduce a single point of chemical attachment on the parent CB molecule would guarantee a high level of control over molecular structures and topologies on the nanoscale. The synthesis of monohydroxylated CB[6] from regular CB[6] in aqueous solution using host–guest interactions to control the oxidation mechanism has been recently reported.[171] Alternatively, a methylene-bridged glycoluril hexamer can be prepared, which can form mono-functionalized cucurbituril derivatives in a subsequent step.[172]

Several concerted intermolecular interactions promote the binding of guests by CB[*n*]s. On the one hand, hydrophobic effects are present in a similar way to

CDs. This contributing effect results from the interplay between the release of "high-energy" water molecules upon inclusion of nonpolar organic residues and concomitant differential dispersion interactions inside the cavity and in bulk water. Ion–dipole interactions between metal or organic cations and the ureido carbonyl rims come into play, while H-bonding interactions prevail less frequently. The preference of CB[n]s to bind cationic species and their reluctance to bind complex anions can be rationalized by the negative electrostatic potential of the carbonyl rims and the inner cavity.[173,174] This feature clearly differentiates CB[n]s from CDs, which preferentially bind to neutral or anionic guests. Higher order CB[n]s also exhibit a degree of flexibility in their binding and they are known to create an induced fit, constrictively binding sterically bulky guests.[175,176] Variation of the number of repeating units and therefore the sizes of both the inner cavity and the portals results in different molecular recognition properties with the CB[n] family. The smallest CB[5] and its homologues (dimethylglycoluril and cyclohexanoglycoluril) can encapsulate a variety of gas molecules, including krypton, xenon, nitrogen, oxygen, argon, nitrous oxide, nitric oxide, carbon monoxide, carbon dioxide, methane, ethylene and ethane, as well as small solvent molecules such as methanol and acetonitrile inside its cavity.[165,167,177–180] They can also bind simultaneously two cations (alkali, alkali earth, NH_4^+, Pb^{2+}, Cd^{2+}) through electrostatic interactions with the ureido carbonyl groups, fully occupying the two portals.[181,182] CB[6] can not only bind alkali metal, alkaline earth, transition metal and lanthanide cations, and several gas molecules, but also a large variety of positively charged and neutral organic guests on account of its larger inner cavity.[163,183–188] CB[6] forms remarkably stable complexes with protonated aminoalkanes, as the hydrophobic oligomethylene chain can partially fill the cavity while the ammonium cation can be simultaneously bound to one of the portals through strong ion–dipole interactions. In particular, CB[6] tightly binds protonated polyamines such as 1,6-diaminohexane or spermine, yielding highly stable 1:1 host–guest complexes (K_a up to 10^{12} M^{-1}). CB[6] also forms moderately stable complexes with protonated bulkier amines such as the 4-methylbenzylammonium ion.[189] CB[7] forms strong 1:1 complexes with positively charged amphiphilic guests, including AD, FC, *p*-xylylene and trimethylsilyl derivatives containing one or two amino groups, as well as viologen derivatives.[190,191] CB[7] exhibits association constants for some selected guests that reach and even surpass that of avidin–biotin interaction ($K_a \approx 10^{15}$ M^{-1}) and represent the strongest non-covalent interactions for a synthetic system.

In 2005, Isaacs and co-workers reported the association constant of the rimantadine/CB[7] complex to be in the order of 10^{12} M^{-1}.[176] Two years later, Kaifer and co-workers reported a $K_a = 3.0 \times 10^{15}$ M^{-1} for the 1,1'-bis-(trimethylammoniomethyl)ferrocene/CB[7] pair.[191] The groups of Kim, Inoue and Gilson reported the record-breaking affinity of $K_a = 5.0 \times 10^{15}$ for the 1-(2-aminoethylamino)adamantine/CB[7].[190] The extremely high affinities of these complexes, with rigid, near-perfect complementary structures, are traceable to a large enthalpic gain, originating from the tight fit of the FC or AD core to the rigid CB cavity, assisted by the entropic gain arising from the dehydration of the CB portals. CB[7] has also been found to interact with diphenylmethane, triphenylmethane and triphenylpyrylium carbocations ($K_a = 2.0 \times 10^5$ to 7.5×10^5) as well as several radicals and neutral molecules (*e.g.* FC, cobaltocene and carborane).[192–194] CB[8] also displays remarkable binding affinities towards positively charged and relatively large guests such as AD derivatives, cyclen and cyclam macrocyles (as well as their doubly charged Cu and Zn complexes) and long alkylammonium aliphatic chains. In contrast to CB[5]–CB[7], the cavity of CB[8] is large enough to accommodate two organic guests simultaneously, thus forming highly stable ternary complexes. In 2001, Kim and co-workers reported the formation of a stable ternary complex between CB[8] and two doubly charged 2,6-bis(4,5-dihydro-1*H*-imidazol-2-yl)naphthalene molecules.[160] This group has also reported the selective formation of a 1:1:1 complex between CB[8], paraquat and 2,6-dihydroxynaphthalene, which results in an enhanced charge-transfer complex within the complex.[195] The supramolecular chemistry of CB[8] has blossomed since the disclosure of its binding properties and a large variety of CB[8] ternary complexes have been recently described and exploited in the preparation of non-covalent constructs such as intermolecular folding systems, vesicles, molecular switches, supramolecular dendrimers and viscoelastic polymeric networks.[183]

Although in its infancy, the supramolecular chemistry of CBs has attracted widespread attention during the past 12 years and their recognition properties have enabled the preparation of a large variety of stimuli-responsive self-organized molecular constructs with interesting properties. Today, there is an increasing interest in the utilization of the unique properties of CBs in the implementation of materials for advanced applications such catalysis, sensing, cell culture or the controlled delivery and release of specific substances. The following section is an overview of the different hydrogel systems which incorporate CB binding motifs and the reported applications of this new type of transient polymeric network.

3.2.2.1 CB Hydrogels

The first hydrogel based on CBs was reported by Kim and co-workers in 2007 and constitutes one of the rare cases of macrocycle-induced gelation in water.[196–198] Only CB[7], and none of the other CB family members, dissolves in warm solutions of dilute mineral acids and forms gels upon cooling. The formation of CB[7] gel fibrils was demonstrated by a combination of AFM, IR, SAXS and single-crystal spectroscopies. These fibrils are held together by extensive C–H···O water/oxonium ion mediated hydrogen bonding. The gels are pH sensitive with optimum formation below pH 2. The system reverts to the solution state upon addition of alkali metal ions, a feature that highlights the importance of the cation complexation with the CB[7] portals in gel formation. The guest-induced stimuli-responsive behaviour of the gel was investigated in the presence of a small amount of 4,4′-diaminostilbene dihydrochloride, both the *cis* and *trans* isomers of which form stable 1:1 inclusion complexes with

CB[7]. A white gel is formed in the presence of small amounts of the *trans* isomer; however, irradiation of the gel with UV light results in the formation of a yellow solution. Heating for 2 h and subsequent cooling regenerates a pale yellow gel, which again forms a fluid solution on irradiation. The breakup of the gels through disruption of the H-bonding network is thought to be greater with the *cis* isomer than with the *trans* isomer, a feature which is the origin of the sol–gel transition. Tam and co-workers have shown that CBs can also be useful compounds to exert control over the gelation of small molecules.[199] The authors found that the addition of small amounts of CB[6] to butan-1-aminium 4-toluenesulfonate resulted in the formation of hydrogels upon cooling. In the absence of CB[6], the organic salt crystallizes as large block-like aggregates. However, in the presence of CB[6], long, thick rod-like fibers form of several micrometers in length. The inclusion of the alkyl chain of butan-1-aminium 4-toluenesulfonate within the CB[6] cavity was demonstrated by NMR spectroscopy. The CB[6] molecule acts as an additive which changes the crystallization process of the salt and induces gel formation.

The previous investigations instigated the new research area of CB hydrogels. The preparation of this new family of aqueous materials at this point of their development is aligned towards the controlled cross-linking of polymer chains through CB host–guest complexes. Two different strategies have been devised to produce CB transient polymeric networks: three-component systems cross-linked by ternary CB[8] inclusion complexes (Figure 3.9a) and two-component systems based on CB[7]/polyamines host–guest pairs (Figure 3.9b); both will be discussed in turn.

Scherman and co-workers reported in 2010 the first example of a supramolecular polymeric hydrogel based on CB host–guest inclusion complexes.[200] By employing a pair of multivalent side-chain functional polymers bearing either viologen or naphthoxy derivatives, the authors have demonstrated that the presence of CB[8] induced supramolecular cross-linking and subsequent gelation (Figure 3.9a). No gelation was observed when all polymeric materials were dissolved together. It is only upon addition of CB[8] that the pendant guest molecules can interact to form a cross-linked network.

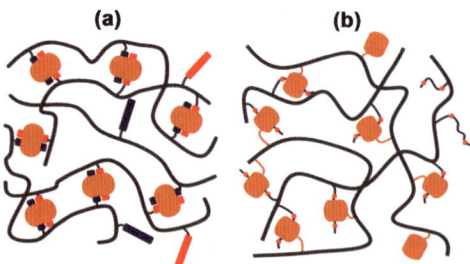

(a) **(b)**

Figure 3.9 Schematic representations of polymeric hydrogels based on cucurbituril inclusion complexes. (a) Hetero-ternary CB[8] binding and (b) binary CB[6] complexes utilising functionalised CB[6] derivatives.

Rheological characterization of the viscoelastic polymeric networks allowed for the determination of the dissociation kinetics of the ternary complex, with $k_d = 1200\,\text{s}^{-1}$. The materials exhibited intermediate mechanical properties at around 5 wt% in water (plateau modulus ~ 350–600 Pa and zero-shear viscosity ~ 5–55 Pa s), for systems with a cross-linked density in the 2.5–10.0% range (percentage of monomer units participating in cross-link formation and determined by the molar equivalent addition of CB[8]). The authors also demonstrated that the facile control over cross-link density through the addition of CB[8] to the system allowed control over the microstructure, as changes in the pore sizes measured by SEM images were proportional to the cross-link density of the network. In a follow-up investigation, Scherman and co-workers prepared extremely high water content (up to 99.75% water), self-assembled polymeric hydrogels derived from renewable cellulosic derivatives.[201,202] The hydrogels had highly tunable mechanical properties, based on the strong and dynamic hetero-ternary complex formation with the macrocyclic host CB[8]. The simple preparation process, their availability from inexpensive renewable resources and the tunability of their mechanical properties are important for biomedical applications. Extremely well-sustained release of bovine serum albumin is observed over the course of 160 days from supramolecular hydrogels containing only 1.5 wt% polymeric constituents. In addition, the bioactivities of the proteins were maintained for up to 50 days, showing its utility as a protein delivery agent.

Very recently, Kim and co-workers have demonstrated hydrogel formation based on the interaction between CB[6] and alkylammonium guests derived from 1,6-diaminohexane (DAH) and spermine (SPM) (Figure 3.9b).[203] On the one hand, the authors prepared CB[6]-containing hyaluronic acid (around 6 mol% of functionalized repeating units) by grafting (allyloxy)$_{12}$CB[6] onto a thiol-functionalized hyaluronic acid through a thiol–ene reaction (although both CB and hyaluronic acid have multiple reactive sites, it is not clear whether the CB molecules are only attached to one or several polymer chains).[204] The mixing of solutions of the CB[6]-containing polymer and a DAH-containing hyaluronic acid (around 50 mol% of functionalized repeating units) produced a hydrogel after 2 min. The addition of excess of SPM to the hydrogel resulted in a phase transition from gel to sol within 10 min, suggesting that the cross-links of the polymer network were indeed due to host–guest interactions. Storage moduli as high as 3.4 kPa were measured for a hydrogel composed of a mixture of spermine and CB[6]-functionalized hyaluronic acid (2 wt%). Cytocompatibility studies demonstrated the high cell viability, enzymatic degradability and negligible cytotoxicity of gels. The most interesting feature of the system is that the presence of free alkylammonium guest in the polymeric network allows for further functionalization of the gel with biorelevant motifs attached to CB[6]. The authors demonstrated the incorporation of functional tags, CB[6] including FITC and rhodamine B isothiocyanate as well as c(RGDyK), a fibronectin motif known to promote cell adhesion. The authors even demonstrated the *in situ* formation of the hydrogel under the skin of mice by sequentially injecting solutions of the complementary polymers CB[6]- and DAH-containing hyaluronic acid. The hydrogel was formed within a few minutes after the injection and was stable for longer than 2 weeks; it was even possible to be modified *in situ* by injection of a solution of FITC-CB[6]. The preparation of a physical hydrogel that can either be injected after formation or generated *in situ* has proven to be of great importance, as these materials may act as effective and modular platforms for a wider variety of biomedical applications, including the delivery of specific substances or the scaffolding of cell structures.

3.3 Hydrogels Based On Non-Covalent Interactions Other Than Host–Guest Inclusion Complexes

Besides host–guest interactions, other interactions such as hydrogen bonding interactions, ionic interactions and metal–ligand interactions can be exploited for hydrogel formation. The following sections describe how this is done.

3.3.1 Hydrogen Bonding Interactions

There has also been much work on the preparation of hydrogels using several natural polysaccharides, including cellulose, guar gum, zanthan gum, starch and dextran, which gel through strong H-bonding interactions between highly hydroxyl-functionalized chains.[205] Rowan and co-workers have utilized cellulose nanowhiskers to prepare gels using a specially designed method that mimics the stimulus-responsive nature of sea cucumber dermis.[206–210] The nanowhisker-based materials gel *via* strong H-bonding interactions through a slow solvent exchange of a strong (*i.e.* water) and a weak (*i.e.* acetone) H-bonding solvent. These materials are strong and stable, and composite materials are formed upon subsequent exchange of the solvent with a weak H-bonding polymer such as poly(vinyl alcohol) from a solution of a nonsolvent (*i.e.* toluene).

Lehn and co-workers described reversible pH-responsive hydrogels formed from the potassium-stabilized association of a bis-guanine telechelic (PEG) oligomer.[211] These oligomers form linear supramolecular polymers; however, the potassium cations promote aggregation of the dimerized guanine end groups to form G-quartets, the cross-linking motif of the network. Reversible binding and release of the potassium ions by cryptand [2.2.2] undergoing deprotonation/protonation promotes gel-to-sol and sol-to-gel transitions.

3.3.2 Ionic Interactions

There are several recent examples of polymeric hydrogels prepared *via* electrostatic interactions between multivalent polymers. Mixing two aqueous solutions of oppositely charged polyelectrolytes generally leads to phase separation.[212] However, the addition of a neutral solvophilic block to the

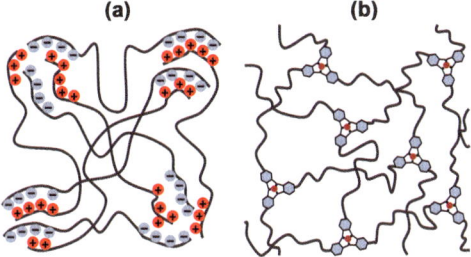

Figure 3.10 Schematic representation of two transient polymeric hydrogels based on (a) ionic interactions and (b) metal–ligand complexes.

polyelectrolyte chain can result in gel formation by preventing macroscopic phase separation. Aida and co-workers reported the preparation of high-water-content (96–98%) non-covalent hydrogels prepared with poly(sodium acrylate) (ASAP)-treated clay nanosheets and ABA-type end-functionalized PEG polymers bearing multivalent guanidinium dendritic end groups.[213] A PEG linear precursor was functionalized at both chain ends with dendrons of various generations (G1–G3) bearing varying numbers of guanidine hydrochloride groups (two, four and eight for G1, G2 and G3 dendrons, respectively). These cationic "binder" materials quickly form cross-links with the anionic silicate-based clay nanosheets to form hydrogels with exceptional mechanical strength and self-healing properties (recovery was observed within 1 min of fracture) that are mouldable into transparent, shape-persistent, freestanding objects. Moreover, the gel materials demonstrated the ability to transport biological activity as a myoglobin-containing hydrogel that retained its catalytic capacity relative to free myoglobin.

Cohen Stuart and co-workers reported a novel class of multi-responsive reversible gels based on the co-assembly of a triblock copolymer having two negatively charged end blocks with a positively charged homopolymer. The authors showed that the hydrogel consisted of a network of interconnected polyelectrolyte complex micelles stabilized by a corona of neutral solvophilic blocks. The gel responded not only to changes in temperature and concentration but also to ionic strength, cationic/anionic composition and, if weak polyelectrolytes are used, pH value.[214] Following a similar strategy, Hawker and co-workers reported the formation of well-defined supramolecular networks by mixing ionic ABA triblock copolymers (Figure 3.10a).[215] Four types of ionic functional groups (sulfonate, carboxylate, ammonium and guanidinium), representing a range of different pK_a values, were incorporated from a parent ABA triblock copolymer containing a central PEG block (10–35 kDa) and varying numbers of reactive sites in the terminal A blocks. As expected, an increase in G'' and a corresponding decrease in G' was observed with increasing amounts of sodium chloride.

3.3.3 Metal–Ligand Interactions

Coordination complexes form cross-links when two or more ligands each donate a non-bonding electron pair to empty orbitals in a transition metal ion.

Because of their high stability and rates of formation, coordination-based cross-links have been proposed to endow certain biological structures with a number of desirable material properties, including triggered self-assembly, increased toughness, self-repair, adhesion, high hardness in the absence of mineralization and mechanical tenability. Tong and co-workers identified one of the major advantages of metal-based systems in describing a poly(acrylic acid)-based hydrogel formed through complexation of the polymer with Fe(III) ions.[216] Simple addition of Fe(III) to a polymer solution forms strong yet reversible hydrogels. Upon irradiation with light in the absence of oxygen, the metal is reduced to Fe(II), which no longer binds the acid-functionalized polymer, thereby disrupting the gel network. Reintroduction of oxygen in the dark reoxidizes the metal ion and the hydrogel reforms in a completely reversible process. Messersmith and co-workers described a strategy for introducing bis- and/or tris-catechol-Fe(III) cross-links into a synthetic polymer network which displayed high elastic moduli and self-healing properties (Figure 3.10b).[217] Tris- and bis-catechol-Fe(III) complexes possess some of the highest known stability constants of metal–ligand chelates (log K_a up to 40).[218] Additionally, the stoichiometry of these complexes can be controlled by pH. Neutral or higher pH is required to stabilize the bis and tris complexes. The authors incorporated 3,4-dihydroxyphenylalanine (DOPA) moieties at the chain ends of a four-arm PEG star polymer (10 kDa) which formed a reversible aqueous polymeric network in the presence of Fe(III) cations. Polymer/FeCl$_3$ mixtures at pH ≈ 5 displayed a viscous response in dynamic oscillatory rheology, whereas the bis- and tris-catechol-Fe(III) cross-linked gels at pH > 8 behaved increasingly elastically. The gels displayed near covalent stiffness at high strain rates, supporting the idea that Fe(III) coordinate bonds can provide significant strength to bulk materials despite their transient nature, given that the pH is high enough to ensure cross-link stability on relevant timescales.

3.4   Conclusions

This chapter has treated the main research that is ongoing in the field of supramolecular polymeric hydrogels. Different from conventional cross-linking methods, the combination of polymer chains and supramolecular cross-links present an interesting platform from which one can modify the polymeric backbone, strength of interaction, directionality, multiple responsiveness and tunable degradability in a modular fashion. Supramolecular cross-links allow for dynamic behaviour: structural error correction, self-healing properties, elasticity and mouldability. Some of these properties are fundamental for the fabrication of performance materials, fault-tolerant products and components encompassing a range of industries, including coatings, electronics, transportation and energy. The use of supramolecular chemistry in the assembly of networked structures allows spatiotemporal control of the viscoelastic properties which could have implications in biomedical applications, particularly cell growth and differentiation. The current developments in this field have provided us with a toolbox which we can utilize to build customizable structures and these advances auger well for the future development of improved supramolecular polymeric hydrogels.

References

1. X. Huang, P. Terech, S. R. Raghavan and R. G. Weiss, *J. Am. Chem. Soc.*, 2005, **127**, 4336–4344.
2. F. L. Buchholz and A. T. Graham, *Modern Superabsorbent Polymer Technology*, Wiley-VCH, Weinheim, 2000.
3. R. Westermeier, *Electrophoresis in Practice*, Wiley-VCH, Weinheim, 2004.
4. S. Mori and H. G. Barth, *Size Exclusion Chromatography*, Springer, Heidelberg, 1999.
5. *Handbook of Bioseparations*, ed. S. Ahuja, Academic Press, San Diego, CA, 2000.
6. *Polymers in Drug Delivery*, ed. I. F. Uchegbu and A. G. Schatzlein, CRC Press, Boca Raton, FL, 2006.
7. Y. Qiu and K. Park, *Adv. Drug Delivery Rev.*, 2001, **53**, 321–339.
8. O. Wichterle and D. Lim, *Nature*, 1960, **185**, 117–118.
9. J. A. Cadée, M. J. A. van Luyn, L. A. Brouwer, J. A. Plantinga, P. B. van Wachem, C. J. de Groot, W. den Otter and W. E. Hennink, *J. Biomed. Mater. Res.*, 2000, **50**, 397–404.
10. H. Park and K. Park, *Pharm. Res.*, 1996, **13**, 1770–1776.
11. N. A. Peppas, P. Bures, W. Leobandung and H. Ichikawa, *Eur. J. Pharm. Biopharm.*, 2000, **50**, 27–46.
12. C. D. Pritchard, T. M. O'Shea, D. J. Siegwart, E. Calo, D. G. Anderson, F. M. Reynolds, J. A. Thomas, J. R. Slotkin, E. J. Woodard and R. Langer, *Biomaterials*, 2011, **32**, 587–597.
13. D.-Y. Teng, Z.-M. Wu, X.-G. Zhang, Y.-X. Wang, C. Zheng, Z. Wang and C.-X. Li, *Polymer*, 2010, **51**, 639–646.
14. H. Zhang, A. Qadeer, D. Mynarcik and W. Chen, *Biomaterials*, 2011, **32**, 890–898.
15. K. P. Koutroumanis, K. Avgoustakis and D. Bikiaris, *Carbohydr. Polym.*, 2010, **82**, 181–188.
16. A. E. Rydholm, C. N. Bowman and K. S. Anseth, *Biomaterials*, 2005, **26**, 4495–4506.
17. C. P. Pathak, A. S. Sawhney and J. A. Hubbell, *J. Am. Chem. Soc.*, 1992, **114**, 8311–8312.
18. M. Kurisawa, J. E. Chung, Y. Y. Yang, S. J. Gao and H. Uyama, *Chem. Commun.*, 2005, 4312–4314.
19. R. Jin, C. Hiemstra, R. Zhong and J. Feijen, *Biomaterials*, 2007, **28**, 2791–2800.
20. C. A. DeForest, B. D. Polizzotti and K. S. Anseth, *Nat. Mater.*, 2009, **8**, 659–664.
21. A. B. W. Brochu, S. L. Craig and W. M. Reichert, *J. Biomed. Mater. Res.*, 2010, **96**, 492–506.

22. A. S. Sawhney, C. P. Pathak and J. A. Hubbell, *Macromolecules*, 1993, **26**, 581–587.

23. B. Jeong, Y. H. Bae, D. S. Lee and S. W. Kim, *Nature*, 1997, **388**, 860–862.

24. L. Aulisa, H. Dong and J. D. Hartgerink, *Biomacromolecules*, 2009, **10**, 2694–2698.

25. L. Yu and J. Ding, *Chem. Soc. Rev.*, 2008, **37**, 1473–1481.

26. M. Guvendiren, H. D. Lu and J. A. Burdick, *Soft Matter*, 2012, **8**, 260–272.

27. C. T. Huynh, M. K. Nguyen and D. S. Lee, *Macromolecules*, 2011, **44**, 6629–6636.

28. A. S. Hoffman, *Adv. Drug Delivery Rev.*, 2002, **54**, 3–12.

29. M. P. Lutolf, *Nat. Mater.*, 2009, **8**, 451–453.

30. M. P. Lutolf and J. A. Hubbell, *Nat. Biotechnol.*, 2005, **23**, 47–55.

31. S. Kiyonaka, K. Sada, I. Yoshimura, S. Shinkai, N. Kato and I. Hamachi, *Nat. Mater.*, 2004, **3**, 58–64.

32. K. Y. Lee and D. J. Mooney, *Chem. Rev.*, 2001, **101**, 1869–1879.

33. J. L. Drury and D. J. Mooney, *Biomaterials*, 2003, **24**, 4337–4351.

34. S. Seiffert and J. Sprakel, *Chem. Soc. Rev.*, 2012, **41**, 909–930.

35. L. A. Estroff and A. D. Hamilton, *Chem. Rev.*, 2004, **104**, 1201–1217.

36. *Molecular Gels*, ed. R. G. Weiss and P. Terech, Springer, Dordrecht, 2005.

37. M. L. Bender and M. Komiyama, *Cyclodextrin Chemistry; Reactivity and Structure Concepts in Organic Chemistry*, Springer, Berlin, 1978.

38. J. Szetjli, *Chem. Rev.*, 1998, **98**, 1743–1753.

39. M. V. Rekharsky and Y. Inoue, *Chem. Rev.*, 1998, **98**, 1875–1917.

40. L. Liu and Q. X. Guo, *J. Inclusion Phenom. Macrocyclic Chem.*, 2002, **42**, 1–14.

41. D. Harries, D. C. Rau and V. A. Parsegian, *J. Am. Chem. Soc.*, 2005, **127**, 2184–2190.

42. M. E. Davis and M. E. Brewster, *Nat. Rev. Drug Discovery*, 2004, **3**, 1023–1035.

43. V. J. Stella, V. M. Rao, E. A. Zannou and V. V. Zia, *Adv. Drug Delivery Rev.*, 1999, **36**, 3–16.

44. S. Li and W. C. Purdy, *Chem. Rev.*, 1992, **92**, 1457–1470.

45. G. Crini and M. J. Morcellet, *J. Sep. Sci.*, 2002, **25**, 789–813.

46. J. Hu, Z. Tao, S. Li and B. Liu, *J. Mater. Sci.*, 2005, **40**, 6057–6061.

47. H.-J. Buschmann and E. Schollmeyer, *J. Cosmet. Sci.*, 2002, **53**, 185–191.

48. H.-J. Buschmann, D. Knittel and E. Schollmeyer, *J. Inclusion Phenom. Macrocyclic Chem.*, 2001, **40**, 169–172.

49. A. R. Hedges, *Chem. Rev.*, 1998, **98**, 2035–2044.

50. F. van de Manakker and T. Vermonden, C. F. van Nostrum and W. E. Hennink, *Biomacromolecules*, 2009, **10**, 3157–3175.

51. F. Yuen and K. C. Tam, *Soft Matter*, 2010, **6**, 4613–4630.

52. K. L. Liu, Z. Zhang and J. Li, *Soft Matter*, 2011, **7**, 11290–11297.

53. J. Li, *NPG Asia Mater.*, 2010, **2**, 112–118.

54. J. Li, *Adv. Polym. Sci.*, 2009, **222**, 175–203.

55. G. Chen and M. Jiang, *Chem. Soc. Rev.*, 2011, **40**, 2254–2266.
56. A. Harada, J. Li and M. Kamachi, *Nature*, 1992, **356**, 325–327.
57. A. Harada and M. Kamachi, *Macromolecules*, 1990, **23**, 2821–2823.
58. A. Harada and M. Kamachi, *J. Chem. Soc., Chem. Commun.*, 1990, 1322–1323.
59. A. Harada, J. Li and M. Kamachi, *Chem. Lett.*, 1993, **22**, 237–240.
60. G. Wenz, B. H. Han and A. Müller, *Chem. Rev.*, 2006, **106**, 782–817.
61. A. Harada, J. Li and M. Kamachi, *Nature*, 1994, **370**, 126–128.
62. A. Harada, J. Li and M. Kamachi, *Macromolecules*, 1993, **26**, 5698–5703.
63. L. H. He, J. Huang, Y. M. Chen and L. P. Liu, *Macromolecules*, 2005, **38**, 3351–3355.
64. A. Harada, M. Okada, J. Li and M. Kamachi, *Macromolecules*, 1995, **28**, 8406–8411.
65. S. Loethen, J.-M. Kim and D. H. Thompson, *Polym. Rev.*, 2007, **47**, 383–418.
66. T. Ooya and N. Yui, *Crit. Rev. Ther. Drug Carrier Syst.*, 1999, **16**, 289–330.
67. J. Araki and K. Ito, *Soft Matter*, 2007, **3**, 1456–1473.
68. J. Li and X. J. Loh, *Adv. Drug Delivery Rev.*, 2008, **60**, 1000–1017.
69. A. Harada, Y. Takashima and H. Yamaguchi, *Chem. Soc. Rev.*, 2009, **38**, 875–882.
70. T. Takata, *Polym. J.*, 2006, **38**, 1–20.
71. H. Tian and Q.-C. Wang, *Chem. Soc. Rev.*, 2005, **35**, 361–374.
72. F. M. Raymo and J. F. Stoddart, *Chem. Rev.*, 1999, **99**, 1643–1663.
73. N. Nakashima, A. Kawabuchi and H. Murakami, *J. Inclusion Phenom. Mol. Recognit. Chem.*, 1998, **32**, 363–373.
74. K. M. Huh, T. Ooya, W. K. Lee, S. Sasaki, I. C. Kwon, S. Y. Jeong and N. Yui, *Macromolecules*, 2001, **34**, 8657–8662.
75. K. M. Huh, Y. W. Cho, H. Chung, I. C. Kwon, S. Y. Jeong, T. Ooya, W. K. Lee, S. Sasaki and N. Yui, *Macromol. Biosci.*, 2004, **4**, 92–99.
76. T. Nakama, T. Ooya and N. Yui, *Polym. J.*, 2004, **36**, 338–344.
77. A. Harada, A. Hashidzume, H. Yamaguchi and Y. Takashima, *Chem. Rev.*, 2009, **109**, 5974–6023.
78. J. Li, X. P. Ni, Z. H. Zhou and K. W. Leong, *J. Am. Chem. Soc.*, 2003, **125**, 1788–1795.
79. J. Li, X. Li, X. Ni, X. Wang, H. Li and K. W. Leong, *Biomaterials*, 2006, **27**, 4132–4140.
80. J. Li, X. Li, Z. Zhou, X. Ni and K. W. Leong, *Macromolecules*, 2001, **34**, 7236–7237.
81. X. Ni, A. Cheng and J. Li, *J. Biomed. Mater. Res., A*, 2009, **88**, 1031–1036.
82. X. Li and J. Li, *J. Biomed. Mater. Res., A*, 2007, **86**, 1055–1061.
83. F. Hirayama and K. Uekama, *Adv. Drug Delivery Rev.*, 1999, **36**, 125–141.
84. J. Li, X. P. Ni and K. W. Leong, *J. Biomed. Mater. Res., A*, 2003, **65**, 196–202.

85. G. Wenz, *Angew. Chem., Int. Ed. Engl.*, 1994, **33**, 803–822.
86. G. Wenz and B. Keller, *Angew. Chem., Int. Ed. Engl.*, 1992, **31**, 197–199.
87. H. S. Choi, T. Ooya, S. C. Lee, S. Sasaki, M. Kurisawa, H. Uyama and N. Yui, *Macromolecules*, 2004, **37**, 6705–6710.
88. Y. K. Joung, T. Ooya, M. Yamaguchi and N. Yui, *Adv. Mater.*, 2007, **19**, 396–400.
89. A. Harada, J. Li and M. Kamachi, *Nature*, 1993, **364**, 516–518.
90. T. Ikeda, T. Ooya and N. Yui, *Macromol. Rapid Commun.*, 2000, **21**, 1257–1262.
91. Y. Yu, C. Chipot, W. Cai and X. Shao, *J. Phys. Chem. B*, 2006, **110**, 6372–6378.
92. X. Guo, A. A. Abdala, B. L. May, S. F. Lincoln, S. A. Khan and R. K. Prud'homme, *Macromolecules*, 2005, **38**, 3037–3040.
93. X. Guo, A. A. Abdala, B. L. May, S. F. Lincoln, S. A. Khan and R. K. Prud'homme, *Polymer*, 2006, **47**, 2976–2983.
94. A. Hashidzume, F. Ito, I. Tomatsu and A. Harada, *Macromol. Rapid Commun.*, 2005, **26**, 1151–1154.
95. A. Hashidzume, I. Tomatsu and A. Harada, *Polymer*, 2006, **47**, 6011–6027.
96. H. S. Choi and N. Yui, *Prog. Polym. Sci.*, 2006, **31**, 121–144.
97. I. Tomatsu, A. Hashidzume and A. Harada, *Macromolecules*, 2005, **38**, 5223–5227.
98. I. Tomatsu, A. Hashidzume and A. Harada, *J. Am. Chem. Soc.*, 2006, **128**, 2226–2227.
99. S. Tamesue, Y. Takashima, H. Yamaguchi, S. Shinkai and A. Harada, *Angew. Chem. Int. Ed.*, 2010, **49**, 7461–7464.
100. X. J. Liao, G. S. Chen, X. X. Liu, W. X. Chen, F. Chen and M. Jiang, *Angew. Chem. Int. Ed.*, 2010, **49**, 4409–4413.
101. I. Tomatsu, A. Hashidzume and A. Harada, *Macromol. Rapid Commun.*, 2006, **27**, 238–241.
102. A. Charlot, A. Heyraud, P. Guenot, M. Rinaudo and R. Auzély-Velty, *Biomacromolecules*, 2006, **7**, 907–913.
103. A. Charlot, R. Auzély-Velty and M. Rinaudo, *J. Phys. Chem. B*, 2003, **107**, 8248–8254.
104. R. Auzély-Velty and M. Rinaudo, *Macromolecules*, 2001, **34**, 3574–3580.
105. R. Auzély-Velty and M. Rinaudo, *Macromolecules*, 2002, **35**, 7955–7962.
106. A. Charlot and R. Auzély-Velty, *Macromolecules*, 2007, **40**, 9555–9563.
107. L. Li, X. Guo, J. Wang, P. Liu, R. K. Prud'homme, B. L. May and S. F. Lincoln, *Macromolecules*, 2008, **41**, 8677–8681.
108. A. Charlot and R. Auzély-Velty, *Macromolecules*, 2007, **40**, 1147–1158.
109. N. A. Semenov, A. Charlot, R. Auzély-Velty and M. Rinaudo, *Rheol. Acta*, 2007, **46**, 541–568.
110. O. Kretschmann, S. W. Choi, M. Miyauchi, I. Tomatsu, A. Harada and H. Ritter, *Angew. Chem. Int. Ed.*, 2006, **45**, 4361–4365.

111. T. Lecourt, J.-M. Mallet and P. Sinaÿ, *Eur. J. Org. Chem.*, 2003, 4553–4560.

112. O. Bistri, K. Mazeau, R. Auzély-Velty and Matthieu Sollogoub, *Chem.–Eur. J*, 2007, **13**, 8847–8857.

113. T. Lecourt, P. Sinaÿ, C. Chassenieux, M. Rinaudo and R. Auzély-Velty, *Macromolecules*, 2004, **37**, 4635–4642.

114. F. van de Manakker, M. van der Pot, T. Vermonden, C. F. van Nostrum and W. E. Hennink, *Macromolecules*, 2008, **41**, 1766–1773.

115. F. van de Manakker, T. Vermonden, N. el Morabit, C. F. van Nostrum and W. E. Hennink, *Langmuir*, 2008, **24**, 12559–12567.

116. K. A. Aamer, H. Sardinha, S. R. Bhatia and G. N. Tew, *Biomaterials*, 2004, **25**, 1087–1093.

117. F. van de Manakker, K. Braeckmans, N. el Morabit, S. C. De Smedt, C. F. van Nostrum and W. E. Hennink, *Adv. Funct. Mater.*, 2009, **19**, 2992–3001.

118. F. van de Manakker, L. M. J. Kroon-Batenburg and T. Vermonden, C. F. van Nostrum and W. E. Hennink, *Soft Matter*, 2010, **6**, 187–194.

119. B. Chen, K. L. Liu, Z. Zhang, X. Ni, S. H. Goh and J. Li, *Chem. Commun.*, 2012, **48**, 5638–5640.

120. C. Koopmans and H. Ritter, *Macromolecules*, 2008, **41**, 7418–7422.

121. A. Sandier, W. Brown and H. Mays, *Langmuir*, 2000, **16**, 1634–1642.

122. C. Amiel and B. Sébille, *Adv. Colloid Interface Sci.*, 1999, **79**, 105–122.

123. N. M. Gosselet, H. Naranjo, E. Renard, C. Amiel and B. Sébille, *Eur. Polym. J.*, 2002, **38**, 649–654.

124. D. Mislovičová, G. Kogan, N. M. Gosselet, B. Sébille and L. Šoltés, *Chem. Biodiversity*, 2007, **4**, 52–57.

125. L. Moine, C. Amiel, W. Brown and P. Guerin, *Polym. Int.*, 2001, **50**, 663–676.

126. N. M. Gosselet, F. Beucler, E. Renard, C. Amiel and B. Sébille, *Colloids Surf., A*, 1999, **155**, 177–188.

127. N. M. Gosselet, C. Borie, C. Amiel and B. Sébille, *J. Dispersion Sci. Technol*, 1998, **19**, 805–820.

128. S. Hvilsted, C. Sánchez and R. Alcalá, *J. Mater. Chem.*, 2009, **19**, 6641–6648.

129. R. Eelkema, M. M. Pollard, N. Katsonis, J. Vicario, D. J. Broer and B. L. Feringa, *J. Am. Chem. Soc.*, 2006, **128**, 14397–11407.

130. H. Yu, T. Iyoda and T. Ikeda, *J. Am. Chem. Soc.*, 2006, **128**, 11010–11011.

131. J. del Barrio, R. M. Tejedor, L. S. Chinelatto, C. Sánchez, M. Piñol and L. Oriol, *Chem. Mater.*, 2010, **22**, 1714–1723.

132. Y. L. Zhao and J. F. Stoddart, *Langmuir*, 2009, **25**, 8442–8446.

133. K. Peng, C. Cui, I. Tomatsu, F. Porta, A. H. Meijer, H. P. Spaink and A. Kros, *Soft Matter*, 2010, **6**, 3778–3783.

134. K. Peng, I. Tomatsu, A. V. Korobko and A. Kros, *Soft Matter*, 2010, **6**, 85–87.

135. K. Peng, I. Tomatsu and A. Kros, *Chem. Commun.*, 2010, **46**, 4094–4096.

136. K. M. Huh, H. Tomita, W. K. Lee, T. Ooya and N. Yui, *Macromol. Rapid Commun.*, 2002, **23**, 179–182.

137. H. S. Choi, K. M. Huh, T. Ooya and N. Yui, *J. Am. Chem. Soc.*, 2003, **125**, 6350–6351.

138. H. S. Choi, T. Ooya, S. Sasaki and N. Yui, *Macromolecules*, 2003, **36**, 5342–5347.

139. H. S. Choi, T. Ooya, K. M. Huh and N. Yui, *Biomacromolecules*, 2005, **6**, 1200–1204.

140. H. S. Choi, A. Takahashi, T. Ooya and N. Yui, *Macromolecules*, 2004, **37**, 10036–10041.

141. N. M. Gosselet, V. Wintgens and C. Amiel, *Macromol. Biosci.*, 2005, **5**, 306–313.

142. G. Volet and C. Amiel, *Eur. Polym. J.*, 2009, **45**, 852–862.

143. A. Harada and S. Takahashi, *J. Chem. Soc., Chem. Commun.*, 1984, 645–646.

144. A. Harada, Y. Hu, S. Yamamoto and S. Takahashi, *J. Chem. Soc., Dalton Trans.*, 1988, 729–732.

145. Y. Odagaki, K. Hirotsu, T. Higuchi, A. Harada and S. Takahashi, *J. Chem. Soc., Perkin Trans.*, 1990, **1**, 1230–1231.

146. B. Gonzalez, I. Cuadrado, B. Alonso, C. M. Casado, M. Moran and A. E. Kaifer, *Organometallics*, 2002, **21**, 3544–3551.

147. A. Mirzoian and A. E. Kaifer, *Chem.–Eur. J.*, 1997, **3**, 1052–1058.

148. J. F. Bergamini, M. Jouini, S. Aeiyach, K. I. Chane-Ching, J. C. Lacroix, J. Tanguy and P. C. Lacaze, *J. Electroanal. Chem.*, 2005, **579**, 125–131.

149. T. Komura, T. Yamaguchi, K. Noda and S. Hayashi, *Electrochim. Acta*, 2002, **47**, 3315–3325.

150. C. A. Nijhuis, J. Huskens and D. N. Reinhoudt, *J. Am. Chem. Soc.*, 2004, **126**, 12266–12267.

151. C. A. Nijhuis, J. K. Sinha, G. Wittstock, J. Huskens, B. J. Ravoo and D. N. Reinhoudt, *Langmuir*, 2006, **22**, 9770–9775.

152. T. Matsue, D. H. Evans, T. Osa and N. Kobayashi, *J. Am. Chem. Soc.*, 1985, **107**, 3411–3417.

153. H.-J. Thiem, M. Brandl and R. Breslow, *J. Am. Chem. Soc.*, 1988, **110**, 8612–8616.

154. R. Isnin, C. Salam and A. E. Kaifer, *J. Org. Chem.*, 1991, **56**, 35–41.

155. A. E. Kaifer, *Acc. Chem. Res.*, 1999, **32**, 62–71.

156. M. Nakahata, Y. Takashima, H. Yamaguchi and A. Harada, *Nat. Commun.*, 2011, **2**, 511.

157. A. Harada, R. Kobayashi, Y. Takashima, A. Hashidzume and H. Yamaguchi, *Nat. Chem.*, 2011, **3**, 34–37.

158. R. Behrend, E. Meyer and F. Rusche, *Justus Liebigs Ann. Chem*, 1905, **339**, 1–37.

159. W. A. Freeman, W. L. Mock and N.-Y. Shih, *J. Am. Chem. Soc.*, 1981, **103**, 7367–7368.

160. J. Kim, I.-S. Jung, S.-Y. Kim, E. Lee, J.-K. Kang, S. Sakamoto, K. Yamaguchi and K. Kim, *J. Am. Chem. Soc.*, 2000, **122**, 540–541.

161. A. I. Day, A. P. Arnold, R. J. Blanch and B. Snushall, *J. Org. Chem.*, 2001, **66**, 8094–8100.

162. A. I. Day, R. J. Blanch, A. P. Arnold, S. Lorenzo, G. R. Lewis and I. Dance, *Angew. Chem. Int. Ed.*, 2002, **41**, 275–277.

163. J. Lagona, P. Mukhopadhyay, S. Chakrabarti and L. Isaacs, *Angew. Chem. Int. Ed.*, 2005, **44**, 4844–4870.

164. D. Bardelang, K. Udachin, D. M. Leek, J. M. Margeson, G. Chan, C. I. Ratcliffe and J. A. Ripmeester, *Cryst. Growth Des.*, 2011, **11**, 5598–5614.

165. A. Flinn, G. C. Hough, J. F. Stoddart and D. J. Williams, *Angew. Chem. Int. Ed.*, 1992, **31**, 1475–1477.

166. S. Sasmal, M. K. Sinha and E. Keinan, *Org. Lett.*, 2004, **6**, 1225–1228.

167. J. Zhao, H.-J. Kim, J. Oh, S.-Y. Kim, J. W. Lee, S. Sakamoto, K. Yamaguchi and K. Kim, *Angew. Chem. Int. Ed.*, 2001, **40**, 4233–4235.

168. S. Y. Jon, N. Selvapalam, D. H. Oh, J.-K. Kang, S.-Y. Kim, Y. J. Jeon, J. W. Lee and K. Kim, K., *J. Am. Chem. Soc.*, 2003, **125**, 10186–10187.

169. F. Wu, L.-H. Wu, X. Xiao, Y.-Q. Zhang, S.-F. Xue, Z. Tao and A. I. Day, *J. Org. Chem.*, **77**, 606–611.

170. N. Selvapalam, Y. H. Ko, K. M. Park, D. Kim and J. Kim, *Chem. Soc. Rev.*, 2007, **36**, 267–279.

171. N. Zhao, G. O. Lloyd and O. A. Scherman, *Chem. Commun.*, 2012, 3070–3072.

172. D. Lucas, T. Minami, G. Iannuzzi, L. Cao, J. B. Wittenberg, P. Anzenbacher and L. Isaacs, *J. Am. Chem. Soc.*, 2011, **133**, 17966–17976.

173. J. W. Lee, S. Samal, N. Selvapalam, H.-J. Kim and K. Kim, *Acc. Chem. Res.*, 2003, **36**, 621–630.

174. F. Biedermann and O. A. Scherman, *J. Phys. Chem. B*, 2012, **116**, 2842–2849.

175. S. Liu, A. D. Shukla, S. Gadde, B. D. Wagner, A. E. Kaifer and L. Isaacs, *Angew. Chem. Int. Ed.*, 2008, **47**, 1–5.

176. S. Liu, C. Ruspic, P. Mukhopadhyay, S. Chakrabarti, P. Zavalij and L. Isaacs, *J. Am. Chem. Soc.*, 2005, **127**, 15959–15967.

177. D. M. Rudkevich, *Angew. Chem. Int. Ed.*, 2004, **43**, 558–571.

178. K. A. Kellersberger, J. D. Anderson, S. M. Ward, K. E. Krakowiak and D. V. Dearden, *J. Am. Chem. Soc.*, 2001, **123**, 11316–11317.

179. G. Huber, F.-X. Legrand, V. Lewin, D. Baumann, M.-P. Heck and P. Berthault, *Chem. Phys. Chem.*, 2011, **12**, 1053–1055.

180. M. Florea and W. M. Nau, *Angew. Chem. Int. Ed.*, 2011, **50**, 9338–9342.

181. H. Zhang, E. S. Paulsen, K. A. Walker, K. E. Krakowiak and D. V. Dearden, *J. Am. Chem. Soc.*, 2003, **125**, 9284–9285.

182. K. Jansen, H.-J. Buschmann, A. Wego, D. Dopp, C. Mayer, H. J. Drexler, H. J. Holdt and E. Schollmeyer, *J. Inclusion Phenom. Macrocyclic Chem.*, 2001, **39**, 357–363.

183. E. Masson, X. Ling, R. Joseph, L. Kyeremeh-Mensah and X. Lu, *RSC Adv.*, 2012, **2**, 1213–1247.

184. R. Hoffmann, W. Knoche, C. Fenn and H.-J. Buschmann, *J. Chem. Soc., Faraday Trans.*, 1994, **90**, 1507–1511.

185. H.-J. Buschmann, E. Cleve and E. Schollmeyer, *Inorg. Chim. Acta*, 1992, **193**, 93–97.

186. H.-J. Buschmann, K. Jansen, C. Meschke and E. Schollmeyer, *J. Solution Chem.*, 1998, **27**, 135–140.

187. H.-J. Buschmann, K. Jansen and E. Schollmeyer, *Inorg. Chem. Commun.*, 2003, **6**, 531–534.

188. X. X. Zhang, K. E. Krakowiak, G. Xue, J. S. Bradshaw and R. M. Izatt, *Ind. Eng. Chem. Res.*, 2000, **39**, 3516–3520.

189. C. Marquez, R. R. Hudgins and W. M. Nau, *J. Am. Chem. Soc.*, 2004, **126**, 5806–5816.

190. S. Moghaddam, C. Yang, M. Rekharsky, Y. H. Ko, K. Kim, Y. Inoue and M. K. Gilson, *J. Am. Chem. Soc.*, 2011, **133**, 3570–3581.

191. M. V. Rekharsky, T. Mori, C. Yang, Y. H. Ko, N. Selvapalam, H. Kim, D. Sobransingh, A. E. Kaifer, S. Liu, L. Isaacs, W. Chen, S. Moghaddam, M. K. Gilson, K. Kim and Y. Inoue, *Proc. Natl. Acad. Sci. U. S. A.*, 2007, **104**, 20737–20742.

192. W. Ong and A. E. Kaifer, *Organometallics*, 2003, **22**, 4181–4183.

193. R. Wang and D. H. Macartney, *Tetrahedron Lett.*, 2008, **49**, 311–314.

194. P. Montes-Navajas, L. Teruel, A. Corma and H. García, *Chem.–Eur. J.*, 2008, **14**, 1762–1768.

195. H.-J. Kim, J. Heo, W. S. Jeon, E. Lee, J. Kim, S. Sakamoto, K. Yamaguchi and K. Kim, *Angew. Chem. Int. Ed.*, 2001, **40**, 1526–1529.

196. I. Hwang, W. S. Jeon, H. J. Kim, D. Kim, H. Kim, N. Selvapalam, N. Fujita, S. Shinkai and K. Kim, *Angew. Chem. Int. Ed.*, 2007, **46**, 210–213.

197. S. R. Haines and R. G. Harrison, *Chem. Commun.*, 2002, 2846–2847.

198. J. A. Foster and J. W. Steed, *Angew. Chem. Int. Ed.*, 2010, **49**, 6718–6724.

199. H. Yang, Y. B. Tan and Y. X. Wang, *Soft Matter*, 2009, **5**, 3511–3516.

200. E. A. Appel, F. Biedermann, U. Rauwald, S. T. Jones, J. M. Zayed and O. A. Scherman, *J. Am. Chem. Soc.*, 2010, **132**, 14251–14260.

201. E. A. Appel, X. J. Loh, S. T. Jones, F. Biedermann, C. A. Dreiss and O. A. Scherman, *J. Am. Chem. Soc.*, 2012, **10**, 134, 11767–11773.

202. E. A. Appel, X. J. Loh, S. T. Jones, C. A. Dreiss and O. A. Scherman, *Biomaterials*, 2012, **33**, 4646–4652.

203. K. M. Park, J.-A. Yang, H. Jung, J. Yeom, J. S. Park, K.-H. Park, A. S. Hoffman, S. K. Hahn and K. Kim, *ACS Nano*, 2012, **6**, 2960–2968.

204. H. Jung, K. M. Park, J. A. Yang, E. J. Oh, D. W. Lee, K. T. Park, S. H. Ryu, S. K. Hahn and K. Kim, *Biomaterials*, 2011, **32**, 7687–7694.

205. Y. Habibi, L. A. Lucia and O. J. Rojas, *Chem. Rev.*, 2010, **110**, 3479–3500.

206. K. Shanmuganathan, J. R. Capadona, S. J. Rowan and C. Weder, *J. Mater. Chem.*, 2010, **20**, 180–186.

207. R. Rusli, K. Shanmuganathan, S. J. Rowan and S. J. Eichhorn, *Biomacromolecules*, 2010, **11**, 762–768.

208. J. R. Capadona, K. Shanmuganathan, S. Triftschuh, S. Seidel, S. J. Rowan and C. Weder, *Biomacromolecules*, 2009, **10**, 712–716.

209. J. R. Capadona, O. V. D. Berg, L. A. Capadona, M. Schroeter, S. J. Rowan and D. J. Tyler, *Nat. Nanotechnol.*, 2007, **2**, 765–769.
210. J. R. Capadona, K. Shanmuganathan, D. J. Tyler, S. J. Rowan and C. Weder, *Science*, 2008, **319**, 1370–1374.
211. A. Ghoussoub and J. M. Lehn, *Chem. Commun.*, 2005, 5763–5765.
212. M. A. Cohen Stuart, B. Hofs, I. K. Voets and A. de Keizer, *Curr. Opin. Colloid Interface Sci.*, 2005, **10**, 30–36.
213. Q. Wang, J. L. Mynar, M. Yoshida, E. Lee, M. Lee, K. Okuro, K. Kinbara and T. Aida, *Nature*, 2010, **463**, 339–343.
214. M. Lemmers, J. Sprakel, I. K. Voets, J. van der Gucht and M. A. Cohen Stuart, *Angew. Chem. Int. Ed.*, 2010, **49**, 708–711.
215. J. N. Hunt, K. E. Feldman, N. A. Lynd, J. Deek, L. M. Campos, J. M. Spruell, B. M. Hernandez, E. J. Kramer and C. J. Hawker, *Adv. Mater.*, 2011, **23**, 2327–2331.
216. F. Peng, G. Li, X. Liu, S. Wu and Z. Tong, *J. Am. Chem. Soc.*, 2008, **130**, 16166–16167.
217. N. Holten-Andersen, M. J. Harrington, H. Birkedal, B. P. Lee, P. B. Messersmith, K. Y. C. Lee and J. H. Waite, *Proc. Natl. Acad. Sci. U. S. A.*, 2011, **108**, 2651–2655.
218. A. Avdeef, S. R. Sofen, T. L. Bregante and K. N. Raymond, *J. Am. Chem. Soc.*, 1978, **100**, 5362–5370.

CHAPTER 4

Synthesis and Properties of Slide-Ring Gels

KAZUAKI KATO* AND KOHZO ITO*

Graduate School of Frontier Sciences, The University of Tokyo, 5-1-5 Kashiwanoha, Kashiwa, Chiba 277-8561, Japan
*Email: kato@molle.k.u-tokyo.ac.jp; kohzo@k.u-tokyo.ac.jp

4.1 Introduction

All gels are classified into two categories: chemical gels and physical gels.[1] When cross-linked polymeric materials are immersed in a good solvent, they absorb the liquid until the swelling force associated with the mixing entropy between the chains and the solvent balances the elastic force of the chains between junction points. These cross-linked polymeric systems containing solvent are called chemical gels. The swelling behaviour of chemical gels was explained in detail by Flory and Rehner.[2] Tanaka discovered the volume phase transition of chemical gels, in which the swelling and shrinking behaviours exhibit discontinuous profiles with hysteresis.[3] This novel discovery regarding cross-linked polymeric materials has attracted considerable interest from researchers in the field of polymer science. As a result, some interesting aspects of chemical gels have been discovered in succession, including the kinetics of the volume phase transition (Tanaka and Fillmore[4]), the frozen or fixed inhomogeneous structure of the chemical gels (Shibayama[5]), and the abnormally small friction behaviour of gels (Gong and Osada[6]). In addition, various new types of gels have been developed. For example, Gong and Osada developed a double network gel having a high modulus up to the sub-megapascal range, with a

Monographs in Supramolecular Chemistry No. 11
Polymeric and Self Assembled Hydrogels: From Fundamental Understanding to Applications
Edited by Xian Jun Loh and Oren A. Scherman
© The Royal Society of Chemistry 2013
Published by the Royal Society of Chemistry, www.rsc.org

failure compressive stress as high as 20 MPa.[7] The double network gel had both soft and hard components so as to avoid fracture, as in the case of biomaterials. Furthermore, Yoshida incorporated a dissipative system into gels to realize a self-oscillating gel device,[8] and also reported comb-type grafted hydrogels which showed a rapid de-swelling response to temperature changes.[9]

Physical gels have noncovalent cross-linking junctions arising from ionic interactions, hydrophobic interactions, hydrogen bonding, microcrystal formation, helix formation, and so on. In general, these noncovalent cross-links are not as strong as the covalent cross-links in chemical gels, and physical gels show a sol–gel transition response to temperature, pH, and solvent. The mechanical behaviour of physical gels is complex because the recombination of cross-linking points occurs on deformation; hence, the affine deformation model is not available in the physical gel. The recombination causes hysteresis in the stress–strain curve of the physical gel, which means that it cannot regain its original shape after deformation as quickly as the chemical gel. Haraguchi *et al.* have recently developed a novel type of physical gel, the nanocomposite gel, which uses clay as the cross-linking junction.[10,11] Polymer chains of *N*-isopropylacrylamide strongly absorb onto the clay surface at both ends, thus bridging different clays. Surprisingly, the nanocomposite gel shows high stretchability, up to 10 times its original length. This suggests that the structures of the cross-linking junctions play important roles in determining the mechanical properties of polymeric materials.

Another recent approach towards a polymer network structure has been developed using polyrotaxane in supramolecular chemistry. Supramolecules and their topological characteristics have attracted considerable interest. A typical example is that of rotaxanes, in which cyclic molecules are threaded on a single polymer chain and trapped by capping the chain with bulky end groups.[12 14] In 1990, Harada and Kamachi reported the first synthesis of pseudo-polyrotaxane in which many α-cyclodextrin (α-CD) molecules were threaded on a single polymer chain of poly(ethylene glycol) (PEG):[15] CDs mixed with PEG in water were threaded onto a PEG self-assembly. Subsequently, in 1992, both ends of the pseudo-polyrotaxane were capped with bulky groups to form polyrotaxane.[16] In recent years, this novel architecture in supramolecular chemistry has attracted great attention as a new approach for developing functional polymeric materials.[14,17]

The first report of a physical hydrogel based on the polyrotaxane architecture was also presented by Harada *et al.*[18] When α-CDs were mixed with long PEG chains at high concentration in water, a sol–gel transition occurred due to hydrogen bonding between the α-CDs threaded on the PEG chains in different pseudo-polyrotaxanes. In addition, Yui and co-workers formed some hydrogels using biodegradable CD polyrotaxane[19] as the cross-linker for use in regenerative medicine.[20] The biodegradable polyrotaxane is subject to hydrolysis at an ester bond between a bulky end group and the polymer axis. Consequently, the erosion time of the biodegradable hydrogel strongly depends on its polyrotaxane content. Furthermore, Takata *et al.* synthesized recyclable cross-linked polyrotaxane gels: topologically networked polyrotaxane capable

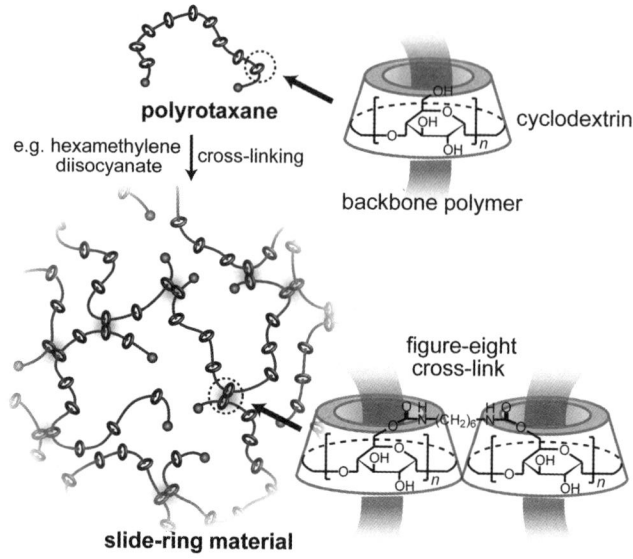

Figure 4.1 Cross-linking of polyrotaxanes by coupling of cyclodextrins inter-molecularly to produce slide-ring materials.

of undergoing reversible assembly and disassembly based on the concept of dynamic covalent bond chemistry.[21] They cross-linked poly(crown ether)s with dumbbell-shaped axle molecules, which showed a reversible cleavage of the disulfide bond. As a result, a novel reversible cross-linking/uncross-linking system that could recycle networked polymeric materials was realized.

We have recently developed another cross-linking structure based on the polyrotaxane architecture.[22] We prepared polyrotaxane sparsely containing α-CDs and subsequently cross-linked the α-CDs on different polyrotaxanes. As a result, the cross-linked junctions of the figure-of-eight shapes were not fixed on the PEG chains and could move freely in the polymer network (Figure 4.1). We refer to this new cross-linked polymer network as a *slide-ring gel*. Such a polymeric material with freely movable cross-links was theoretically considered as a sliding gel by de Gennes in 1999.[23] We describe herein highlights of the synthesis, properties, and applications of slide-ring gels, including recent synthetic breakthroughs and findings regarding their viscoelastic properties. It is noteworthy that the concept of freely movable cross-links is not limited only to slide-ring gels containing solvents, but can also be applied to slide-ring polymeric materials without solvents. This may bring about a paradigm shift in cross-linked polymeric materials, since their initial discovery by Goodyear.

4.2 Synthesis of Slide-Ring Materials

As mentioned above, the slide-ring gel was obtained by intermolecular cross-linking of polyrotaxanes *via* covalent bonds between CDs. Similar to the original, the precursor polyrotaxane consisted of PEG and α-CD. The molecular weight of the PEG was much larger, with a molecular weight of *ca.* 20 000, compared to the original (*ca.* 3000), because polymers having larger molecular weights form sparse inclusion complexes with α-CD. As the result of end-capping of the inclusion complex, a sparse polyrotaxane was obtained with 28% surface coverage, which is a measure of the CD packing density along the main chain polymer. Then, intermolecular cross-linking of the captured α-CDs yielded a transparent gel. A control experiment with the mixture of end-capped PEG and α-CD instead of polyrotaxane showed no gelation under the same conditions. Additionally, treatment with a strong base liquefied the gel, as the result of the elimination reaction of bulky groups at both ends of the PEG. This was the first experimental evidence that indicated the slidability of the cross-links.

The originally synthesized PEG-α-CD slide-ring material has been studied intensively in terms of physical properties. Simultaneously, various material designs have been demonstrated through modification of the cyclic components. This rapid progress of study has been supported by the established synthesis of the PEG-based polyrotaxanes.[16] As will be discussed, all the fundamental studies, until very recently, have been performed with the first-generation slide-ring material synthesized from PEG-based polyrotaxanes. Our recent progress in polyrotaxane synthesis has enabled diverse polyrotaxanes with various backbone polymers. This allows us to produce the next generation of slide-ring gels and discover additional features of these materials.

Generally, a polymer network that is cross-linked covalently is well known to have large areas of inhomogeneity, as shown in Figure 4.2. As the result of uneven cross-linking reactions, partial chains, which are the chains between neighbouring cross-links, have large distributions in length. Such structural inhomogeneity considerably reduces the mechanical strength of the network, because the tensile stress is concentrated in the shorter partial chains when the network is stretched. On the other hand, topological cross-links can avoid the concentration of stress. Even when the shorter partial chains are subjected to large stress, the stress is distributed to the adjacent partial chains by the movement of cross-links along polymer chains like a moving pulley, as shown in Figure 4.3(a). It is noteworthy that the equalization of tension can take place

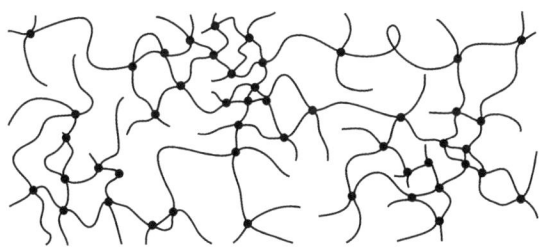

Figure 4.2 Inhomogeneity of polymer network cross-linked covalently as the result of unavoidably uneven cross-linking reactions.

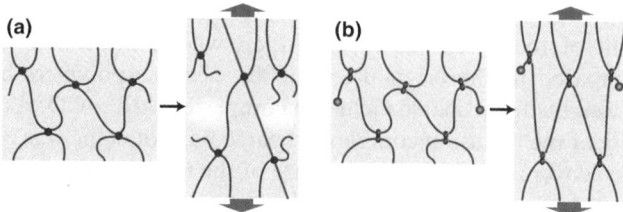

Figure 4.3   (a) Stress concentration of covalently cross-linked network under deformation and (b) the pulley effect of a topological network to equalize the
tensions in each chain.

not only in a single polymer chain but also among adjacent chains that are
interlocked by figure-of-eight cross-links. We refer to such an effect that arises
from the sliding of cross-links as the "pulley effect".[22] The pulley effect is a
concept that was created originally to explain the abnormal stretch properties
of a swollen gel. The concept represents the characteristics of slide-ring materials, and therefore we can expect various unique properties that are different
from those in a chemical gel, as will be discussed later.

4.3   Homogeneous and Isotropic Networks under Deformation

As mentioned above, covalently cross-linked networks have inherently large
inhomogeneous structures due to the non-random distribution of their cross-
links.[5] This inhomogeneity increases with swelling or deformations. Since the
cross-linking junctions are permanently fixed on the polymer chains, the network cannot adjust its cross-link distribution or polymer length in the network
when its environment changes. Experimentally, such inhomogeneities under
deformations are detected clearly by small-angle neutron scattering (SANS) as
abnormal butterfly patterns, *i.e.* prolate patterns that are *parallel* to the
stretching direction.[24,25] A slide-ring gel, however, shows normal butterfly
patterns, *i.e.* prolate patterns that are *perpendicular* to the stretching direction.[26] The normal butterfly pattern is generally observed in homogeneous
materials, such as uncross-linked polymer films and polymer solutions, due to
the orientation of the polymer chains along the elongation or flow direction.[27]
Thus, the normal butterfly pattern of a slide-ring gel indicates that the inhomogeneity caused by non-random cross-linking reactions decreases upon
deformation. This abnormality can be interpreted by the flowing of polymer
chains, like solutions, through the cross-links, *i.e.* the pulley effect.

   Another proof of the pulley effect is observed in the isotropic nature of slide-
ring gels. Generally, cross-linked materials exhibit anisotropic orientations of
chain segments under deformation.[28] The orientation anisotropy is the origin of
the rubber elasticity, as has been demonstrated by accumulated studies of the
photoelasticity. This principle is applied generally to existing rubbery materials,
and it is known as the stress-optic law.[29] However, a small-angle X-ray

scattering (SAXS) study implies that this principle might not be applicable to slide-ring materials. Chemically cross-linked gels, when they are stretched horizontally, generally exhibit vertically elliptic patterns in the two-dimensional SAXS profiles. This is the result of the anisotropic orientation of chain segments: partial chains horizontal to the stretching direction are elongated, whereas vertical chains are compressed, as schematically shown in Figure 4.3(a). On the other hand, the slide-ring gels exhibit an isotropic scattering even when the gels are stretched to 1.5 times, as if the gels were not deformed.[30] The isotropic scattering that arises from the isotropic orientation of chain segments can be also interpreted by the pulley effect, as shown in Figure 4.3(b). The compressed partial chains can flow in the direction of extension through the cross-link, changing the counter lengths of partial chains. As a result, the orientation anisotropy of chain segments caused by deformation is relaxed.

Since the orientation anisotropy is closely correlated to the rubber elasticity, this relaxation suggests the possibility of stress relaxation of slide-ring gels. At the same time, the fact that the materials do not flow like liquids indicates elastic body characteristics. Therefore, the elasticity of the slide-ring gel might be different from the rubber elasticity, accompanying the deviation from the stress-optic law. The answer to the question of the possible elasticity is described later, based on our very recent findings of a viscoelastic relaxation that is observed in a new class of slide-ring gels.

4.4 Abnormal Mechanical Properties and Theoretical Descriptions

The mechanical properties of slide-ring gels are quite different from those of conventional gels. Noncovalently cross-linked gels show a J-shaped stress–strain curve with large hysteresis. This large hysteresis arises from the recombination of cross-links under deformation. Covalently cross-linked gels, on the other hand, exhibit no hysteresis because of the practically infinite lifetime of the cross-links, and the stress–strain curve is S-shaped. This S-shaped curve is well explained theoretically.[31,32] Unlike these conventional gels, the slide-ring gel exhibits a J-shaped curve without a hysteresis loop, as shown in Figure 4.4.[33] This peculiar behaviour can be explained qualitatively by a theoretical model, the so-called free junction model, in which three Gaussian chains at a cross-link are able to slide towards one another.[34] Figure 4.5 shows theoretically obtained stress–strain curves of a chemically cross-linked gel and a slide-ring gel in a low extension region. The slide-ring gel shows a concave-up stress–strain curve that agrees qualitatively with the J-shape in the experimental results shown in Figure 4.4, whereas the chemical gel is concave-down.

Many biomaterials, such as mammalian skin, vessels, and tissues, show J-shaped stress–strain curves, which usually provide toughness and no elastic instabilities, among other advantages.[35,36] The J-shaped stress–strain curve affords toughness because its low shear modulus drastically reduces the energy released in the fracture, driving crack propagation. In addition, the material

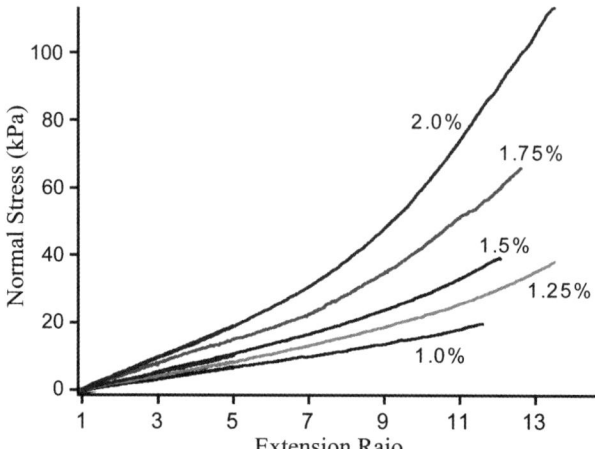

Figure 4.4   (a) Anisotropic orientation of chain segments under deformation observed in a covalently cross-linked network; (b) isotropic orientation observed in slide-ring gels due to the pulley effect.

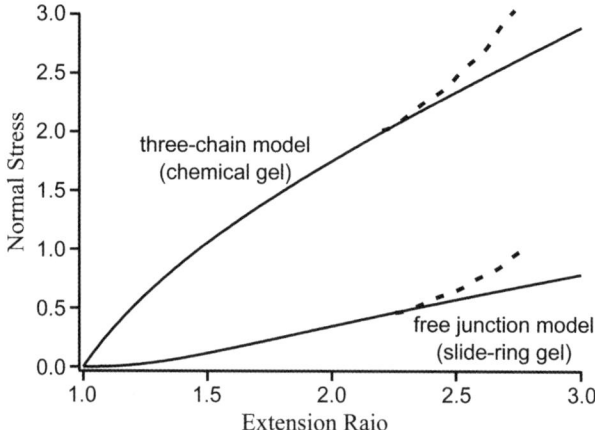

Figure 4.5   J-shaped stress–strain curves without hysteresis observed in slide-ring gels. Numbers indicate the cross-linker concentrations.

becomes stiffer as the extension ratio approaches the fracture point. The slide-ring materials show J-shaped stress–strain curves that are similar to the curves shown by biomaterials. This indicates that the slide-ring materials are suitable for use as substitutes for various types of biomaterials. If artificial arteries were made of the slide-ring materials, they may integrate better with native tissue than fixed cross-linked materials.

   Recently, systematic biaxial strain testing elucidated the details of the stress–strain behaviour of slide-ring gels.[37] Conventional gels and elastomers with covalent cross-links generally show large strain-coupling; for example, elongation of materials affects the stresses not only on the vertical face to the direction of elongation but also on the horizontal faces.[38,39] This is because all chains in a network are connected and therefore deformations of each chain are strongly coupled with each other. However, in the case of slide-ring gels, strain-coupling is significantly small. The stress–strain behaviour of slide-ring gels shows good agreement with the neo-Hookean model, which describes the behaviour of ideal rubber elasticity.[40] Additionally, the abnormally small coupling is within the result of volume conservation. Thus, this behaviour demonstrates a possible mechanism that can decouple the strains in different directions, *i.e.* the pulley effect.

4.5 Functionalization by Modification of the Cyclic Components

As the details of the pulley effect are revealed, chemical modifications to the cyclic components of slide-ring materials have also helped develop new application studies. By taking an advantage of CDs bearing hydroxyl groups, various functional groups have been introduced on the cyclic component to obtain new materials with different properties. Photo-responsive changes in the volume of a slide-ring gel have been realized by the modification of CDs with azobenzene derivatives.[41] The magnitude of volume change is larger than that of azobenzene-modified gels with covalent cross-links, because of the pulley effect which fully produces the macroscopic effect of the proven *trans–cis* photoisomerization. Mesogen-modified slide-ring materials are expected to respond much faster than existing liquid crystalline polymers. Various fundamental properties have been investigated precisely for a promising precursor of liquid-crystalline materials, a mesogen-modified polyrotaxane.[42] Additionally, modifications with polymers are also possible. Slide-ring materials that have CDs modified with polycaprolactone are transparent and flexible amorphous films, even without solvents.[43] Subsequent studies revealed that this kind of material shows remarkable scratch resistance; the materials are currently being applied as coating materials for automobiles and cell phones. This property is considered to be a result of the slidable grafted chains.

All these studies draw on the common design of attaching the new mobility mode to proven functional groups, to innovate new materials that excel in these functions compared to existing polymer materials. Thus, the diversity of the chemical modification of the cyclic component is quite important. Fortunately, the most readily available slide-ring material has α-CD as the cyclic component, on which various modifications have performed.

4.6 Significance of the Diversification of the Backbone Polymers

In contrast to the diversity of the cyclic components, however, the backbone polymer of the slide-ring material is uniform. Since the cross-linking manner is the most unique feature, the effect of the backbone polymer has not attracted the interest of researchers. As shown schematically in Figure 4.1, CD is represented as a ring and the backbone polymer is a mere string. This coarse graining, which is a successful technique in polymer physics, has contributed to the progress in understanding the abnormal properties of slide-ring materials. However, various material properties must depend on the chemical nature of the string, as is experimentally observed in existing polymer materials. Thus, also in the case of slide-ring materials, it is not possible to design materials without considering the diversity of the string. Moreover, it is easy to assume that a difference in the string influences different slidability: thicker strings weaken the ability and bulky side chains of strings disable the sliding. Therefore, the diversification of backbone polymers is key to the material design and extraction of the latent potential of slide-ring materials.

Although many studies have elucidated the features of slide-ring materials since the original synthesis from PEG-based polyrotaxane in 2001,[22] almost all studies to date have dealt with the single-species precursor polyrotaxane. The main reason for this is the difficulty of synthesizing diverse polyrotaxanes, although inclusion complexes have a wide range of backbones.[14] The end-capping reactions of inclusion complexes always compete against dissociation, and the end groups must be sufficiently bulky to prevent unravelling. The thicker polymer chains select larger CDs, and therefore bulkier capping reagents are needed. Many polymers are known to form inclusion complexes with γ-CD selectively.[14] Besides, unlike water-soluble PEG, hydrophobic polymers require inhomogeneous complexation with the water-soluble CDs and specific techniques to achieve the subsequent end-capping. Another reason is that the diversification appears to be unpromising, because no-one has demonstrated the properties of diversified materials. Very recently, we established a versatile synthesis of diverse polyrotaxanes with various backbone polymers, as mentioned in Chapter 8. This synthetic breakthrough enabled us to produce various slide-ring materials, resulting in the elucidation of the new properties of these materials, as mentioned in Chapter 9. In the next chapter, we introduce a successful example of material design based on the diversity of backbone polymers.

4.7 Organic–Inorganic Hybrid Slide-Ring Materials

Inorganic polymers represented by silicone have characteristic properties such as thermostability, insulation, and gas permeability. By hybridization techniques with organic components, we can obtain materials that exhibit properties derived from both the inorganic and organic components. Such organic–inorganic hybrid materials rely on mixing techniques to avoid macroscopic separation of both components, though each component has quite different affinity. Three representative methods of hybridization are shown in Figure 4.6. For example, in the case of Figure 4.6(a), organic monomers attached at both the ends of silicone are polymerized in the presence of hydrophilic organic monomers, yielding a network composed of silicone and hydrophilic organic

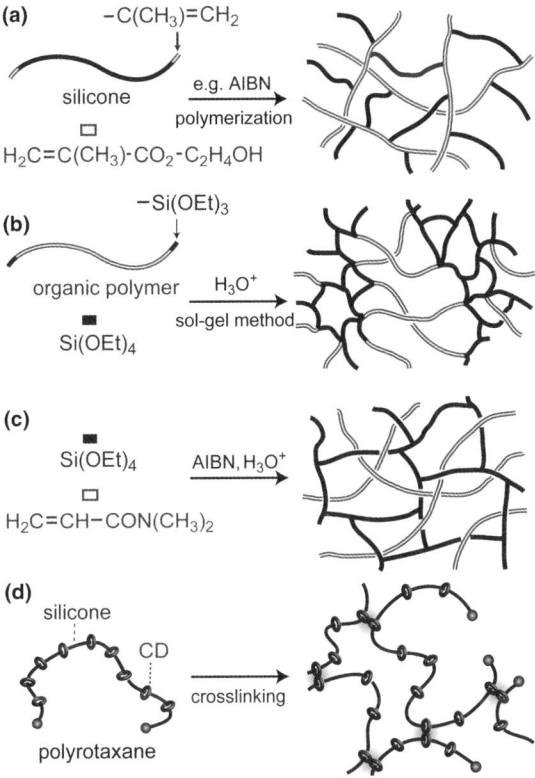

Figure 4.6 (a–c) Three major techniques for the hybridization of organic and inorganic components; (d) a newly introduced method based on an organic–inorganic polyrotaxane and subsequent cross-linking.

polymer.[44] This product is the material of soft contact lenses with high oxygen permeability. In any of the methods shown in Figure 4.6(a)–4.6(c) the bonding or entangling of each component during the polymerization processes is the key for avoiding macroscopic phase separation.[45,46]

Polyrotaxanes are molecules composed of a backbone polymer and topologically bound cyclic molecules, and therefore neither component can separate from the other, even without chemical bonds or strong interactions. Thus, polyrotaxanes can be cross-linked without the macroscopic phase separation of each component to yield a homogeneous network. This feature of polyrotaxanes inspired a new class of hybrid materials, shown in Figure 4.6(d), based on organic–inorganic polyrotaxanes. Replacing the backbone polymer of slide-ring materials with silicone, for example, results in a hybrid material in which each component never separates from the other macroscopically, but microscopically. Also, the slidability enables the variation of the relative position of each component. This unique structural feature is expected to change various physical properties of the material.

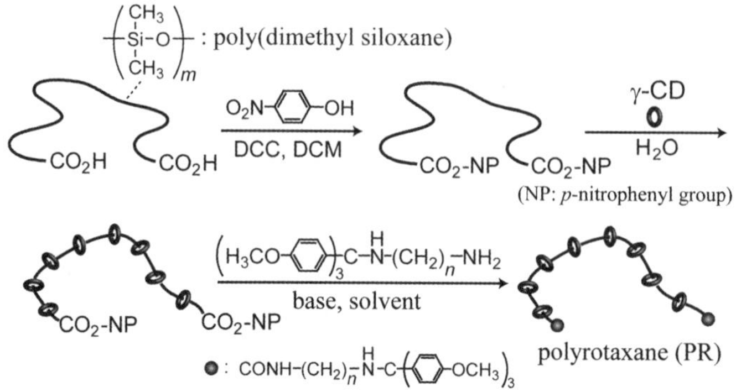

Figure 4.7 Synthesis of silicone-based polyrotaxanes. Production yield and filling ratio of the polyrotaxanes strongly depend on the conditions of the capping reaction and the numbers of methylene spacers in the capping reagent.

n of spacer	base[a]	solvent	condition	yield (%)[b]	filling ratio (%)[c]
2	EDIPA	MeCN	RT, 48 h	15	2
4	EDIPA	MeCN	RT, 72 h	21	2
4	EDIPA	DMF	RT, 72 h	< 1	-
4	EDIPA	DCM	RT, 72 h	< 1	-
6	None	MeCN	RT, 72 h	30	2
6	None	MeCN	50 °C, 72 h	53	7

[a] EDIPA: ethyldiisopropylamine. [b] Yields based on PDMS. [c] Estimated from TGA based on a definition that 100% means 3:2 molar ratio between the monomer unit and γ-CD.

Silicone-based polyrotaxane, which is a key precursor of the hybrid material, is synthesized by the process shown in Figure 4.7.[47] Important points of this synthesis are: (1) pretreatment of the polymer ends with an active ester; (2) adoption of a quite bulky moiety for capping; and (3) strong dependence on the reaction solvent. Since silicone is substantially thicker than PEG, complex formation does not occur with small cyclic molecules such as α- and β-CD, but rather with γ-CD.[48] Thus, proven capping reagents for the PEG-α-CD polyrotaxane, such as adamantane and dinitrobenzene derivatives, are too small for the PDMS-γ-CD polyrotaxane and result in unravelling. Bulky *p*-methoxytrityl derivatives prevent unravelling, whereas non-substituted trityl groups are inadequate. The bulky capping reagents successfully yield polyrotaxanes only when the ends of the silicone are activated with *p*-nitrophenyl ester in advance. In the case of PEG-based polyrotaxane, the carboxyl ends of PEG react with bulky amines in the presence of condensation agents such as BOP.[49] In contrast, however, silicone-based polyrotaxanes are not obtained through condensation using BOP. We speculate that γ-CD covers the end groups of the polymer to prevent the formation of the active complex among the carboxylic acid, amine, and BOP. The significant effect of the methylene spacer on the yield of polyrotaxane indicates that the reaction site is crowded. In addition to the bulky reagents and active polymer ends, suitable solvents are needed for the capping reaction. Reactions in DMF or dichloromethane produce no polyrotaxane. A majority of the silicone is extracted by dichloromethane within five minutes. Capping reactions must be faster than unravelling of the inclusion complexes. Since DMF and dichloromethane are, respectively, good solvents for γ-CD and silicone, the unravelling occurs before the end-capping reaction. On the other hand, acetonitrile is a good solvent only for the capping reagents, and not for γ-CD or silicone. The same reactions in acetone yield polyrotaxanes because of the similar affinity to acetonitrile. These solvents might be ideal for the capping reaction, because they accelerate the end-capping reaction and decelerate the unravelling.

In this way, we must consider various points for the synthesis of polyrotaxanes. The poor solubility of the obtained polyrotaxanes presents another problem in terms of characterization and the subsequent cross-linking reaction to obtain the slide-ring materials. Particularly in the case of silicone-based products, the polyrotaxane is not soluble in any general solvents, except for DMF in the presence of LiCl. The poor solubility mainly comes from the radically different solubility of each component, unlike PEG-α-CD polyrotaxanes. Also, the ratio between both components dramatically changes the solubility of the polyrotaxane. A polyrotaxane with a 2% filling ratio does not dissolve in DMF/LiCl, whereas one with a 7% filling ratio is soluble. This may be because the ratio of insoluble silicone increases. The fact that the silicone-based polyrotaxane with a 7% filling ratio dissolves in DMF/LiCl is interesting, because silicone itself is not soluble in the solvent. The solubility of fully-covered polyrotaxanes, with a 100% filling ratio, is determined by the solubility of the cyclic components, and therefore polyrotaxanes are often used to dissolve insoluble polymers.[50] We intensively investigate now the structure and dynamics using SANS and SAXS combined with electron microscopy and quasi-elastic light scattering.

The solution of silicone-based polyrotaxane in DMF/LiCl is cured by cross-linking agents reactive to the cyclic components, yielding transparent and flexible gels as shown in Figure 4.8. Since the polyrotaxane formation between silicone and γ-CD already prevents macroscopic phase separation, slide-ring gels of the cross-linked polyrotaxane also exhibit no phase separation.

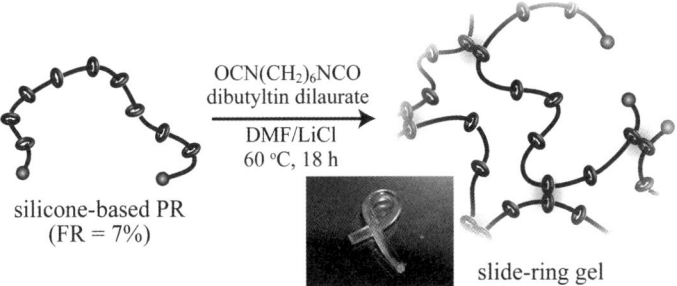

silicone-based PR
(FR = 7%)

OCN(CH$_2$)$_6$NCO
dibutyltin dilaurate

DMF/LiCl
60 °C, 18 h

slide-ring gel

Figure 4.8 Cross-linking reaction of the silicone-based polyrotaxane to yield a slide-ring gel.

Although the obtained gel is transparent, microscopic domains or phase separations are possible when CDs aggregate within the network. Combined with the slidability, it might be possible to vary and control such sub-micron structures and the resulting physical properties of the hybrid materials.

4.8 A Breakthrough Synthesis of Polyrotaxanes for the Diversification of Backbones

The synthesis of the above-mentioned silicone-based polyrotaxanes and the subsequent cross-linking is the first example which realized a slide-ring material based on backbone polymers other than PEG. This method, however, requires the synthesis of an excessive amount of bulky capping reagents. As mentioned above, the general synthesis of polyrotaxane consists of inclusion complex formation and end-capping of the complex. The complex formation is accelerated by excess CDs, and the excess CDs that do not form the complex are wasted. We reused the excess CDs for the subsequent capping reactions, yielding polyrotaxanes consisting of various CDs and backbone polymers as shown in Figure 4.9.[51] The differences between this method and the one shown in Figure 4.7 are the omission of the isolation process of the inclusion complex and the use of tertiary amines only without bulky capping regents. Adding the tertiary amine to the inclusion complex between CDs and linear polymers modified with the active ester triggers a transesterification reaction between the active ends of the polymer and excess CDs. As the result, polyrotaxanes

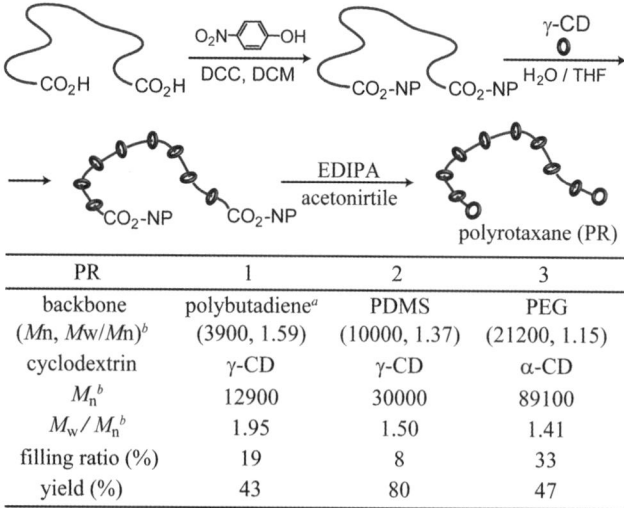

PR	1	2	3
backbone	polybutadiene[a]	PDMS	PEG
(Mn, Mw/Mn)[b]	(3900, 1.59)	(10000, 1.37)	(21200, 1.15)
cyclodextrin	γ-CD	γ-CD	α-CD
$M_n{}^b$	12900	30000	89100
$M_w / M_n{}^b$	1.95	1.50	1.41
filling ratio (%)	19	8	33
yield (%)	43	80	47

[a] trans:cis:vinyl = 42:37:21. [b] Measured by size exclusion chromatography.

Figure 4.9   Versatile synthesis of diverse polyrotaxanes. The single method produces three different polyrotaxanes with different backbones without additional bulky capping reagents.

end-capped with CDs are obtained in relatively high yields. The possible side reaction between polymer ends and CDs that form the inclusion complex is suppressed by the excess amount of the free CDs.

The crucial advantage of this method is that no additional end-capping reagent is required. Generally, according to the thickness and solubility of the backbone polymers and the size of the CDs, we must synthesize the suitable reagents and optimize the conditions of the capping reaction.[14,52] On the other hand, by a single common method, three different polyrotaxanes are synthesized from different linear polymers and different CDs. A polyrotaxane consisting of polybutadiene and γ-CD was the first synthesized by this method. As we describe later, this novel polyrotaxane enabled us to discern an unknown crucial feature of slide-ring materials. This synthesis is a breakthrough in polyrotaxane synthesis because of its versatility. Currently, we can obtain other novel polyrotaxanes by this method, such as polyalkene-γ-CD, poly(propylene glycol) (PPG)-CDs, and PEG-γ-CD.

4.9 Viscoelastic Relaxation Reflecting the Sliding Dynamics and the Mobility of the CDs

It is well known that common chemical gels and rubbers have finite equilibrium moduli (in the long time limit) arising from the entropic elasticity of the polymer conformation between fixed cross-links. In contrast, entangled polymer chains have no finite equilibrium modulus, like liquids, since the release or reptation of entangled polymers eventually relaxes the chain deformation responsible for the entropic elasticity. The slide-ring gel has a similar sliding motion to polymer chains at the cross-linking junction of the entangled polymer system. The experimental results show that the slide-ring gel has a Young's modulus that is much lower than the chemical gel of the same cross-linking density and finite equilibrium in the low-frequency limit.[53] This is because the polymer chain can slide but is not released by the topological restriction, since the stopper of polyrotaxane at both ends hinders the perfect dissociation of the axis polymer chain from the cross-links. What, then, determines the low Young's modulus of the slide-ring gel?

The experimental results show that a number of CDs remain uncross-linked in the slide-ring gel.[54] These CDs have a sliding motion along an axis polymer in polyrotaxanes, yielding the alignment entropy. These free cyclic molecules play a dominant role on the Young's modulus in the slide-ring gel. Figure 4.10 schematically shows the molecular mechanism of the elasticity in the slide-ring gel. When the slide-ring gel is stretched along the horizontal axis, polymer chains are deformed to an anisotropic conformation on a short timescale. As time passes, the axis polymer relaxes the polymer deformation by sliding at the cross-links, *via* the pulley effect.[22] The SAXS results on uniaxial deformation indicate the relaxation of the conformational anisotropy in the slide-ring gel, as mentioned before.[30] However, free cyclic molecules cannot pass through the cross-links consisting of other rings of the same size, but the axis polymer

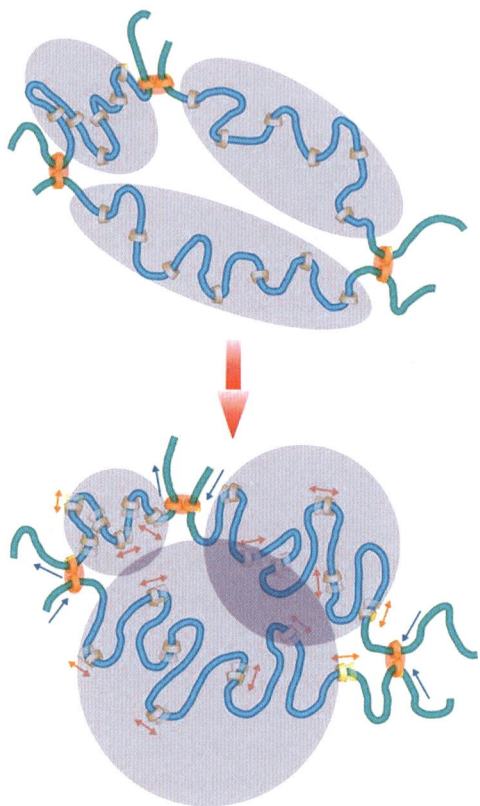

Figure 4.10   The relaxation of the conformation anisotropy of polymer chains by the
sliding of the axis polymer at the cross-linking junctions. Then free cyclic
molecules form a heterogeneous density distribution since they cannot
pass through cross-links consisting of other rings with the same size.

chains can easily do so. As a result, the axis polymer sliding forms a heterogeneous density distribution of free cyclic molecules as axis polymer chains
relax the conformation anisotropy. The heterogeneous distribution reduces
the alignment entropy of the free molecules dramatically, which leads to an
entropic elasticity completely different from that due to the polymer conformation. The finite equilibrium, low-modulus characteristics of the slide-ring
gel should come from the novel kind of entropic elasticity due to the alignment
of free cyclic molecules on cross-linked polyrotaxanes.

Figure 4.11 shows the frequency dependences of storage moduli E' and the
loss moduli E'' obtained by dynamic viscoelastic measurements of polybutadiene-based slide-ring gels that are prepared by cross-linking of the above-
mentioned polybutadiene polyrotaxane solution in DMSO.[55] It is obvious that
two different plateaus of E' exist with a viscoelastic relaxation, whereas existing
cross-linked materials generally show no relaxation in this frequency range.
This relaxation is not like the well-known glass transition that also shows a

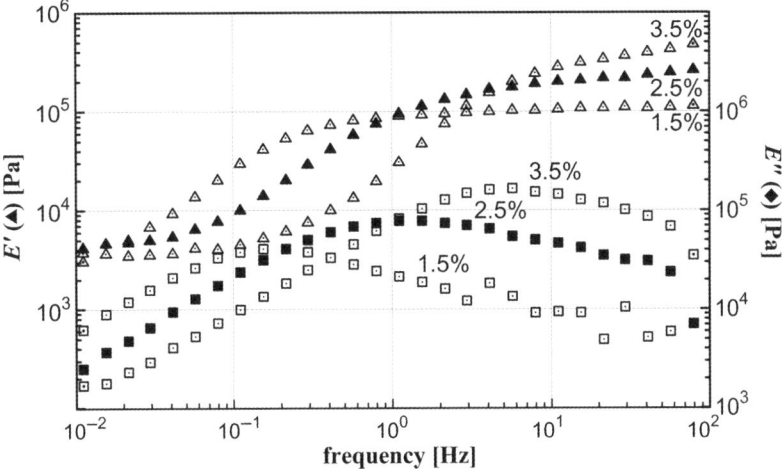

Figure 4.11   Viscoelastic relaxation of polybutadiene-based slide-ring gels. Frequency dependences of elastic modulus E' and E'' measured at 30 °C in DMSO are shown. Numbers indicate the concentrations of cross-linker.

viscoelastic relaxation from the glassy state, which generally has E' of 10^9 Pa, to the rubbery state, with 10^6 Pa. Initially, to elucidate the relaxation mechanism, the number-average molar mass of the network strands M_x is estimated from each plateau modulus E_r and the network density ρ (mass per unit volume) by the following equation, with the assumption of an affine network:[56]

$$E_r = \frac{3\rho RT}{M_x} \tag{4.1}$$

where R is the gas constant and T is the absolute temperature. The values of E' at 80 Hz or 0.01 Hz were substituted for E_r. The values of E' at higher frequency give values of several thousand for M_x, whereas those at lower frequency result in several hundred thousand for M_x. The molar mass of the precursor polyrotaxane is about 25 000 and it has about ten CDs. Thus, the former M_x value gives a picture that one polyrotaxane has several cross-link junctions, whereas the latter would mean several tens of polyrotaxanes form a single partial chain of the network. Therefore, the former picture is more probable, meaning that the plateau at higher frequency is assigned to be the rubbery plateau and that the lower one indicates a previously unknown state.

In addition to the consideration from the plateau moduli, the relaxation dynamics depending on the cross-linking density gives us clearer pictures of the molecular origin of this relaxation. Although detailed discussion is omitted here, the relaxation time shows a proportional relation to the cube of M_x, as shown in Figure 4.12.[55] From the analogy with reptation theory[28] that draws the cubic power dependence on the average molar mass of polymer chain M rather than M_x, the dynamics of this relaxation would correspond to a local

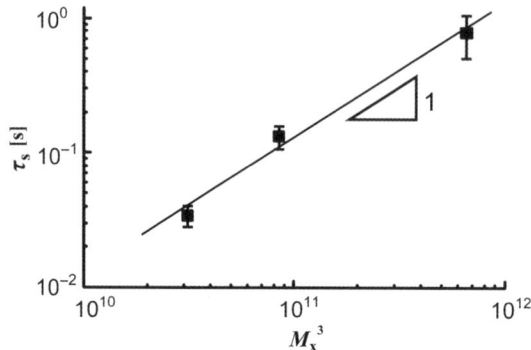

Figure 4.12   Cubic power dependence on the average molecular weight of partial chain M_x to the relaxation time τ.

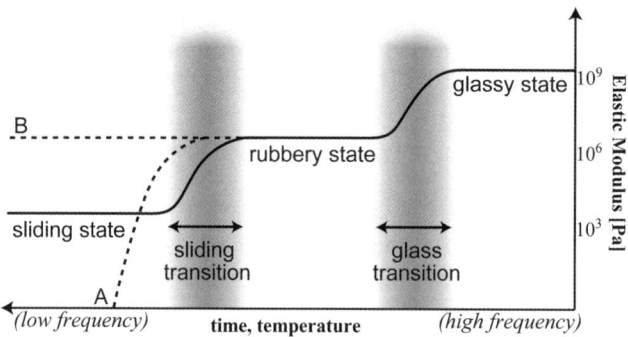

Figure 4.13   Schematic overview of viscoelastic behaviour of slide-ring materials, comparing general amorphous polymer (*dashed line A*) and cross-linked rubber (*dashed line B*).

diffusion of partial chains of the network. The partial chains can diffuse beyond the figure-of-eight cross-links just like entangled polymer chains, though the overall diffusion of polymer chains is inhibited by the bulky end-groups. Such a picture of reptation-like local diffusion of partial chains is consistent with the dependence of relaxation time on M_x rather than M. Consequently, the viscoelastic relaxation indicates a peculiar transition between one state where the cross-links behave as though fixed and another where they slide freely.

   Figure 4.13 shows a rough comparison of the viscoelastic behaviour between general polymers and slide-ring materials. It is well known that amorphous long polymers generally exhibit the glass state, glass transition, rubbery state and macroscopic flow.[56] In the glass state, only local vibrations of covalent bonds are permitted. During the glass transition, conformational changes begin and then micro-Brownian motion becomes dominant in the rubbery state. Micro-Brownian motion is considered to be a cooperative segment motion consisting of several monomer units and is an origin of rubber elasticity. On longer timescales or at higher temperature, materials flow, reflecting the overall diffusion of polymers (dashed line A in the figure), whereas flow is prevented by the cross-linking of polymers, showing a longer rubbery plateau (dashed line B). On the other hand, slide-ring materials exhibit another relaxation: the partial chains flow locally through the cross-links, independent of the overall diffusion that is not allowed because of the end-capping. Thus, this relaxation represents the nature of the topological cross-linking that has been called the pulley effect, and the timescale of relaxation indicates the dynamics of sliding in the slide-ring gels.

Notably, another plateau exists on longer timescales. Combined with the previously described experimental fact of the isotropic orientation of chain segments under deformation, the novel plateau is thought to indicate an origin of elasticity other than the micro-Brownian motion. The local diffusion that reduces the orientation anisotropy of chain segments must accompany the heterogeneous distribution of uncross-linked CDs. The reason for this is that the number of uncross-linked CDs between the cross-links remains constant, independent of the fluctuation of partial chain lengths. Thus, the heterogeneous distribution of uncross-linked CDs can generate a new type of entropy elasticity that yields macroscopic stress.[53] That is, the lower plateau modulus arises from the entropy of CDs, not the backbone polymer. This novel elasticity gives an interpretation to the abnormal softness of slide-ring materials that is observed as, for example, the gradual initial gradient of the J-shaped stress–strain curve. Through this viscoelastic study, we found that the arrangement entropy of uncross-linked CDs is another characteristic that affects the mechanical properties of slide-ring materials to the same degree as the sliding of the backbone.

4.10 Conclusions

Since the first slide-ring material was created in 2001, studies in polymer physics have contributed greatly to elucidating the peculiar physical properties that can be summarized in the pulley effect. Synthetic organic chemistry has advanced the functionalization of these materials by diverse chemical modifications of the cyclic components. However, the uniformity of the backbone has prevented the diversification of the physical properties. Our recent studies on slide-ring materials with various backbone polymers have certainly diversified their mechanical properties and revealed previously unknown features of slide-ring materials, such as a viscoelastic relaxation, the dynamics of the sliding, and the contribution of the cyclic components to the elasticity.

Since its discovery by Goodyear in 1839, cross-linking has been considered to be fixed at the polymer chains. Rubber elasticity was well explained by the fixed junction model by assuming the affine deformation of the polymer chains. In contrast, the slide-ring gel has movable cross-linking and free cyclic molecules. The pulley effect due to the sliding of the axis polymer at cross-linking junctions relaxes the stress due to polymer deformation similar to the reptation of an entangled polymer system. Instead of the relaxation of polymer deformation by sliding, free cyclic molecules form a heterogeneous density distribution, which results in a substantial entropy loss. In other words, an entropy loss of the axis polymer deformation transfers to that of the heterogeneous distribution of free cyclic molecules. This means that the pulley effect yields a dynamic coupling between the conformational entropy of chains and the alignment entropy of rings. As a result, the sliding elasticity appears in the sliding state, where axis polymer chains and cyclic molecules are actively sliding. In addition, the slide-ring gel should show a sliding transition between the rubber and sliding states. These aspects indicate that the slide-ring gel demonstrates unique mechanical characteristics in its dynamics.

Furthermore, the diversity of the backbone polymer enabled us to design a new hybrid material. As the history of polymer rheology illustrates, understanding of material features has developed remarkably with every synthetic breakthrough, which has produced, for example, monodisperse polymers, branch polymers, and macrocyclic polymers. Similarly, the diversification of backbones and cyclic components would show us new features of the slide-ring material, and the understanding of the material must be accelerated by close collaborative research between organic chemists and polymer physicists. At the same time, the accumulated knowledge of supramolecular chemistry is necessary to design slide-ring materials, because the slidability, which is the clearest characteristic of these materials, is synonymous with the asynchronous motion between host molecules and guest polymers. We believe that the promising future of slide-ring material lies in a new field of study combining polymer physics, synthetic organic chemistry, and supramolecular chemistry.

References

1. *Gels Handbook*, ed. Y. Osada and K. Kajiwara, Elsevier Academic Press, Amsterdam, 2000.
2. P. J. Flory and J. Rehner, Jr., *J. Chem. Phys.*, 1943, **11**, 512; *ibid.*, 1943, **11**, 521.
3. T. Tanaka, *Phys. Rev. Lett.*, 1978, **40**, 820.
4. T. Tanaka and D. J. Fillmore, *J. Chem. Phys.*, 1979, **70**, 1214.
5. M. Shibayama, *Macromol. Chem. Phys.*, 1998, **199**, 1.
6. J. P. Gong, M. Higa, Y. Iwasaki, Y. Katsuyama and Y. Osada, *J. Phys. Chem.*, 1997, **101**, 5487.
7. J. P. Gong, Y. Katsuyama, T. Kurokawa and Y. Osada, *Adv. Mater.*, 2003, **15**, 1155.
8. R. Yoshida, T. Takahashi, T. Yamaguchi and H. Ichijo, *J. Am. Chem. Soc.*, 1996, **118**, 5134.
9. R. Yoshida, K. Uchida, Y. Kaneko, K. Sakai, A. Kikuchi, Y. Sakurai and T. Okano, *Nature*, 1995, **374**, 240.
10. K. Haraguchi and T. Takehisa, *Adv. Mater.*, 2002, **16**, 1120.

11. K. Haraguchi and H.-J. Li, *Angew. Chem. Int. Ed.*, 2005, **44**, 6500.
12. F. H. Huang and H. W. Gibson, *Prog. Polym. Sci.*, 2005, **30**, 982.
13. G. Wenz, B. H. Han and A. Muller, *Chem. Rev.*, 2006, **106**, 782.
14. A. Harada, A. Hashidzume, H. Yamaguchi and Y. Takashima, *Chem. Rev.*, 2009, **109**, 5974.
15. A. Harada and M. Kamachi, *Macromolecules*, 1990, **23**, 2821.
16. A. Harada, J. Li and M. Kamachi, *Nature*, 1992, **356**, 325.
17. J. Araki and K. Ito, *Soft Matter*, 2007, **3**, 1456.
18. J. Li, A. Harada and M. Kamachi, *Polym. J.*, 1994, **26**, 1019.
19. T. Ooya and N. Yui, *Macromol. Chem. Phys.*, 1998, **199**, 2311.
20. J. Watanabe, T. Ooya and N. Yui, *J. Artif. Organs*, 2000, **3**, 136.
21. N. Kihara, K. Hinoue and T. Takata, *Macromolecules*, 2005, **38**, 223.
22. Y. Okumura and K. Ito, *Adv. Mater.*, 2001, **13**, 485.
23. P. G. de Gennes, *Physica A*, 1999, **271**, 231.
24. J. Bastide and L. Leibler, *Macromolecules*, 1988, **21**, 2647.
25. M. Shibayama, K. Kawakubo, F. Ikkai and M. Imai, *Macromolecules*, 1998, **31**, 2586.
26. T. Karino, Y. Okumura, C. Zhao, T. Kataoka, K. Ito and M. Shibayama, *Macromolecules*, 2005, **38**, 6161.
27. M. Shibayama, H. Kurokawa, S. Nomura, S. Roy, R. S. Stein and W. L. Wu, *Macromolecules*, 1990, **23**, 1438.
28. M. Doi and S. F. Edwards, *The Theory of Polymer Dynamics*, Clarendon Press, Oxford, 1986.
29. H. Janeschitz-Kriegl, *Polymer Melt Rheology and Flow Birefringence*, Springer, Berlin, 1983.
30. Y. Shinohara, K. Kayashima, Y. Okumura, C. Zhao, K. Ito and Y. Amemiya, *Macromolecules*, 2006, **39**, 7386.
31. L. R. G. Treloar, *The Physics of Rubber Elasticity*, Oxford University Press, Oxford, 3rd edn., 1975.
32. J. E. Mark and B. Erman, *Rubber Elasticity: A Molecular Primer*, Cambridge University Press, Cambridge, 2nd edn., 2007.
33. Y. Okumura and K. Ito, *Nippon Gomu Kyokaishi*, 2003, **76**, 31.
34. K. Ito, *Polym. J.*, 2007, **39**, 488.
35. J. F. V. Vincent, *Structural Biomaterials*, Princeton University Press, Princeton, NJ, 1990.
36. S. Vogel, *Comparative Biomechanics: Life's Physical World*, Princeton University Press, Princeton, NJ, 2003.
37. Y. Bitoh, N. Akuzawa, K. Urayama, T. Takigawa, M. Kidowaki and K. Ito, *Macromolecules*, 2011, **44**, 8661.
38. T. Kawamura, K. Urayama and S. Kohjiya, *Macromolecules*, 2001, **34**, 8252.
39. T. Kawamura, K. Urayama and S. Kohjiya, *J. Polym. Sci., Part B: Polym. Phys.*, 2002, **40**, 2780.
40. P. J. Flory, *Principles of Polymer Chemistry*, Cornell University Press, Ithaca, NY, 1953.
41. T. Sakai, H. Murayama, S. Nagano, Y. Takeoka, M. Kidowaki, K. Ito and T. Seki, *Adv. Mater.*, 2007, **19**, 2023.

42. M. Kidowaki, T. Nakajima, J. Araki, A. Inomata, H. Ishibashi and K. Ito, *Macromolecules*, 2007, **40**, 6859.
43. J. Araki, T. Kataoka and K. Ito, *Soft Matter*, 2008, **4**, 245.
44. P. C. Nicolson and J. Vogt, *Biomaterials*, 2001, **22**, 3273.
45. T. Ogoshi and Y. Chujo, *Compos. Interfaces*, 2005, **11**, 539.
46. R. Tamaki, K. Naka and Y. Chujo, *Polym. J.*, 1998, **30**, 60.
47. K. Kato, K. Inoue, M. Kidowaki and K. Ito, *Macromolecules*, 2009, **42**, 7129.
48. H. Okumura, Y. Kawaguchi and A. Harada, *Macromolecules*, 2001, **34**, 6338.
49. J. Araki and K. Ito, *Macromolecules*, 2005, **38**, 7524.
50. S. Brochsztain and M. J. Politi, *Langmuir*, 1999, **15**, 4486.
51. K. Kato, H. Komatsu and K. Ito, *Macromolecules*, 2010, **43**, 8799.
52. G. Wenz, B. H. Han and A. Müller, *Chem. Rev.*, 2006, **106**, 782.
53. K. Ito, *Polym. J.*, 2012, **44**, 38.
54. A. Bando, K. Kato, Y. Sakai, H. Yokoyama and K. Ito, to be submitted.
55. K. Kato and K. Ito, *Soft Matter*, 2011, **7**, 8737.
56. M. Rubinstein and R. Colby, *Polymer Physics*, Oxford University Press, New York, 2003.

CHAPTER 5

Peptide and Protein Hydrogels

LAWRENCE J. DOOLING AND DAVID A. TIRRELL*

California Institute of Technology, Division of Chemistry and Chemical
Engineering, 1200 East California Boulevard, MC 210-41, Pasadena,
CA 91125, USA
*Email: tirrell@caltech.edu

5.1   Introduction

Proteins are a fascinating class of macromolecules from both functional and
structural perspectives. They catalyze the reactions that sustain life, bind
ligands with high affinity and specificity, and mediate interactions among
biomolecules in complex cellular milieux. Proteins also assemble into higher-
order structures that are responsible for the mechanical integrity of cells and
tissues. Their diverse functional and structural properties have made proteins
important building blocks in the development of new biomaterials.[1–4]

Hydrogels are physically or chemically cross-linked polymer networks with
high water content.[5,6] Their formation requires a balance between the forces
driving the association of polymer chains and those mediating solvation of the
network. Given that proteins have evolved to fold and function in aqueous
environments and that many proteins self-assemble into larger structures, they
would seem to be ideal candidates for use as hydrogel precursors. Indeed,
proteins, and more broadly peptides of all sizes, are widely used for this
purpose.

The development of peptide and protein hydrogels has been a cross-
disciplinary effort combining the knowledge of protein structure and function
from biology with the synthesis and characterization tools of macromolecular

Monographs in Supramolecular Chemistry No. 11
Polymeric and Self Assembled Hydrogels: From Fundamental Understanding to Applications
Edited by Xian Jun Loh and Oren A. Scherman
© The Royal Society of Chemistry 2013
Published by the Royal Society of Chemistry, www.rsc.org

chemistry and materials science. The primary motivation driving this field is the need for implantable scaffolds for soft-tissue engineering,[7,8] benign methods for encapsulation of cells and biomolecules,[9,10] matrices for *in vitro* cell culture,[11] and injectable delivery vehicles for therapeutics.[12,13] Moreover, peptide hydrogels show promise as templates for biomineralization[14] and inorganic nanostructures.[15] The ability of proteins to undergo biochemical and structural changes in response to pH and temperature changes, ligand binding, light, and mechanical force also suggests potential applications for peptide hydrogels as biosensors and stimulus-responsive materials.[16,17] Finally, peptide hydrogels provide simplified systems for studying and engineering the assembly of biological molecules as well as inspiration for the development of self-assembling synthetic structures.

We identify four classes of peptide hydrogels: (1) hydrogels from self-assembling oligopeptides, (2) hydrogels from recombinant proteins, (3) hydrogels that are hybrid materials combining peptides and proteins with synthetic polymers, and (4) hydrogels from naturally sourced proteins and proteoglycans such as collagen. Despite the importance of this last class of hydrogels in cell culture and tissue engineering applications, there are only limited opportunities for engineering and rational design of its macromolecular components, and detailed characterization can be challenging because of material heterogeneity. Therefore, after a brief background on peptide and protein structure, we will limit our discussion to examples from the first three classes of peptide hydrogels before concluding with future directions and challenges in this field.

5.2   Peptide and Protein Structure

Peptides are polymers formed by the condensation of amino acids. Short polymer chains with degrees of polymerization less than approximately 25 amino acids are generally described as oligopeptides, while longer chains are known as polypeptides. The term protein will be reserved for polypeptides that have been synthesized by ribosomal translation.

5.2.1   Peptide and Protein Synthesis

The most basic level of protein structure is the linear sequence of amino acids, also known as the primary structure. The protein sequence is genetically encoded in DNA, or more specifically in the portion of the DNA that is transcribed into messenger RNA. Protein synthesis occurs on the ribosome, which translates the messenger RNA and catalyzes the formation of amide bonds between amino acids and the growing polypeptide chain. Through genetic engineering, it is possible to produce natural or artificial proteins in a wide variety of host organisms, typically *Escherichia coli* but occasionally yeast and higher eukaryotes. Recombinant protein production yields monodisperse polymers with precise sequence control that is not possible with synthetic polymerization methods. Oligopeptides and short polypeptides are synthesized *in vitro* using solid-phase peptide synthesis (SPPS), with an upper limit of approximately 50 amino acids.[18] In this method, the peptide chain remains tethered to a resin support as each residue is added by amide coupling of a protected amino acid and its subsequent deprotection. Longer sequences can be synthesized by combining oligopeptides using native chemical ligation.[19]

With a few important exceptions,[20,21] nature incorporates only 20 different amino acids into proteins. These are known as the canonical or natural amino acids. While the number of ways in which these monomers can be combined is nearly infinite, the diversity of the canonical amino acid side chains is somewhat limited from the viewpoint of synthetic chemistry. From the perspective of investigators who wish to design new peptide and protein biomaterials, it is advantageous to augment the canonical amino acids with new monomers that carry more diverse functionality, including reactive groups for bioorthogonal chemistry,[22,23] halogens,[24-26] and photo-reactive moieties.[27] Furthermore, natural proteins contain exclusively L-amino acids, which can influence protein structure and biomaterial assembly.[28] Non-canonical amino acids can be incorporated into synthetic oligopeptides, assuming that their side chains do not interfere with the coupling and deprotection reactions or that suitable protecting groups are available. Numerous strategies have also been described for the co-translational incorporation of non-canonical amino acids in a residue-specific[29] or site-specific manner.[30]

5.2.2 Higher-Order Structure and Hydrogel Assembly

Protein secondary structure describes the local conformation of the peptide backbone. Different conformations arise from the rotational degrees of freedom of the N–C_α and C_α–C bonds. The most common secondary structures are the α-helix and β-sheet, which are characterized by extensive hydrogen bonding between amine and carbonyl groups. Other secondary structures include turns and loops, β-hairpins, β-spirals, and polyproline helices. As discussed in the next section, regular secondary structure is a common feature in peptide and protein hydrogels and provides a basis for classifying these materials.

Tertiary structure is the spatial arrangement of all of the atoms in a protein. It may also be referred to as the folded state of a protein and is closely related to function. Many proteins also have a quaternary structure that describes the assembly of multiple polypeptide subunits. Typically these subunits associate noncovalently through hydrogen bonding, hydrophobic interactions, and salt bridges. Alternatively they may be bound covalently through disulfide bond formation or enzymatic cross-linking. The intermolecular interactions that are responsible for protein quaternary structure closely resemble those that facilitate the self-assembly of peptide and protein hydrogels.

In addition to classification based on the type of peptide precursor (*i.e.* oligopeptides, recombinant proteins, *etc.*) or secondary structures (*i.e.* β-sheets, α-helices, *etc.*), peptide hydrogels can also be described by the mechanism of

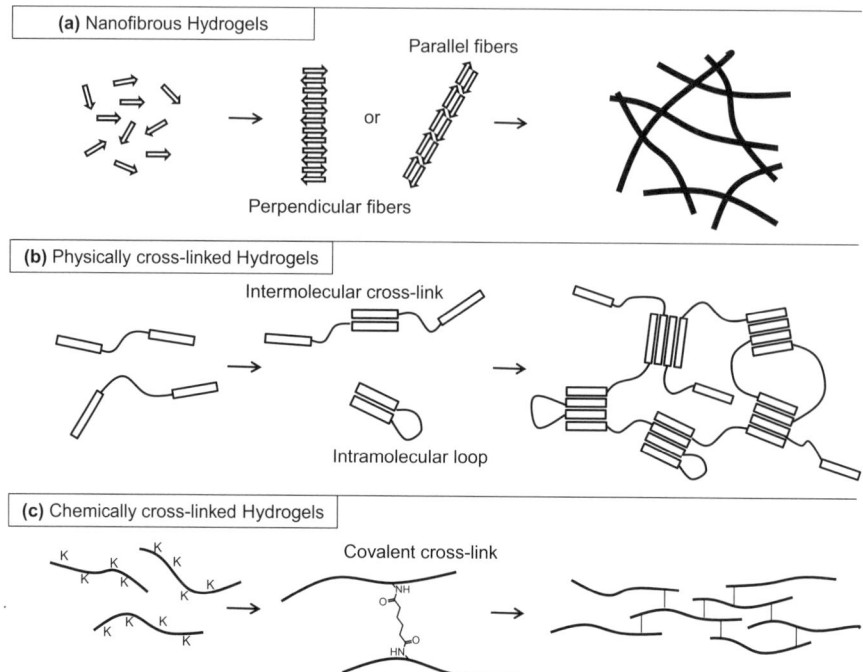

Figure 5.1 Three modes of assembly of peptide hydrogels. (a) Oligopeptide precursors assemble into perpendicular or parallel nanofibers that entangle to
form physically cross-linked gels. (b) Artificial proteins and peptide–
polymer hybrids contain physical cross-linking domains separated by
water-soluble linkers. Noncovalent association of these domains gives rise
to intermolecular cross-links and intramolecular loops, which further
assemble into hydrogel networks. (c) Hydrogels can also be formed by
chemical cross-linking of reactive amino acid side chains such as the
ε-amine of lysine (K).

network assembly. Figure 5.1 depicts three such mechanisms for forming
peptide hydrogels. The first is the gelation of nanofibrous peptide assemblies
and is observed predominantly with oligopeptide precursors. Gel formation
in such systems involves supramolecular assembly of precursors into nanofibers, -tapes, -tubes, or -ribbons that entangle and form nonspecific physical
cross-links. The second mode of assembly is observed in longer multidomain
proteins and peptide–polymer hybrids. Network assembly results from the
noncovalent association of physical cross-linking domains that are separated
by water-soluble linker regions. The final assembly mode is chemical cross-
linking and involves covalent bond formation at reactive amino acid side chains
such as lysine and cysteine. It should be noted, however, that these assembly
mechanisms are not mutually exclusive. For example, nanofibrous hydrogels
from oligopeptides may be chemically cross-linked for increased stability and
mechanical strength.[31,32]

5.3   Hydrogels from Oligopeptides

In a typical strategy for preparing hydrogels from synthetic oligopeptides, precursors self-assemble noncovalently into fibrous nanostructures that form gels through physical cross-linking and entanglements. The key to gel formation is the tuning of molecular and supramolecular interactions to produce a self-supporting network that does not precipitate and is capable of immobilizing water. The noncovalent interactions between self-assembling oligopeptides often mimic the secondary structures responsible for local protein conformation. However, because these interactions generally occur intermolecularly, they also represent a form of quaternary structure.

Most examples of oligopeptide hydrogels utilize the β-sheet motif to produce supramolecular fibers, with only a few recent examples describing the gelation of α-helical peptides and collagen mimetic polyproline helices (CMPs). This is slightly counterintuitive, given that certain natural analogs to α-helical coiled coils and CMPs, keratin and collagen respectively, are components of the hydrogel-like cytoskeleton and extracellular matrix. In contrast, the closest natural analogs of β-sheet oligopeptides are amyloids, protein aggregates with prominent roles in diseases such as Alzheimer's. Another example of hydrogel formation from oligopeptide building blocks is the supramolecular assembly and gelation of peptide amphiphiles (PAs).[33] Although these materials also exhibit some secondary structure, their self-assembly and nanostructures are sufficiently different to warrant separate classification.

5.3.1   β-Sheet Peptide Hydrogels

Zhang *et al.* first observed the gelation of a self-complementary oligopeptide while investigating the EAK16 sequence, Ac-(AEAEAKAK)$_2$-NH$_2$, derived from the yeast Z-DNA binding protein zuotin.[34] The pattern of alternating alanine (A) and glutamic acid (E) or lysine (K) residues promotes the formation of β-strands in which the hydrophobic and charged side-chains are segregated to opposite sides of the peptide backbone. Hydrogen bonding between peptides leads to the formation of intermolecular β-sheets containing a hydrophobic Ala face and a charged Glu/Lys face. Upon addition of salt, the β-sheets further assemble into filaments with diameters of 10–20 nm as a result of hydrophobic bonding between Ala faces and ionic interactions between Glu/Lys faces. The filaments entangle to produce physical hydrogels.

The self-assembly and gelation of EAK16 prompted the development of the rationally designed oligopeptide RAD16 or Ac-(RARADADA)$_2$-NH$_2$. The replacement of Glu with aspartic acid (D) and Lys with arginine (R) preserves both the alternating pattern of hydrophobic and charged amino acids and the self-complementarity of the acidic and basic residues.[35] As a result, RAD16 also forms intermolecular β-sheets that assemble into filaments and hydrogels. The EAK16 and RAD16 peptides support the attachment of a variety of cell types *in vitro* and form the basis for the commercially available PuraMatrix™ hydrogel system for cell culture and encapsulation.[36]

Several additional peptide hydrogels are also based on sequences that form β-sheets. For example, Aggeli *et al.* have investigated the secondary structure and self-assembly of two peptides derived from lysozyme and the trans-membrane protein IsK as a function of the solvent dielectric constant and hydrogen bonding ability.[37] The resulting phase diagram led to the *de novo* design of DN1 (Ac-QQRFQWQFEQQ-NH$_2$), an 11-residue peptide that forms antiparallel, intermolecular β-sheets referred to as β-tapes (Q = glutamine, F = phenylalanine, W = tryptophan). Hydrophobic and π–π interactions between aromatic side chains may cause two β-tapes to associate along their hydrophobic faces to form a β-ribbon. TEM reveals helical and twisting behavior in β-tapes and ribbons as well as higher-order self-assembly into fibrils and fibers.[38,39]

An important consideration in many hydrogel applications is the ability to control the self-assembly process.[40] For example, the EAK16 and RAD16 family of peptides requires physiological salt concentrations to form hydrogels. In water, electrostatic repulsions between like charges on opposing peptides result in a significant energy barrier for self-assembly. Ions lower this barrier and promote peptide gelation by forming an electric double layer that screens the repulsive charges and permits β-sheet assembly.[41,42] Using the self-complementary oligopeptide FEK16 or (FEFEFKFK)$_2$, Messersmith and co-workers have demonstrated controlled assembly of hydrogels by photo-chemical and thermal release of CaCl$_2$ from light- and temperature-sensitive liposomes.[43] Yu and co-workers have described an alternative approach for controlling self-assembly that involves two complementary peptide sequences, Ac-WKVKVKVKVK-NH$_2$ and Ac-EWEVEVEVEV-NH$_2$ (V = valine). Under appropriate conditions, the individual peptides remain in solution because of ionic self-repulsion but assemble into nanofibers and hydrogels upon mixing due to the complementary acidic and basic residues.[44]

The Pochan and Schneider groups have developed an elegant approach to controlling peptide gelation in which the intramolecular folding of a random coil into a β-hairpin is a prerequisite for self-assembly and hydrogel formation (Figure 5.2). Their MAX peptides are derivatives of the rationally designed sequence VKVKVKV^{D}PPTKVKVKVKVK-NH$_2$.[45] Whereas the alternat-ing pattern of hydrophobic valine residues and charged lysine residues is a common feature of peptide hydrogels due to its tendency to form β-sheets, the V^{D}PPT (valine-D-proline-proline-threonine) sequence in the center of this peptide is unique in that it adopts a type II′ turn structure. Deprotonation of the charged lysine side chains at pH 9 eliminates electrostatic repulsions and triggers intramolecular folding to produce a β-hairpin with a hydrophobic valine face and a hydrophilic lysine face. Lateral assembly into an extended β-sheet and facial assembly to bury valine residues produces fibrils that further associate into physical hydrogels that are shear-thinning and reversible. Modifications to the primary sequence that alter its hydrophobicity or charge state result in different folding and self-assembly in response to changes in temperature,[46] ionic strength,[47] and pH.[48] Light may also serve as a trigger for β-hairpin formation in a MAX peptide incorporating a photocaged cysteine

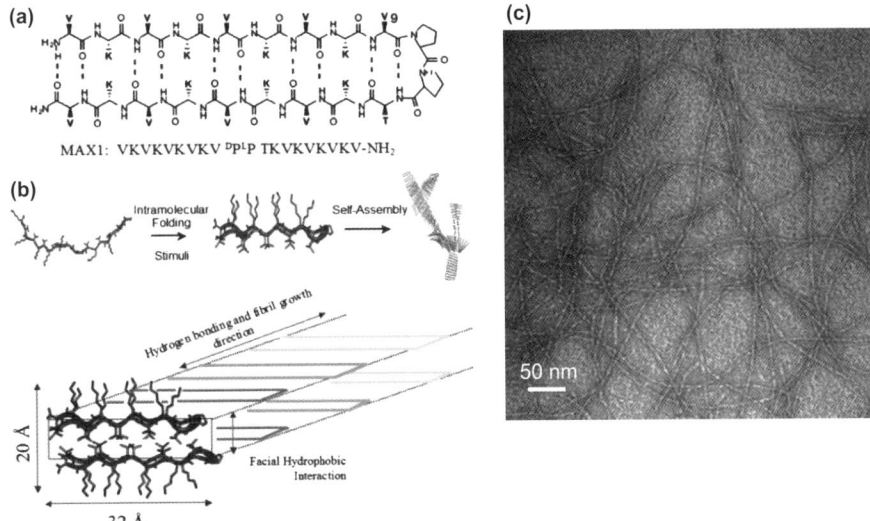

MAX1: VKVKVKVKV ᴰᴾᴸᴾ TKVKVKVKV-NH₂

Figure 5.2 Hydrogel formation from the MAX β-hairpin oligopeptide. (a) The MAX peptide contains alternating Val and Lys residues with the potential to form β-hairpins. (b) A stimulus triggers folding of a random coil to a β-hairpin, which can then further assemble into fibrils to form a hydrogel. The proposed mechanism for fibril self-assembly involves burial of the hydrophobic Val face and lateral association of hairpins through intermolecular hydrogen bonding. (c) Fibrils can be observed in a diluted hydrogel by transmission electron microscopy.
(Reproduced from Ozbas *et al.*[47] by permission of the American Chemical Society.)

residue.[49] While the replacement of a valine residue with an α-carboxy-2-nitrobenzyl-protected cysteine (CNB-Cys) prevents intramolecular folding, cleavage of the protecting group by UV irradiation results in peptide self-assembly and gelation. Thus, in addition to demonstrating a novel method for controlling hydrogel formation, this work also highlights the utility of incorporating two non-natural amino acids, D-Pro and CNB-Cys, into hydrogel precursors.

The examples considered thus far involve the gelation of oligopeptides with lengths of 10–20 amino acids. An interesting question is how small such precursor peptides can be while still retaining the ability to self-assemble into hydrogels. A number of research groups are addressing this question in their investigations of low molecular weight hydrogelators (LMWG).[50] In these systems, fibrous hydrogels assemble from short peptide sequences, typically dipeptides and tripeptides, conjugated to large aromatic groups such as fluorenylmethoxycarbonyl (Fmoc). Triggers for LMWG self-assembly include acidification to protonate the C-terminus,[51,52] temperature change,[53] solvent change,[54] and enzymatic activity.[55,56] The best studied example of a LMWG is the FmocPhePhe dipeptide. Spectroscopic measurements indicating β-sheet formation and π–π stacking have led to a model for self-assembly in which anti-parallel FmocFF β-sheets form nanotubes 3 nm in diameter.[57] The keys to nanotube formation appear to be the π–π interactions between Fmoc groups on adjacent β-sheets and the twisting of β-sheets that leads to tube closure. Functionalization of LMWG systems with the FmocRGD peptide results in hydrogel matrices that support cell attachment *in vitro*.[58] These systems offer the potential to form hydrogels from relatively inexpensive and easily synthesized starting materials.

5.3.2   α-Helical Coiled-Coil Oligopeptide Hydrogels

When compared to β-sheets, α-helical structures have played a less prominent role in the development of hydrogels from synthetic oligopeptides. While there are numerous examples of gelating β-sheet peptides, similar α-helical or coiled-coil systems have required more careful design to assemble into hydrogels. As a result, the earliest examples of coiled-coil hydrogels were self-assembling recombinant proteins[59] and protein–polymer hybrids,[60] and only recently have research groups reported the gelation of synthetic oligopeptide coiled coils.[61–63]

Coiled-coil peptides are not true α-helices but are closely related.[64] The name derives from the fact that two helices or coils wrap around one another to form a dimer, although higher-order oligomers are also possible. Rather than the expected 3.6 residues per turn for an α-helix, coiled coils complete 3.5 residues per turn. A common description of coiled-coil peptide sequences uses an *abc-defg* heptad repeat notation, with seven residues representing two turns of the helix (Figure 5.3a). The residues at positions *a* and *d* are nonpolar and form a hydrophobic face to facilitate contact with the other coil (or coils in the case of multimers). Positions *e* and *f* are frequently charged residues while *b*, *c*, and *g* are polar. Coiled coils are important motifs in natural proteins, including

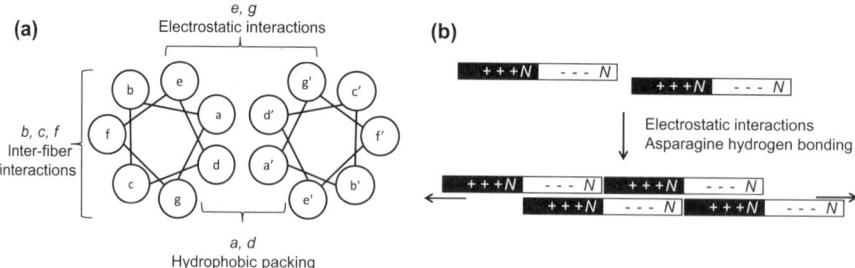

Figure 5.3   Coiled-coil peptides and fibers. (a) Helical wheel representation of a coiled-coil dimer. Dimer formation is driven by hydrophobic association of the *a* and *d* residues and is often stabilized by electrostatic attractions at the *e* and *g* positions. The *b*, *c*, and *f* positions have important roles in controlling interactions between dimers that govern fiber thickening and hydrogel formation. (b) Staggered dimerization leading to fiber formation can be enforced by electrostatic interactions (*e* and *g*) and asparagine hydrogen bonding on the hydrophobic face.

transcription factors that must dimerize to bind DNA, and fibrous structural proteins such as keratin. The preponderance of leucine at the *d* position of the heptad repeat has given rise to the term "leucine zipper" to describe certain coiled coils.[65]

The first step toward coiled-coil peptide hydrogels involves the self-assembly of oligopeptides into longer fibers. Kajava and co-workers have formalized rules for fiber formation and applied them to the *de novo* design of homopentamer coils.[66] Their design employs the staggered assembly of coils in the axial direction to produce fibers at acidic pH. Zimenkov *et al.* have adopted some of these rules to form fibers from dimeric coiled coils using a precursor peptide containing six heptad repeats.[67] Electrostatic interactions at the *e* and *g* positions and a hydrogen bond between asparagine residues on the hydrophobic face enforce parallel dimerization between three N-terminal heptads of one strand and three C-terminal heptads of another (Figure 5.3b). The result of this staggered arrangement is a dimer with "sticky ends" that promotes coiled-coil extension into nanofibers that thicken through lateral association. Using a similar strategy, the Woolfson group has developed heterodimeric coiled-coil peptides that self-assemble into fibers upon mixing.[68,69] Finally, Hartgerink and co-workers have demonstrated the formation of uniform fibers from blunt-ended homodimers with increasing peptide concentrations and acidification.[62] A recent review provides additional examples of coiled-coil fiber formation.[70]

Although there are now several examples of fibers produced by extending coiled-coil peptides, only a few have led to hydrogel formation. The Woolfson group has engineered their self-assembling fiber system to form hydrogels by controlling fiber thickening.[61] They accomplish this by careful choice of amino acids at the *b*, *c*, and *f* positions of the coiled-coil heptad. These residues have a minimal effect on coil dimerization but mediate the lateral association between multiple dimers in solution. Whereas strong electrostatic interactions result in thick fibers that settle out of solution,[71] weaker interactions can create cross-link points for a physical hydrogel. When all three residues are glutamine, weak gels form but melt as the temperature increases due to a loss of hydrogen bonding. Conversely, peptides with alanine at all three positions form gels that strengthen with increasing temperature, an observation that is consistent with physical cross-linking mediated by hydrophobic interactions.

The blunt-ended coiled-coil dimers of the Hartgerink group also form reversible, self-supporting gels when sufficiently concentrated.[62] The proposed mechanism of gel assembly involves protonation of glutamic acids in the *e* position to reduce electrostatic repulsions. Dexter and co-workers have described the similar pH-dependent gelation of a coiled-coil peptide that features an expanded hydrophobic face at the *a*, *d*, *e*, and *g* positions and likely assembles as a homohexamer.[63] Charged residues at the *b*, *c*, and *f* positions appear to affect the assembly of coiled-coil fibers into hydrogels, with an overall charge of +1 or –1 required for gelation. Thus, while examples of hydrogels from coiled-coil oligopeptides remain rare compared to β-sheet hydrogels, these recent reports indicate that design rules for this system are beginning to emerge.

5.3.3   Collagen Mimetic Peptide Hydrogels

An emerging class of peptide hydrogels is based on the triple-helical structure of collagen and its assembly into higher-order fibrous structures. With at least 28 isoforms, collagens are the most abundant proteins in vertebrates and are responsible for much of the mechanical integrity of the extracellular matrix.[72] While naturally derived sources of collagens are common materials for cell culture and tissue engineering, a synthetic substitute would be more homogeneous and eliminate the potential for contamination from animal tissues. To realize this goal, several research groups have explored the molecular and supramolecular structures of collagen mimetic peptides, or CMPs.[73,74]

Early efforts in this field sought to analyze the assembly and stability of triple-helical CMPs as a model system for collagen. The most common CMP sequences are repeats of the Gly-Pro-Pro (GPP) or Gly-Pro-Hyp (GPO) tri-peptides, where Hyp (O) is hydroxyproline. These primary sequences adopt a left-handed polyproline type II helical structure, and three such helices associate as a right-handed superhelix. The requirement for Gly in CMPs stems from the steric hindrance encountered by larger side chains in the center of the helix, while Pro residues promote the preorganization of helical strands and Hyp contributes to stability through solvation or stereoelectronic effects.[72] While researchers have gained important structural information from these first-generation CMPs, evidence for their higher-order assembly into fibers and gels has been limited.[75]

As with coiled-coil peptides, progress toward higher-order CMP structures requires the extension of triple helices in the axial direction followed by their lateral and head-to-tail association into fibers. Kotch and Raines have accomplished axial extension using disulfide bonds to create cysteine-knotted trimers with single- and double-stranded overhangs at the termini.[76] The overhangs associate with the appropriate strands on opposing molecules to extend the triple helix into nanofibers. The Koide group has also developed a cysteine knot method for producing long triple-helical CMPs that form gels.[77,78] Chaikof, Conticello and co-workers have described the formation of the most collagen-like synthetic fibers to date using a zwitterionic CMP, $(PRG)_4(POG)_4(EOG)_4$.[79] Electrostatic interactions between Arg and Glu side chains produce triple helices that further assemble upon thermal annealing into 70 nm wide fibers with periodic banding patterns similar to those in native collagen. Alternative strategies to produce long triple helices involve chain extension of telechelic CMPs functionalized with thiols and thioesters for native chemical ligation,[80] ligands for metal-ion coordination,[81,82] and aromatic rings for π–π stacking and cation–π interactions.[83,84]

The Hartgerink group has reported the most comprehensive study to date on the hierarchical assembly of CMPs into hydrogels (Figure 5.4).[85] Adopting an electrostatic sticky-end approach similar to that of Chaikof and Conticello, they have demonstrated triple helix formation by a zwitterionic CMP, $(PKG)_4(POG)_4(DOG)_4$. They hypothesize that replacement of Arg with Lys and of Glu with Asp results in more effective salt bridges and improved higher-order

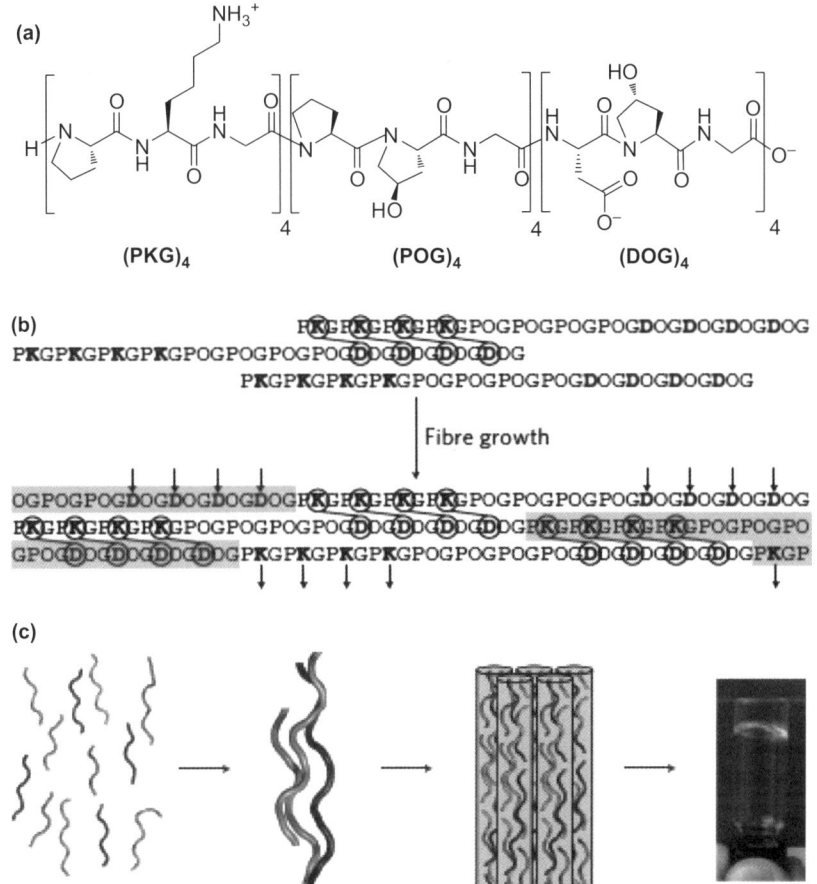

Figure 5.4 Hierarchical assembly of collagen-mimetic peptides. (a) The CMP from O'Leary *et al.*[85] contains Lys and Asp residues that stabilize triple helix formation through salt bridges. (b) "Sticky-ended" CMP triple helices form fibers in the axial direction. Charged residues that do not participate in intrahelical salt bridges may promote lateral association of helices. (c) The consequence of this design is the formation of long triple-helical CMP fibers that further assemble into lateral bundles and entangled hydrogel networks.
(Reproduced from O'Leary *et al.*[85] by permission of Macmillan Publishers.)

assembly. Nanofibers of width 4–5 nm are consistent with bundling of multiple triple helices that further assemble into collagenase-degradable hydrogels.

A future challenge in the field of CMP hydrogels is the development of properly assembling precursor peptides that lack hydroxyproline. While Hyp is important for triple helix stability, it has limited researchers to assembling long polypeptide chains from short, synthetic oligopeptides rather than recombinant proteins. Using a combination of electrostatic stabilization and cysteine knots, Krishna and Kiick have demonstrated the formation of triple helices and fibrillar structures from CMPs lacking Hyp.[86] Engineering recombinant organisms capable of incorporating Hyp during protein synthesis[87] or modifying proline residues post-translationally[88] would offer alternative solutions.

5.3.4 Peptide Amphiphile Hydrogels

Hydrogel formation in many β-sheet and coiled-coil systems is due in part to the amphiphilic nature of the precursor peptides. Self-assembly in such materials results from the burial of apolar regions and a combination of electrostatic attractions, hydrogen bonding, and solvation in polar regions. These characteristics are also specific design features of a class of materials known as peptide amphiphiles (PAs). For example, Tirrell and co-workers have synthesized oligopeptides conjugated to long alkyl chains at the N-terminus and demonstrated that these hydrophobic tails produce monolayers at an air/water interface.[89] The PA monolayers are useful for displaying cell adhesion molecules and orienting peptides to promote intermolecular interactions such as triple helix formation in CMPs.[90,91] The Zhang group has also developed amphiphilic molecules composed completely of amino acids. These PAs associate in water to form nanovesicles and nanotubes by burying hydrophobic residues.[92]

The Stupp group has thoroughly investigated the self-assembly and gelation of peptide amphiphiles, as well as numerous applications for these materials. They have synthesized PAs containing a 16-carbon alkyl tail covalently bonded to the N-terminus of a hydrophilic peptide.[93,94] An example of this type of molecule is their phosphoserine ($S^{(PO4)}$)-containing PA, $CH_3(CH_2)_{15}O$-CCCCGGGS$^{(PO4)}$RGD (C = cysteine). Since the hydrophobic tail is slightly narrower than the peptide head, PAs tend to form cylindrical micelles in water. Upon slow acidification or addition of calcium ions, PAs self-assemble into nanofibers that gel at sufficiently high concentrations. Oxidation of cysteine residues creates reversible disulfide cross-links that stabilize PA nanostructures. Alternatively, UV-mediated cross-linking can occur between hydrophobic tails that contain diacetylene moieties rather than simple alkyl chains.[32] Applications of PA hydrogels include their use as tissue engineering scaffolds and drug delivery vehicles.[95] The Stupp group's original peptide amphiphiles incorporated phosphoserine residues that act as templates for the mineralization of hydroxyapatite, as observed in bone formation.[93] Nanofiber and hydrogel formation have proven robust to significant sequence variation in the peptide region, permitting incorporation of a variety of biologically active peptide domains, including those with considerable hydrophobic content. Examples include PAs displaying the cell-binding peptides RGD and IKVAV, growth factor-mimicking epitopes, and heparin-binding peptides (I = isoleucine).[94,96,97]

Deming and co-workers have produced rapidly recovering hydrogels from a different class of peptide amphiphiles.[98,99] Using *N*-carboxyanhydride (NCA) polymerization, they prepared long diblock copolypeptides of the form $K_{160}L_{40}$ and $K_{160}V_{40}$ with very low polydispersity. The hydrophobic blocks adopt α-helical and β-sheet structures when constituted from leucine (L) and valine, respectively. In the case of polyleucine helical hydrophobic blocks, the proposed model for self-assembly involves perpendicular alignment of α-helices in a twisted fibril that is surrounded by a polyelectrolyte "brush".[100] This is in contrast to other self-assembling helical peptides in which alignment occurs parallel to the helical axis. Gelation occurs above a concentration threshold that is dependent on the length of the hydrophilic block as well as the length and structure of the hydrophobic block. Triblock ($K_xL_yK_z$) and pentablock ($K_xL_yK_zL_yK_x$) copolypeptides also self-assemble and form gels, with pentablocks having the potential for intermolecular cross-links between fibrils.[101]

5.4 Hydrogels from Recombinant Proteins

Recombinant proteins are another class of building blocks for engineered hydrogels. Whereas oligopeptides prepared by SPPS are limited in length and polypeptides prepared NCA polymerization are limited in sequence diversity, recombinant expression exploits the ability of a host organism to synthesize full-length proteins of almost any sequence from DNA templates. One result of this is a fundamental difference in the mechanisms for hydrogel formation by oligopeptides *versus* recombinant proteins. As discussed in the previous section, short oligopeptides self-assemble into nanofibers that entangle to form hydrogels. Proteins, on the other hand, can be long enough to form hydrogel networks without preassembling into fibers. Instead, network formation is driven by noncovalent interactions between physical cross-linking domains or by covalent bonds between chemically cross-linked residues.

Protein engineering offers the potential for unprecedented control over hydrogel structure and functionality.[2,3] Two important categories of recombinant proteins for hydrogels are self-assembling artificial proteins and biomimetic proteins based on elastins and silks. Recent advances in protein engineering have also led to the development of a third category that combines assembly domains with full length, functional proteins to form multifunctional hydrogels.[102]

5.4.1 Self-Assembling, Multidomain Artificial Proteins

Our laboratory has investigated recombinant artificial proteins as precursors for self-assembling hydrogels based on the association of coiled-coil domains.[59] Telechelic triblock proteins denoted AC_xA contain two leucine-zipper domains (A) flanking a random coil polyelectrolyte chain, $(AG)_3PEG$, repeated x times (C_x). Variation in the association of zipper domains in response to temperature and pH changes results in a reversible sol–gel transition. At high temperature or high pH, solutions of AC_xA behave as viscous liquids. Decreasing the temperature to refold the helices, or lowering the pH to near-neutral values, leads to strong association of leucine-zipper domains as tetramers. If the polyelectrolyte linker region remains solvated, coiled-coil tetramers can act as physical cross-links in a hydrogel network. Using genetically engineered endblocks, Kopeček and co-workers have demonstrated that the oligomerization state, self-assembly behavior, and pH and temperature responsiveness of coiled-coil hydrogels can be tuned by altering the amino acid sequence.[103,104]

Hydrogels assembled from AC_xA triblocks are unexpectedly soft and erode rapidly in open solutions, thus limiting their potential applications.[105,106] This behavior is due to the tendency of AC_xA to form non-productive loops that do not contribute to the elasticity of the network. The network is transient and rapid exchange of peptide strands between coiled-coil aggregates allows zippers to disengage and dissolve in the surrounding medium. Our group has demonstrated three strategies that overcome these limitations and produce stiffer hydrogels with tunable erosion rates. The first is to stabilize leucine-zipper aggregates through disulfide bond formation (Figure 5.5a).[106] By the judicious placement of cysteine residues in the zipper domains, it is possible to preferentially stabilize intermolecular aggregates while allowing exchange of looped strands. In an alternative strategy, longer polyelectrolyte linkers suppress loops that arise to avoid energy penalties associated with stretching shorter chains (Figure 5.5b).[105] Unlike typical elastic materials that become softer upon increasing the molecular weight between cross-links, AC_xA hydrogels can become stiffer when the linker region is extended. Finally, redesigned triblock proteins of the form AC_xP and PC_xP (Figure 5.5c), where P is a zipper domain derived from the cartilage oligomeric matrix protein, are stiffer and exhibit slower erosion rates than AC_xA.[107] In the case of PC_xP, these observations arise from the higher aggregation number and parallel alignment of the P zippers. While A zippers can adopt an antiparallel orientation as part of tetrameric coiled coils, P zippers associate exclusively in a parallel orientation in pentameric aggregates. This results in a lower tendency for PC_xP to form intramolecular loops, as the linker region must stretch to permit the proper parallel alignment. In the case of AC_xP triblocks, intramolecular loops rarely form due to the preference of the A and P zippers to aggregate as homo-oligomers rather than hetero-oligomers.

The ability to tune the assembly and erosion of telechelic leucine-zipper hydrogels suggests a potential application as materials for the encapsulation and controlled release of cells or biomolecules. In this respect, desirable features include shear-thinning behavior and the rapid recovery of elastic strength upon cessation of shear. These properties would permit the delivery of the hydrogel and its cargo *via* a minimally invasive injection. Rheological measurements of PC_xP hydrogels indicate a decrease in the elastic modulus of three orders of magnitude at high strain rates (Figure 5.6a).[108] Hydrogels also recover their elastic strength within seconds after large-amplitude oscillatory strain and form self-supporting structures when injected through narrow gauge needles (Figure 5.6a,b). Shear-thinning appears to be due to yielding behavior within the gel, with shear banding potentially protecting cells and biomolecules from high shear rates. The nonlinear rheology of PC_xP hydrogels is largely unaffected by

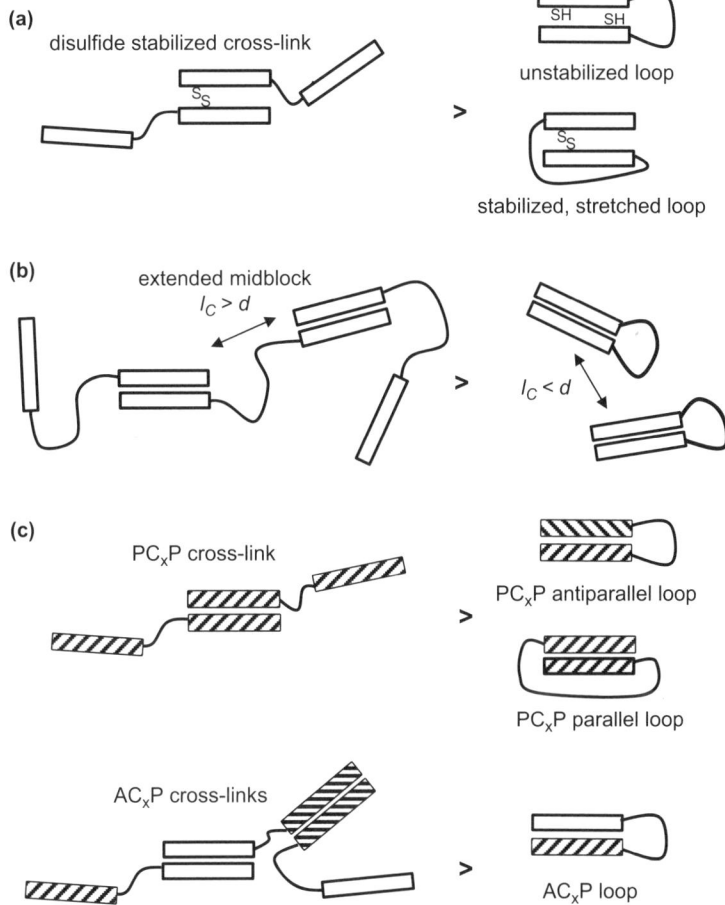

Figure 5.5 Stabilization of self-assembling artificial proteins to control hydrogel stiffness and erosion rate. Cross-links are depicted as dimers for clarity, although tetrameric and pentameric coiled-coils predominate in these systems. (a) Disulfide formation between cysteine residues in the AC$_x$A domain stabilizes intermolecular cross-links. The protein design precludes stabilization of antiparallel intramolecular loops, while parallel loops rarely form due to chain stretching. (b) Cross-linking is favored by an extended midblock C with a mean end-to-end distance (l_C) greater than the average distance between proteins (d). In contrast, shorter midblocks form loops to avoid energy penalties associated with chain stretching. (c) Proteins of the form PC$_x$P contain P coil domains that aggregate exclusively in a parallel orientation. Cross-linking is preferred to loop formation, which requires chain stretching. Proteins of the form AC$_x$P do not form intramolecular loops because the A and P domains (white rectangles and striped rectangles, respectively) do not form hetero-oligomers.

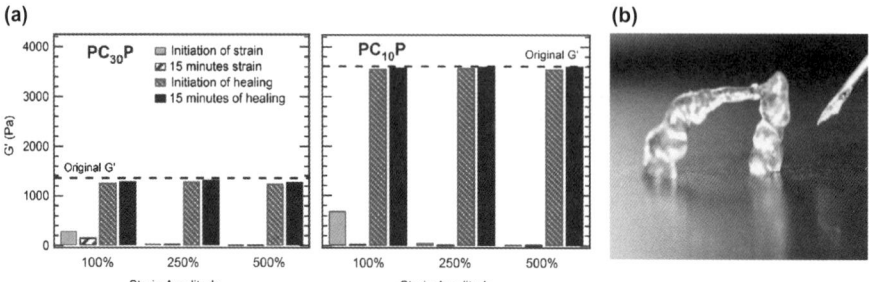

Figure 5.6   Shear-thinning and elastic recovery of PC$_x$P hydrogels. (a) Hydrogels experience a decrease in their shear storage modulus (G') of three orders of magnitude at the onset of large-amplitude oscillatory strain but recover their elastic strength within seconds. Recovery was independent of the length of the midblock in PC$_{10}$P and PC$_{30}$P hydrogels. (b) PC$_x$P forms a self-supporting gel upon injection through a 22 gauge needle. (Reproduced from Olsen *et al.*[108] by permission of the American Chemical Society.)

the length of the polyelectrolyte linker region, suggesting that it should be possible to incorporate biologically active sequences into shear-thinning telechelic proteins.[109]

Telechelic proteins containing collagen-mimetic endblock sequences also produce shear-thinning, thermoreversible hydrogels.[110,111] The triblock architecture of these proteins consists of nine repeats of a Pro-Gly-Pro sequence flanking a central random coil. Recombinant production in yeast yields gram-per-liter quantities of the secreted product.[112] In these materials, endblock aggregation occurs through formation of triple helices. Hydrogels assembled from CMP telechelic triblocks have shown promise in the controlled release of model proteins.[113] Chemical cross-linking of lysine residues in the random coil midblock using glutaraldehyde produces shape-memory hydrogel networks that are dependent on the thermoreversibility of the endblock triple helices.[114]

Many peptide hydrogels require temporary exposure to non-physiological pH, temperature, or ionic strength in order to self-assemble. These conditions may result in unacceptable levels of cell death or denaturation of encapsulated cargo. To overcome these problems, Heilshorn and co-workers have introduced the mixing-induced, two-component hydrogel (MITCH) system in which gelation can occur only upon mixing of two protein solutions.[115] The two components of MITCH gels are artificial proteins consisting of several repeats of either the WW or proline-rich domains separated by random coil linker regions. Physical cross-linking between proteins containing the WW domain, named for its conserved tryptophan residues, and proteins containing the proline-rich domain derives from the noncovalent association between these sequences as found in natural proteins. By varying the stoichiometry of the two components as well as the frequency and binding strength of the cross-linking domains, it is possible to tune the sol–gel transition and viscoelastic properties of the system in a predictable manner.[115,116] The transient nature of the network results in shear-thinning protein hydrogels that display rapid self-healing within minutes after injection.

5.4.2 Biomimetic Recombinant Proteins: Elastins and Silks

Elastins are important components of the extracellular matrix and represent attractive targets for biomedical engineering applications. The desirable mechanical and chemical properties of elastins can be recapitulated in elastin-like polypeptides (ELPs) containing repeats of the VPGXG pentamer, where the "guest residue" X can be any amino acid except proline. Elastin-like polypeptides exhibit inverse temperature behavior in that they are soluble at low temperatures but aggregate in coacervate phases at higher temperatures. The transition from a soluble to an aggregated state occurs at the lower critical solution temperature (LCST), where it has been proposed that ELPs transition from a random coil conformation to a β-spiral. The LCST is dependent on the length of the sequence, the ELP concentration, and the hydrophobicity of the guest residue, with more hydrophilic sequences exhibiting higher LCSTs.[117,118] While highly repetitive ELPs can be produced synthetically, recombinant engineering permits access to more complex protein sequences, including multiblock architectures and bioactive domains.[119,120] One of the most common engineering strategies is to vary the ELP guest residue to incorporate reactive moieties for chemical cross-linking or to alter the LCST behavior of the protein. Temperature cycling above and below the LCST offers a facile method of purification of recombinant ELPs.[121]

Elastin hydrogels have been formed by both physical and chemical cross-linking methods. Conticello and co-workers have described the physical cross-linking of triblock ELPs based on inverse temperature transitions.[122,123] The mechanism for gel formation is similar to that of telechelic leucine-zipper and CMP hydrogels. Triblocks of the form BAB contain hydrophobic endblock elastin sequences (B) flanking a hydrophilic elastin midblock (A). At temperatures above their LCST, endblocks undergo microphase separation while the hydrophilic midblock (which has a higher LCST) remains solvated. This results in hydrogels in which the aggregated hydrophobic blocks serve as thermoreversible, noncovalent cross-linkers. Chilkoti and co-workers have analyzed the rheological behavior of an ELP above its LCST. Although their design did not include explicit cross-linking domains, they observed gel-like behavior in the coacervate with potential applications in cartilage tissue engineering.[124]

Several research groups have employed chemical cross-linking to form hydrogels with mechanical properties that more closely resemble those of native elastin. The most common strategy uses small multifunctional cross-linkers that form covalent bonds with reactive amino acids, typically lysines occupying the ELP guest site or in distinct cross-linking domains.[125–131] Other strategies include enzymatic cross-linking,[132] UV or visible light photo-cross-linking,[27,133] and γ-irradiation.[134]

Hydrogels from silk proteins represent another class of potentially useful biomimetic materials.[135] The best studied silk proteins are those from the Chinese silkworm *Bombyx mori* and from the dragline of the spider *Nephila clavipes*. Fibers spun from silk proteins derive their impressive mechanical strength and extensibility from repetitive Gly- and Ala-rich β-sheet crystalline domains separated by amorphous hydrophilic linker regions. Kaplan and co-workers have demonstrated that solutions of naturally derived silkworm fibroin doped with poly(ethylene glycol) undergo a sol–gel transition over the course of days due to the coalescence of hydrophobic regions.[136–138] Hydrogels produced from silk fibroin are candidates for tissue engineering and cell encapsulation applications.[139,140]

In contrast to extensively engineered ELPs, silk protein engineering has generally focused on producing close facsimiles of the natural silks. In particular, there has been strong interest in developing recombinant sources of dragline spider silk since spiders, unlike silkworms, are not easily farmed. Several research groups have demonstrated progress in this area by producing spider silks in bacterial,[141] plant,[142] mammalian,[143] and silkworm hosts.[144] While significant attention has been directed toward spinning recombinant silks into fibers, only a few studies have examined hydrogel formation from these proteins.[145,146] Further protein engineering could offer the ability to tune the mechanical properties of silk and silk hydrogels or to introduce new bioactive domains.[147,148] An alternative approach to silk-like hydrogels involves genetically engineered block copolymers of silk-like GAGAGS (S = serine) peptides and elastin-like VPGXG peptides.[149,150] Solutions of these proteins spontaneously form swollen hydrogels that display the crystalline hydrophobic domains of silks and the elasticity and responsiveness of ELPs. These materials have shown promise as delivery vehicles for gene therapy.[151]

5.4.3 Multifunctional Protein Hydrogels

Many proposed applications of peptide hydrogels in tissue engineering and drug delivery require materials that display simple binding motifs and assemble or disassemble in response to environmental cues. Both oligopeptides and recombinant proteins have shown promise in meeting these requirements. However, protein engineering also offers the ability to develop hydrogels with much more advanced functionality. In this regard, emerging research has demonstrated the potential of engineering multifunctional protein hydrogels and their applications as sensors and catalytic materials.

Banta and co-workers have developed multifunctional protein hydrogels that combine the self-assembly behavior of leucine zippers with the useful functional properties of fluorescent proteins and enzymes. In their initial demonstration of this approach, they showed that the incorporation of green fluorescent protein in the midblock region of AC_xA does not inhibit protein folding or gel formation.[152] Using this method, they produced multicolor hydrogels and probed gel structure in these systems by Förster resonance energy transfer (FRET). A similar design that incorporated an oxidase enzyme and electron conductors was used to create bioelectrocatalytic hydrogels capable of reducing molecular oxygen to water.[153] The electron-conducting hydrogel network consisted of triblock leucine-zipper proteins with osmium bis-bipyridine complexes bound to histidine residues. Enzymatic activity was derived from a chimeric protein containing a leucine-zipper domain and the small laccase (SLAC) polyphenol oxidase from *Streptomyces coelicolor*. The zipper domain mediates incorporation of the chimera into the hydrogel network, while dimerization of the SLAC enzyme provides additional cross-linking and is required for catalytic activity. This novel class of materials has potential applications in biofuel cells and oxygen sensors. Other examples of enzymatic hydrogels incorporate an aldo-keto reductase and an organophosphate hydrolase, suggesting the broad applicability of this method.[154,155]

In similar work, Gallivan and co-workers have engineered a chimeric calmodulin protein fused to a leucine-zipper domain.[156] Calmodulin is an important regulatory protein that undergoes a conformational change in the presence of calcium ions to bind partner proteins. A solution of the chimeric calmodulin-zipper protein and a telechelic cross-linker containing calmodulin-binding endblocks forms a hydrogel network in the presence of Ca^{2+}. The network is reversible upon chelation of calcium.

5.5 Hydrogels from Peptide–Polymer and Protein–Polymer Hybrids

As discussed earlier in this chapter, a common design for recombinant protein hydrogels is a telechelic triblock architecture with physical cross-linking domains separated by a polyelectrolyte linker. Hydrogel formation results from an appropriate balance between endblock aggregation and linker solubility. While genetically encoded linkers offer clear advantages such as monodispersity, sequence diversity, and bioactivity, synthetic polymers can also fill this role in hydrogels formed from peptide–polymer and protein–polymer hybrids. From a synthesis and cost perspective, hydrogels in which the bulk of the dry weight derives from synthetic macromolecules may be more desirable than hydrogels from oligopeptides or recombinant proteins. Furthermore, whereas ribosomal translation limits recombinant proteins to linear architectures, synthetic methods offer access to branched architectures such as graft, star, and dendritic polymers.

In hybrid hydrogels, peptides and proteins may serve several roles. Most commonly, they are responsible for chemical or physical cross-linking of the prepolymer solution. The mechanisms for cross-linking and self-assembly are analogous to those described for recombinant protein hydrogels. Peptides and proteins may also confer biological activities such as cell binding and enzymatic degradability to otherwise inert polymers. Finally, proteins embedded in hybrid hydrogel networks may exhibit enzymatic activity or ligand binding for applications in stimulus-responsive materials and biosensors. We present several examples of hybrid hydrogels as an introduction to this field. A more detailed account of protein– and peptide–polymer materials is provided by Krishna and Kiick.[157]

5.5.1 Peptides as Physical and Chemical Cross-Linkers

Kopeček and co-workers have pioneered the development of physically cross-linked hybrid hydrogels from proteins and synthetic polymers.[60] Their design features recombinant coiled-coil proteins attached to a copolymer of N-(2-hydroxypropyl)methacrylamide and (N',N'-dicarboxymethylaminopropyl)-methacrylamide. The attachment is mediated by Ni^{2+} coordination by the polyhistidine tag on the protein and the iminodiacetate side chains of the copolymer. As in triblock leucine-zipper recombinant proteins, coiled-coil aggregation leads to hydrogel formation. A second generation design features two copolymers covalently grafted with coiled coils that associate as hetero-dimers.[158] Reversible hydrogels form upon mixing the two copolymers at neutral pH at concentrations as low as 0.1 wt%.

Shen and co-workers have further demonstrated the cross-linking potential of coiled coils in a bottom-up approach to fabricating hydrogels for tissue engineering.[159] Using photolithography, they encapsulate cells in 400 µm star-shaped microgels by photopolymerization of poly(ethylene glycol) (PEG) diacrylate. Subsequent attachment of cysteine-containing coiled coils to the residual acrylate groups facilitates reversible assembly of microgels into porous macroscopic scaffolds. The high porosity and pore interconnectivity of these scaffolds and the short length scales of the microgels may aid nutrient transport to the encapsulated cells.

The Yu group has investigated hydrogel formation from four-arm PEG stars functionalized with collagen-mimetic peptides.[160] The association of CMPs as triple helices forms a network that is reversible by heating above the melting temperature and addition of competitor CMPs. They exploit this reversible behavior to form gradients in gel stiffness through local injection of a hot CMP solution. As the solution diffuses from the injection site, the thermal gradient melts the triple helical cross-links, which do not reform upon cooling due to the high concentration of competitor peptides. They have also demonstrated the use of CMPs to modify natural collagens noncovalently with PEG and growth factors through a strand invasion mechanism.[161,162] These systems highlight the potential of hybrid hydrogels containing synthetic polymers, peptides, and naturally derived proteins.

Covalent cross-linking offers another route to peptide–polymer hydrogels (Figure 5.7). Hubbell and co-workers have pioneered the development of peptide- and protein-functionalized PEG hydrogels using photopolymerization of acrylated precursors and Michael-type addition reactions.[163–165] Both methods have been used to introduce cell-binding ligands and protease-sensitive peptide sequences for enzymatic degradation by plasmin and matrix metalloproteases.[166,167] Anseth and co-workers have extended this method for cell encapsulation under mild conditions using the bioorthogonal strain-promoted azide–alkyne cycloaddition (SPAAC). To accomplish this, they have

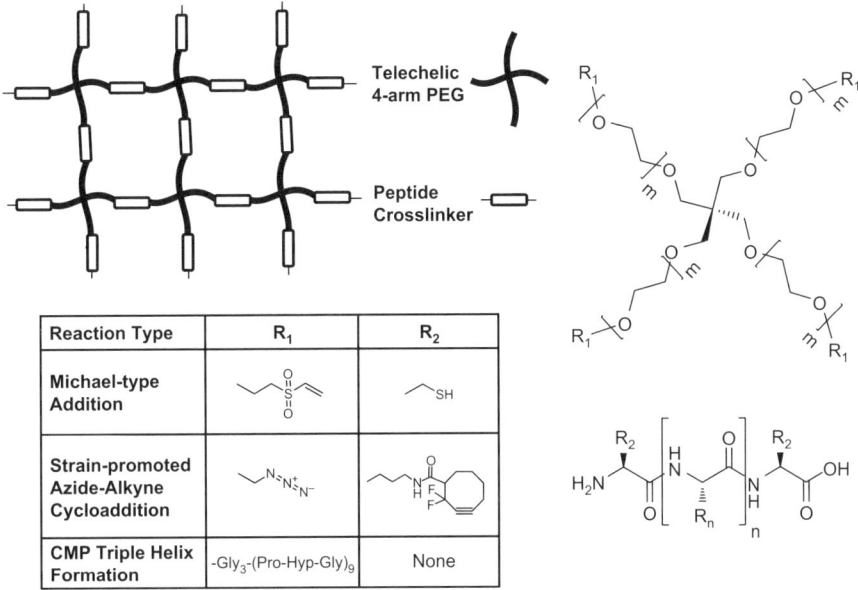

Reaction Type	R₁	R₂
Michael-type Addition		SH
Strain-promoted Azide-Alkyne Cycloaddition	N=N⁺=N⁻	
CMP Triple Helix Formation	-Gly₃-(Pro-Hyp-Gly)₉	None

Figure 5.7 Chemical and physical cross-linking of four-arm PEG stars. Telechelic four-arm PEG is functionalized with vinyl sulfone and azide groups for Michael-type addition and strain-promoted azide–alkyne cycloaddition, respectively. Chemical cross-linking occurs through difunctional peptides that contain terminal cysteines or cyclooctynes. Alternatively, physical hydrogels are formed by functionalizing four-arm PEG with collagen-mimetic peptides that associate noncovalently as triple helices.

cross-linked tetraazide-functionalized four-arm PEG stars with peptides containing terminal cyclooctynes.[22] Their peptide cross-linker also contains an alkene moiety for subsequent photopatterning of cell-binding ligands by the thiol–ene reaction and a nitrobenzyl ether moiety for controlled photodegradation.[168]

5.5.2 Synthetic Hydrogels Containing Functional Protein Domains

In parallel with the development of self-assembling, multifunctional hydrogels from recombinant proteins, several groups have demonstrated the potential of incorporating functional proteins such as enzymes and antibodies into synthetic polymer scaffolds. For example, two groups have produced calcium-responsive materials through the covalent incorporation of calmodulin (CaM) into cross-linked polyacrylamide (PA) and PEG scaffolds.[169,170] Both networks undergo contraction in the presence of Ca^{2+} and swelling upon its chelation. In the case of Ehrick *et al.*, contraction of the PA hydrogel results from binding between CaM and a similarly immobilized CaM ligand. This material has been used to create chemically tunable microlenses.[171] In the system developed by Murphy *et al.*, conformational changes in CaM drive contraction of the PEG scaffold. This system has been used for the controlled release of encapsulated growth factors.[172]

Other examples of functional protein–polymer hybrids include hydrogels that swell and contract due to antigen–antibody binding[173] and catalytic activity of grafted enzymes.[174] In addition, there are several examples in which growth factors are covalently cross-linked to a polymer scaffold for tissue engineering applications.[175–177] Although the protein components do not necessarily contribute to hydrogel assembly or function, these systems still represent an important class of hybrid materials.

5.6 Future Directions and Challenges

In the past 20 years, the field of peptide and protein hydrogels has grown from early observations of self-assembly of oligopeptides to the design of advanced materials with well-controlled biological activity and mechanical properties. Hydrogels are now routinely produced from oligopeptides, recombinant proteins, and peptide–polymer hybrids. Together with hydrogels from naturally sourced biomolecules and synthetic polymers, these materials provide scientists, engineers, and clinicians with a multitude of options to address problems in medicine and basic biology. In this regard, an important challenge will be to match the most appropriate materials with each application, particularly when transitioning hydrogels into clinical settings. Other challenges include assessing the immune response to peptide and protein hydrogels and producing hydrogel precursors in sufficient quantity and purity for their intended applications. These and other challenges will guide the design of future generations of engineered hydrogels.

5.6.1 Immune Response to Peptide and Protein Hydrogels

Peptide and protein hydrogels are frequently touted as biocompatible on the basis of the fact that their precursors resemble natural biopolymers and should be susceptible to enzymatic or hydrolytic degradation. However, implantation of hydrogels still has the potential to generate undesirable immune responses. While more thorough investigations will be required, initial *in vivo* and *in vitro* studies to analyze the immunogenicity of peptide hydrogels and their precursors have been encouraging. For example, using the RAD16 and EAK16 peptides, Holmes *et al.* detected no inflammatory response after intramuscular injection in rats and no measureable antibody response when the peptides were conjugated to bovine serum albumin and injected in rabbits and goats.[178] Similarly, myocardial injection of RAD16 hydrogels in mice did not generate significant inflammation,[179] while the MAX β-hairpin hydrogels of the Schneider and Pochan groups did not elicit an inflammatory response from macrophages in an *in vitro* assay.[180] Promising results have also been reported for elastin-like and silk-elastin-like materials.[181,182]

The Collier group has further investigated the immunogenicity of both coiled-coil and β-sheet self-assembling oligopeptides.[183–185] They have demonstrated in a mouse model that β-sheet oligopeptides stimulate antibody production only when displaying a strongly immunogenic epitope.[184] For example, undecorated β-sheet fibrils and fibrils displaying an RGD motif do not elicit an immune response, whereas fibrils that display a 17-mer sequence from chicken egg albumin stimulate high titer antibody production. These findings suggest that self-assembling peptides may be safe for implantation or injection if strongly immunogenic epitopes are avoided. On the other hand, the ability of β-sheet fibrils to enhance the immunogenicity of selected epitopes indicates that self-assembling peptides may be useful as well-defined adjuvants for vaccine delivery and immunotherapy.

In separate work, Collier and co-workers have demonstrated that, like β-sheet fibrils, undecorated coiled coils do not elicit a measureable immune response.[185] However, triblock materials consisting of coiled-coil endblocks separated by a PEG spacer do stimulate a moderate level of antibody production. They suggest that higher molecular weight oligomers may present better targets to the immune system, a hypothesis that is consistent with the increased immunogenicity of protein aggregates.[186] Although this study was conducted with peptide concentrations below the gel point, it still highlights a key challenge for peptide hydrogels. It may not be possible to predict the immunogenicity of hydrogel networks based on the immune response to precursor peptides and proteins. Further investigation with implanted or injected gels will be required.

5.6.2 New Methods for Peptide Synthesis

While solid-phase peptide synthesis has been used extensively for hydrogel applications, there are several limitations to this method that could hinder efforts to scale up peptide production to clinically and industrially useful quantities. The foremost limitation is the inverse relationship between peptide length and overall yield that precludes the synthesis of polypeptides that contain more than approximately 50 amino acids. Low overall yields and the poor atom economy of reactions involving large protecting groups and coupling reagents also result in significant amounts of wasted starting material. Pattabiraman and Bode have recently reviewed these and other challenges as well as promising new methods of amide bond formation and chemoselective ligation.[18] The development of new reagents and catalysts should be closely followed by the peptide hydrogels field.

Recombinant protein production offers an alternative to SPPS, even for small oligopeptides. Riley *et al.* have recently described the bioproduction of β-sheet peptides that self-assemble into hydrogels.[187] To accomplish this, highly repetitive polypeptides are cleaved into oligopeptides at precise positions using cyanogen bromide. This method should be applicable to nearly any peptide sequence, assuming the appropriate selection of chemical or enzymatic cleavage agents. Like SPPS, there are also drawbacks to recombinant protein production. Each new recombinant protein requires cloning to produce the template DNA as well as optimization of expression and purification. Protein purification frequently involves an affinity chromatography step and the removal of lipopolysaccharide endotoxins for therapeutic applications. These factors must be considered when choosing between scaling up SPPS or bio-production of oligopeptides by fermentation.

5.6.3   Spatially Patterned Hydrogels and Epitopes Beyond RGD

Owing to their similar mechanical properties and ability to display bioactive domains, peptide and protein hydrogels are excellent candidates to replace the extracellular matrix (ECM) in tissue engineering scaffolds and *in vitro* cell culture matrices. While the homogeneity and well-defined nature of peptide and protein hydrogels are typically advantageous properties, the natural ECM contains mixtures of physical and chemical signals that vary in each tissue. These signals are often arranged in gradients or spatial patterns that define the cellular microenvironment in such a way as to direct specific cell and tissue behaviors. Hydrogels that mimic these patterns will likely be key to recreating the morphogenic events observed during development and tissue repair.[8,188] A number of research groups are developing strategies to accomplish this.[189] The West, Shoichet, and Anseth groups have created spatial patterns of cell-binding peptides and protein growth factors within hydrogels using three-dimensional photolithography.[168,190,191] With these patterns it is possible to direct cell behaviors such as spreading and migration. Straley *et al.* have patterned elastin-like protein hydrogels using spatially controlled enzymatic degradation.[192] Other potential strategies for hydrogel patterning include layer-by-layer assembly using 3D printers[193,194] and microfluidic approaches.[195]

In addition to spatially patterned hydrogels, more realistic cellular micro-environments can be obtained by incorporating biologically active domains beyond the standard RGD cell-binding sequence. While this tripeptide motif is frequently used to demonstrate the cytocompatibility of hydrogels, it binds only a subset of integrin receptors and lacks the spatial context normally provided by neighboring domains in fibronectin, vitronectin, laminin, or collagen.[196,197] More complex cell-binding domains can be engineered into hydrogels assembled from recombinant proteins. For example, Fong has demonstrated accelerated *in vitro* wound healing on elastin-like protein films containing the full-length fibronectin type III domains 9 and 10.[198] Wound closure on these materials occurs more rapidly than on materials containing only the RGD sequence and approaches the rate observed on fibronectin. It is also possible to include more biological complexity in hydrogels from oligo-peptides. Kokkoli and co-workers have developed hydrogels from peptide amphiphiles that contain both the RGD cell-binding domain and the PHRSN synergy sequence separated by the approximate distance observed in fibro-nectin (H = histidine, N = asparagine).[199,200] Endothelial cells cultured on these materials exhibit high levels of spreading, extracellular matrix deposition, and cytoskeletal organization. Together, spatial patterning and more complex epitopes will allow researchers and clinicians to engineer cellular micro-environments that more closely resemble real tissues without compromising the well-defined nature of peptide and protein hydrogels.

Acknowledgments

Work on protein hydrogels at Caltech is supported by NIH grant U01 DK089533-01.

References

1. S. Gomes, I. B. Leonor, J. F. Mano, R. L. Reis and D. L. Kaplan, *Prog. Polym. Sci.*, 2012, **37**, 1.
2. S. A. Maskarinec and D. A. Tirrell, *Curr. Opin. Biotechnol.*, 2005, **16**, 422.
3. J. C. M. van Hest and D. A. Tirrell, *Chem. Commun.*, 2001, 1897.
4. T. O. Yeates and J. E. Padilla, *Curr. Opin. Struct. Biol.*, 2002, **12**, 464.
5. A. M. Lomas and N. A. Peppas, in *Encyclopedia of Controlled Drug Delivery*, ed. E. Mathiowitz, Wiley, New York, 1999, p. 397.
6. B. V. Slaughter, S. S. Khurshid, O. Z. Fisher, A. Khademhosseini and N. A. Peppas, *Adv. Mater.*, 2009, **21**, 3307.
7. J. L. Drury and D. J. Mooney, *Biomaterials*, 2003, **24**, 4337.
8. M. P. Lutolf and J. A. Hubbell, *Nat. Biotechnol.*, 2005, **23**, 47.
9. G. D. Nicodemus and S. J. Bryant, *Tissue Eng., Part B*, 2008, **14**, 149.
10. J. K. Tessmar and A. M. Göpferich, *Adv. Drug Delivery Rev.*, 2007, **59**, 274.
11. M. W. Tibbitt and K. S. Anseth, *Biotech. Bioeng.*, 2009, **103**, 655.
12. M. K. Nguyen and D. S. Lee, *Macromol. Biosci.*, 2010, **10**, 563.
13. L. Yu and J. Ding, *Chem. Soc. Rev.*, 2008, **37**, 1473.
14. A. Mata, Y. Geng, K. J. Henrikson, C. Aparicio, S. R. Stock, R. L. Satcher and S. I. Stupp, *Biomaterials*, 2010, **31**, 6004.
15. C. Y. Khripin, D. Pristinski, D. R. Dunphy, C. J. Brinker and B. Kaehr, *ACS Nano*, 2011, **5**, 1401.
16. D. W. P. M. Löwik, E. H. P. Leunissen, M. van den Heuvel, M. B. Hansen and J. C. M. van Hest, *Chem. Soc. Rev.*, 2010, **39**, 3394.
17. R. V. Ulijn, N. Bibi, V. Jayawarna, P. D. Thornton, S. J. Todd, R. J. Mart, A. M. Smith and J. E. Gough, *Mater. Today*, 2007, **10**, 40.
18. V. R. Pattabiraman and J. W. Bode, *Nature*, 2011, **480**, 471.
19. P. Dawson, T. Muir, I. Clark-Lewis and S. Kent, *Science*, 1994, **266**, 776.
20. J. F. Atkins and R. Gesteland, *Science*, 2002, **296**, 1409.
21. A. Böck, K. Forchhammer, J. Heider, W. Leinfelder, G. Sawers, B. Veprek and F. Zinoni, *Mol. Microbiol.*, 1991, **5**, 515.
22. C. A. DeForest, B. D. Polizzotti and K. S. Anseth, *Nat. Mater.*, 2009, **8**, 659.
23. C. M. Nimmo and M. S. Shoichet, *Bioconjugate Chem.*, 2011, **22**, 2199.
24. S. K. Holmgren, L. E. Bretscher, K. M. Taylor and R. T. Raines, *Chem. Biol.*, 1999, **6**, 63.

25. Y. Tang, G. Ghirlanda, W. A. Petka, T. Nakajima, W. F. DeGrado and D. A. Tirrell, *Angew. Chem. Int. Ed.*, 2001, **40**, 1494.
26. N. C. Yoder and K. Kumar, *Chem. Soc. Rev.*, 2002, **31**, 335.
27. I. S. Carrico, S. A. Maskarinec, S. C. Heilshorn, M. L. Mock, J. C. Liu, P. J. Nowatzki, C. Franck, G. Ravichandran and D. A. Tirrell, *J. Am. Chem. Soc.*, 2007, **129**, 4874.
28. D. K. Smith, *Chem. Soc. Rev.*, 2009, **38**, 684.
29. J. A. Johnson, Y. Y. Lu, J. A. Van Deventer and D. A. Tirrell, *Curr. Opin. Chem. Biol.*, 2010, **14**, 774.
30. C. C. Liu and P. G. Schultz, *Ann. Rev. Biochem.*, 2010, **79**, 413.
31. E. L. Bakota, L. Aulisa, K. M. Galler and J. D. Hartgerink, *Biomacromolecules*, 2011, **12**, 82.
32. L. Hsu, G. L. Cvetanovich and S. I. Stupp, *J. Am. Chem. Soc.*, 2008, **130**, 3892.
33. T. Aida, E. W. Meijer and S. I. Stupp, *Science*, 2012, **335**, 813.
34. S. Zhang, T. Holmes, C. Lockshin and A. Rich, *Proc. Natl. Acad. Sci. U. S. A.*, 1993, **90**, 3334.
35. S. Zhang, T. C. Holmes, C. M. DiPersio, R. O. Hynes, X. Su and A. Rich, *Biomaterials*, 1995, **16**, 1385.
36. S. Zhang, R. Ellis-Behnke, X. Zhao and L. Spirio, in *Scaffolding in Tissue Engineering*, ed. P. X. Ma and J. Elisseeff, CRC Press, Boca Raton, FL, 2005, p. 217.
37. A. Aggeli, M. Bell, N. Boden, J. N. Keen, P. F. Knowles, T. C. B. McLeish, M. Pitkeathly and S. E. Radford, *Nature*, 1997, **386**, 259.
38. A. Aggeli, I. A. Nyrkova, M. Bell, R. Harding, L. Carrick, T. C. B. McLeish, A. N. Semenov and N. Boden, *Proc. Natl. Acad. Sci. U. S. A.*, 2001, **98**, 11857.
39. C. W. G. Fishwick, A. J. Beevers, L. M. Carrick, C. D. Whitehouse, A. Aggeli and N. Boden, *Nano Lett.*, 2003, **3**, 1475.
40. J. Kopeček and J. Yang, *Acta Biomater.*, 2009, **5**, 805.
41. M. R. Caplan, P. N. Moore, S. Zhang, R. D. Kamm and D. A. Lauffenburger, *Biomacromolecules*, 2000, **1**, 627.
42. M. R. Caplan, E. M. Schwartzfarb, S. Zhang, R. D. Kamm and D. A. Lauffenburger, *Biomaterials*, 2002, **23**, 219.
43. J. H. Collier, B. H. Hu, J. W. Ruberti, J. Zhang, P. Shum, D. H. Thompson and P. B. Messersmith, *J. Am. Chem. Soc.*, 2001, **123**, 9463.
44. S. Ramachandran, P. Flynn, Y. Tseng and Y. B. Yu, *Chem. Mater.*, 2005, **17**, 6583.
45. J. P. Schneider, D. J. Pochan, B. Ozbas, K. Rajagopal, L. Pakstis and J. Kretsinger, *J. Am. Chem. Soc.*, 2002, **124**, 15030.
46. D. J. Pochan, J. P. Schneider, J. Kretsinger, B. Ozbas, K. Rajagopal and L. Haines, *J. Am. Chem. Soc.*, 2003, **125**, 11802.
47. B. Ozbas, J. Kretsinger, K. Rajagopal, J. P. Schneider and D. J. Pochan, *Macromolecules*, 2004, **37**, 7331.
48. K. Rajagopal, M. S. Lamm, L. A. Haines-Butterick, D. J. Pochan and J. P. Schneider, *Biomacromolecules*, 2009, **10**, 2619.

49. L. A. Haines, K. Rajagopal, B. Ozbas, D. A. Salick, D. J. Pochan and J. P. Schneider, *J. Am. Chem. Soc.*, 2005, **127**, 17025.
50. D. J. Adams, *Macromol. Biosci.*, 2011, **11**, 160.
51. V. Jayawarna, M. Ali, T. A. Jowitt, A. F. Miller, A. Saiani, J. E. Gough and R. V. Ulijn, *Adv. Mater.*, 2006, **18**, 611.
52. Y. Zhang, H. Gu, Z. Yang and B. Xu, *J. Am. Chem. Soc.*, 2003, **125**, 13680.
53. Z. Yang, H. Gu, Y. Zhang, L. Wang and B. Xu, *Chem. Commun.*, 2004, 208.
54. A. Mahler, M. Reches, M. Rechter, S. Cohen and E. Gazit, *Adv. Mater.*, 2006, **18**, 1365.
55. S. Toledano, R. J. Williams, V. Jayawarna and R. V. Ulijn, *J. Am. Chem. Soc.*, 2006, **128**, 1070.
56. Z. Yang, H. Gu, D. Fu, P. Gao, J. K. Lam and B. Xu, *Adv. Mater.*, 2004, **16**, 1440.
57. A. M. Smith, R. J. Williams, C. Tang, P. Coppo, R. F. Collins, M. L. Turner, A. Saiani and R. V. Ulijn, *Adv. Mater.*, 2008, **20**, 37.
58. M. Zhou, A. M. Smith, A. K. Das, N. W. Hodson, R. F. Collins, R. V. Ulijn and J. E. Gough, *Biomaterials*, 2009, **30**, 2523.
59. W. A. Petka, J. L. Harden, K. P. McGrath, D. Wirtz and D. A. Tirrell, *Science*, 1998, **281**, 389.
60. C. Wang, R. J. Stewart and J. Kopeček, *Nature*, 1999, **397**, 417.
61. E. F. Banwell, E. S. Abelardo, D. J. Adams, M. A. Birchall, A. Corrigan, A. M. Donald, M. Kirkland, L. C. Serpell, M. F. Butler and D. N. Woolfson, *Nat. Mater.*, 2009, **8**, 596.
62. H. Dong, S. E. Paramonov and J. D. Hartgerink, *J. Am. Chem. Soc.*, 2008, **130**, 13691.
63. N. L. Fletcher, C. V. Lockett and A. F. Dexter, *Soft Matter*, 2011, **7**, 10210.
64. J. M. Mason and K. M. Arndt, *ChemBioChem*, 2004, **5**, 170.
65. W. Landschulz, P. Johnson and S. McKnight, *Science*, 1988, **240**, 1759.
66. S. A. Potekhin, T. N. Melnik, V. Popov, N. F. Lanina, A. A. Vazina, P. Rigler, A. S. Verdini, G. Corradin and A. V. Kajava, *Chem. Biol.*, 2001, **8**, 1025.
67. Y. Zimenkov, V. P. Conticello, L. Guo and P. Thiyagarajan, *Tetrahedron*, 2004, **60**, 7237.
68. M. J. Pandya, G. M. Spooner, M. Sunde, J. R. Thorpe, A. Rodger and D. N. Woolfson, *Biochemistry*, 2000, **39**, 8728.
69. M. G. Ryadnov and D. N. Woolfson, *Nat. Mater.*, 2003, **2**, 329.
70. D. N. Woolfson, *Biopolymers*, 2010, **94**, 118.
71. D. Papapostolou, A. M. Smith, E. D. T. Atkins, S. J. Oliver, M. G. Ryadnov, L. C. Serpell and D. N. Woolfson, *Proc. Natl. Acad. Sci. U. S. A.*, 2007, **104**, 10853.
72. M. D. Shoulders and R. T. Raines, *Annu. Rev. Biochem.*, 2009, **78**, 929.
73. J. A. Fallas, L. E. R. O'Leary and J. D. Hartgerink, *Chem. Soc. Rev.*, 2010, **39**, 3510.

74. S. M. Yu, Y. Li and D. Kim, *Soft Matter*, 2011, **7**, 7927.
75. K. Kar, P. Amin, M. A. Bryan, A. V. Persikov, A. Mohs, Y. H. Wang and B. Brodsky, *J. Biol. Chem.*, 2006, **281**, 33283.
76. F. W. Kotch and R. T. Raines, *Proc. Natl. Acad. Sci. U. S. A.*, 2006, **103**, 3028.
77. T. Koide, D. L. Homma, S. Asada and K. Kitagawa, *Bioorg. Med. Chem. Lett.*, 2005, **15**, 5230.
78. C. M. Yamazaki, S. Asada, K. Kitagawa and T. Koide, *Peptide Sci.*, 2008, **90**, 816.
79. S. Rele, Y. Song, R. P. Apkarian, Z. Qu, V. P. Conticello and E. L. Chaikof, *J. Am. Chem. Soc.*, 2007, **129**, 14780.
80. S. E. Paramonov, V. Gauba and J. D. Hartgerink, *Macromolecules*, 2005, **38**, 7555.
81. M. M. Pires and J. Chmielewski, *J. Am. Chem. Soc.*, 2009, **131**, 2706.
82. W. Hsu, Y.-L. Chen and J.-C. Horng, *Langmuir*, 2012, **28**, 3194.
83. M. A. Cejas, W. A. Kinney, C. Chen, G. C. Leo, B. A. Tounge, J. G. Vinter, P. P. Joshi and B. E. Maryanoff, *J. Am. Chem. Soc.*, 2007, **129**, 2202.
84. C.-C. Chen, W. Hsu, T.-C. Kao and J.-C. Horng, *Biochemistry*, 2011, **50**, 2381.
85. L. E. R. O'Leary, J. A. Fallas, E. L. Bakota, M. K. Kang and J. D. Hartgerink, *Nat. Chem.*, 2011, **3**, 821.
86. O. D. Krishna and K. L. Kiick, *Biomacromolecules*, 2009, **10**, 2626.
87. D. D. Buechter, D. N. Paolella, B. S. Leslie, M. S. Brown, K. A. Mehos and E. A. Gruskin, *J. Biol. Chem.*, 2003, **278**, 645.
88. D. M. Pinkas, S. Ding, R. T. Raines and A. E. Barron, *ACS Chem. Biol.*, 2011, **6**, 320.
89. P. Berndt, G. B. Fields and M. Tirrell, *J. Am. Chem. Soc.*, 1995, **117**, 9515.
90. M. A. Biesalski, A. Knaebel, R. Tu and M. Tirrell, *Biomaterials*, 2006, **27**, 1259.
91. Y.-C. Yu, P. Berndt, M. Tirrell and G. B. Fields, *J. Am. Chem. Soc.*, 1996, **118**, 12515.
92. S. Vauthey, S. Santoso, H. Gong, N. Watson and S. Zhang, *Proc. Natl. Acad. Sci. U. S. A.*, 2002, **99**, 5355.
93. J. D. Hartgerink, E. Beniash and S. I. Stupp, *Science*, 2001, **294**, 1684.
94. J. D. Hartgerink, E. Beniash and S. I. Stupp, *Proc. Natl. Acad. Sci. U. S. A.*, 2002, **99**, 5133.
95. J. B. Matson and S. I. Stupp, *Chem. Commun.*, 2012, **48**, 26.
96. K. Rajangam, H. A. Behanna, M. J. Hui, X. Han, J. F. Hulvat, J. W. Lomasney and S. I. Stupp, *Nano Lett.*, 2006, **6**, 2086.
97. M. J. Webber, J. Tongers, C. J. Newcomb, K.-T. Marquardt, J. Bauersachs, D. W. Losordo and S. I. Stupp, *Proc. Natl. Acad. Sci. U. S. A.*, 2011, **108**, 13438.
98. A. P. Nowak, V. Breedveld, L. Pakstis, B. Ozbas, D. J. Pine, D. Pochan and T. J. Deming, *Nature*, 2002, **417**, 424.

99. A. P. Nowak, V. Breedveld, D. J. Pine and T. J. Deming, *J. Am. Chem. Soc.*, 2003, **125**, 15666.
100. T. J. Deming, *Soft Matter*, 2005, **1**, 28.
101. Z. Li and T. J. Deming, *Soft Matter*, 2010, **6**, 2546.
102. S. Banta, I. R. Wheeldon and M. Blenner, *Annu. Rev. Biomed. Eng.*, 2010, **12**, 167.
103. C. Xu, V. Breedveld and J. Kopeček, *Biomacromolecules*, 2005, **6**, 1739.
104. C. Xu and J. Kopeček, *Pharm. Res.*, 2008, **25**, 674.
105. W. Shen, J. A. Kornfield and D. A. Tirrell, *Soft Matter*, 2007, **3**, 99.
106. W. Shen, R. G. H. Lammertink, J. K. Sakata, J. A. Kornfield and D. A. Tirrell, *Macromolecules*, 2005, **38**, 3909.
107. W. Shen, K. Zhang, J. A. Kornfield and D. A. Tirrell, *Nat. Mater.*, 2006, **5**, 153.
108. B. D. Olsen, J. A. Kornfield and D. A. Tirrell, *Macromolecules*, 2010, **43**, 9094.
109. L. Mi, S. Fischer, B. Chung, S. Sundelacruz and J. L. Harden, *Biomacromolecules*, 2006, **7**, 38.
110. P. J. Skrzeszewska, F. A. de Wolf, M. W. T. Werten, A. P. H. A. Moers, M. A. Cohen Stuart and J. van der Gucht, *Soft Matter*, 2009, **5**, 2057.
111. P. J. Skrzeszewska, J. Sprakel, F. A. de Wolf, R. Fokkink, M. A. Cohen Stuart and J. van der Gucht, *Macromolecules*, 2010, **43**, 3542.
112. M. W. T. Werten, H. Teles, A. P. H. A. Moers, E. J. H. Wolbert, J. Sprakel, G. Eggink and F. A. de Wolf, *Biomacromolecules*, 2009, **10**, 1106.
113. H. Teles, T. Vermonden, G. Eggink, W. E. Hennink and F. A. de Wolf, *J. Controlled Release*, 2010, **147**, 298.
114. P. J. Skrzeszewska, L. N. Jong, F. A. de Wolf, M. A. Cohen Stuart and J. van der Gucht, *Biomacromolecules*, 2011, **12**, 2285.
115. C. T. S. Wong Po Foo, J. S. Lee, W. Mulyasasmita, A. Parisi-Amon and S. C. Heilshorn, *Proc. Natl. Acad. Sci. U. S. A.*, 2009, **106**, 22067.
116. W. Mulyasasmita, J. S. Lee and S. C. Heilshorn, *Biomacromolecules*, 2011, **12**, 3406.
117. D. E. Meyer and A. Chilkoti, *Biomacromolecules*, 2004, **5**, 846.
118. D. W. Urry, *J. Phys. Chem. B*, 1997, **101**, 11007.
119. R. A. McMillan, T. A. T. Lee and V. P. Conticello, *Macromolecules*, 1999, **32**, 3643.
120. A. Panitch, T. Yamaoka, M. J. Fournier, T. L. Mason and D. A. Tirrell, *Macromolecules*, 1999, **32**, 1701.
121. D. T. McPherson, J. Xu and D. W. Urry, *Protein Expression Purif.*, 1996, **7**, 51.
122. E. R. Wright and V. P. Conticello, *Adv. Drug Delivery Rev.*, 2002, **54**, 1057.
123. E. R. Wright, R. A. McMillan, A. Cooper, R. P. Apkarian and V. P. Conticello, *Adv. Funct. Mater.*, 2002, **12**, 149.
124. H. Betre, L. A. Setton, D. E. Meyer and A. Chilkoti, *Biomacromolecules*, 2002, **3**, 910.
125. K. Di Zio and D. A. Tirrell, *Macromolecules*, 2003, **36**, 1553.

126. P. J. Nowatzki and D. A. Tirrell, *Biomaterials*, 2004, **25**, 1261.
127. E. R. Welsh and D. A. Tirrell, *Biomacromolecules*, 2000, **1**, 23.
128. R. A. McMillan and V. P. Conticello, *Macromolecules*, 2000, **33**, 4809.
129. R. E. Sallach, W. Cui, J. Wen, A. Martinez, V. P. Conticello and E. L. Chaikof, *Biomaterials*, 2009, **30**, 409.
130. D. W. Lim, D. L. Nettles, L. A. Setton and A. Chilkoti, *Biomacromolecules*, 2007, **8**, 1463.
131. K. Trabbic-Carlson, L. A. Setton and A. Chilkoti, *Biomacromolecules*, 2003, **4**, 572.
132. M. K. McHale, L. A. Setton and A. Chilkoti, *Tissue Eng.*, 2005, **11**, 1768.
133. K. Nagapudi, W. T. Brinkman, J. E. Leisen, L. Huang, R. A. McMillan, R. P. Apkarian, V. P. Conticello and E. L. Chaikof, *Macromolecules*, 2002, **35**, 1730.
134. J. Lee, C. W. Macosko and D. W. Urry, *Macromolecules*, 2001, **34**, 4114.
135. C. Vepari and D. L. Kaplan, *Prog. Polym. Sci.*, 2007, **32**, 991.
136. H.-J. Jin and D. L. Kaplan, *Nature*, 2003, **424**, 1057.
137. U.-J. Kim, J. Park, C. Li, H.-J. Jin, R. Valluzzi and D. L. Kaplan, *Biomacromolecules*, 2004, **5**, 786.
138. A. Matsumoto, J. Chen, A. L. Collette, U.-J. Kim, G. H. Altman, P. Cebe and D. L. Kaplan, *J. Phys. Chem. B*, 2006, **110**, 21630.
139. P.-H. G. Chao, S. Yodmuang, X. Wang, L. Sun, D. L. Kaplan and G. Vunjak-Novakovic, *J. Biomed. Mater. Res., B*, 2010, **95**, 84.
140. X. Wang, J. A. Kluge, G. G. Leisk and D. L. Kaplan, *Biomaterials*, 2008, **29**, 1054.
141. X.-X. Xia, Z.-G. Qian, C. S. Ki, Y. H. Park, D. L. Kaplan and S. Y. Lee, *Proc. Natl. Acad. Sci. U. S. A.*, 2010, **107**, 14059.
142. J. Scheller, K.-H. Gührs, F. Grosse and U. Conrad, *Nat. Biotechnol.*, 2001, **19**, 573.
143. A. Lazaris, S. Arcidiacono, Y. Huang, J.-F. Zhou, F. Duguay, N. Chretien, E. A. Welsh, J. W. Soares and C. N. Karatzas, *Science*, 2002, **295**, 472.
144. F. Teulé, Y.-G. Miao, B.-H. Sohn, Y.-S. Kim, J. J. Hull, M. J. Fraser, R. V. Lewis and D. L. Jarvis, *Proc. Natl. Acad. Sci. U. S. A.*, 2012, **109**, 923.
145. S. Rammensee, D. Huemmerich, K. D. Hermanson, T. Scheibel and A. R. Bausch, *Appl. Phys. A: Mater. Sci. Process.*, 2006, **82**, 261.
146. K. Schacht and T. Scheibel, *Biomacromolecules*, 2011, **12**, 2488.
147. E. Bini, C. W. P. Foo, J. Huang, V. Karageorgiou, B. Kitchel and D. L. Kaplan, *Biomacromolecules*, 2006, **7**, 3139.
148. S. Yanagisawa, Z. Zhu, I. Kobayashi, K. Uchino, Y. Tamada, T. Tamura and T. Asakura, *Biomacromolecules*, 2007, **8**, 3487.
149. Z. Megeed, J. Cappello and H. Ghandehari, *Adv. Drug Delivery Rev.*, 2002, **54**, 1075.
150. A. Nagarsekar, J. Crissman, M. Crissman, F. Ferrari, J. Cappello and H. Ghandehari, *Biomacromolecules*, 2003, **4**, 602.
151. K. Greish, K. Araki, D. Li, B. W. O'Malley, R. Dandu, J. Frandsen, J. Cappello and H. Ghandehari, *Biomacromolecules*, 2009, **10**, 2183.

152. I. R. Wheeldon, S. Calabrese Barton and S. Banta, *Biomacromolecules*, 2007, **8**, 2990.
153. I. R. Wheeldon, J. W. Gallaway, S. C. Barton and S. Banta, *Proc. Natl. Acad. Sci. U. S. A.*, 2008, **105**, 15275.
154. H. D. Lu, I. R. Wheeldon and S. Banta, *Protein Eng., Des. Sel.*, 2010, **23**, 559.
155. I. R. Wheeldon, E. Campbell and S. Banta, *J. Mol. Biol.*, 2009, **392**, 129.
156. S. Topp, V. Prasad, G. C. Cianci, E. R. Weeks and J. P. Gallivan, *J. Am. Chem. Soc.*, 2006, **128**, 13994.
157. O. D. Krishna and K. L. Kiick, *Biopolymers*, 2010, **94**, 32.
158. J. Yang, C. Xu, C. Wang and J. Kopeček, *Biomacromolecules*, 2006, **7**, 1187.
159. B. Liu, Y. Liu, A. K. Lewis and W. Shen, *Biomaterials*, 2010, **31**, 4918.
160. P. J. Stahl, N. H. Romano, D. Wirtz and S. M. Yu, *Biomacromolecules*, 2010, **11**, 2336.
161. A. Y. Wang, S. Leong, Y.-C. Liang, R. C. C. Huang, C. S. Chen and S. M. Yu, *Biomacromolecules*, 2008, **9**, 2929.
162. A. Y. Wang, X. Mo, C. S. Chen and S. M. Yu, *J. Am. Chem. Soc.*, 2005, **127**, 4130.
163. S. Halstenberg, A. Panitch, S. Rizzi, H. Hall and J. A. Hubbell, *Biomacromolecules*, 2002, **3**, 710.
164. D. L. Hern and J. A. Hubbell, *J. Biomed. Mater. Res.*, 1998, **39**, 266.
165. M. P. Lutolf and J. A. Hubbell, *Biomacromolecules*, 2003, **4**, 713.
166. M. P. Lutolf, J. L. Lauer-Fields, H. G. Schmoekel, A. T. Metters, F. E. Weber, G. B. Fields and J. A. Hubbell, *Proc. Natl. Acad. Sci. U. S. A.*, 2003, **100**, 5413.
167. J. L. West and J. A. Hubbell, *Macromolecules*, 1998, **32**, 241.
168. C. A. DeForest and K. S. Anseth, *Nat. Chem.*, 2011, **3**, 925.
169. J. D. Ehrick, S. K. Deo, T. W. Browning, L. G. Bachas, M. J. Madou and S. Daunert, *Nat. Mater.*, 2005, **4**, 298.
170. W. L. Murphy, W. S. Dillmore, J. Modica and M. Mrksich, *Angew. Chem. Int. Ed.*, 2007, **46**, 3066.
171. J. D. Ehrick, S. Stokes, S. Bachas-Daunert, E. A. Moschou, S. K. Deo, L. G. Bachas and S. Daunert, *Adv. Mater.*, 2007, **19**, 4024.
172. W. J. King, J. S. Mohammed and W. L. Murphy, *Soft Matter*, 2009, **5**, 2399.
173. T. Miyata, N. Asami and T. Uragami, *Nature*, 1999, **399**, 766.
174. W. Yuan, J. Yang, P. Kopečková and J. Kopeček, *J. Am. Chem. Soc.*, 2008, **130**, 15760.
175. A. H. Zisch, M. P. Lutolf, M. Ehrbar, G. P. Raeber, S. C. Rizzi, N. Davies, H. Schmökel, D. Bezuidenhout, V. Djonov, P. Zilla and J. A. Hubbell, *FASEB J.*, 2003, **17**, 2260.
176. J. J. Moon, S.-H. Lee and J. L. West, *Biomacromolecules*, 2007, **8**, 42.
177. J. E. Saik, D. J. Gould, E. M. Watkins, M. E. Dickinson and J. L. West, *Acta Biomater.*, 2011, **7**, 133.

178. T. C. Holmes, S. de Lacalle, X. Su, G. Liu, A. Rich and S. Zhang, *Proc. Natl. Acad. Sci. U. S. A.*, 2000, **97**, 6728.
179. M. E. Davis, J. P. M. Motion, D. A. Narmoneva, T. Takahashi, D. Hakuno, R. D. Kamm, S. Zhang and R. T. Lee, *Circulation*, 2005, **111**, 442.
180. L. A. Haines-Butterick, D. A. Salick, D. J. Pochan and J. P. Schneider, *Biomaterials*, 2008, **29**, 4164.
181. J. Cappello, J. W. Crissman, M. Crissman, F. A. Ferrari, G. Textor, O. Wallis, J. R. Whitledge, X. Zhou, D. Burman, L. Aukerman and E. R. Stedronsky, *J. Controlled Release*, 1998, **53**, 105.
182. D. W. Urry, T. M. Parker, M. C. Reid and D. C. Gowda, *J. Bioact. Compat. Polym.*, 1991, **6**, 263.
183. J. S. Rudra, T. Sun, K. C. Bird, M. D. Daniels, J. Z. Gasiorowski, A. S. Chong and J. H. Collier, *ACS Nano*, 2012, **6**, 1557.
184. J. S. Rudra, Y. F. Tian, J. P. Jung and J. H. Collier, *Proc. Natl. Acad. Sci. U. S. A.*, 2010, **107**, 622.
185. J. S. Rudra, P. K. Tripathi, D. A. Hildeman, J. P. Jung and J. H. Collier, *Biomaterials*, 2010, **31**, 8475.
186. A. Rosenberg, *AAPS J.*, 2006, **8**, E501.
187. J. M. Riley, A. Aggeli, R. J. Koopmans and M. J. McPherson, *Biotechnol. Bioeng.*, 2009, **103**, 241.
188. E. S. Place, N. D. Evans and M. M. Stevens, *Nat. Mater.*, 2009, **8**, 457.
189. S. Khetan and J. A. Burdick, *Soft Matter*, 2011, **7**, 830.
190. M. S. Hahn, J. S. Miller and J. L. West, *Adv. Mater.*, 2006, **18**, 2679.
191. R. G. Wylie, S. Ahsan, Y. Aizawa, K. L. Maxwell, C. M. Morshead and M. S. Shoichet, *Nat. Mater.*, 2011, **10**, 799.
192. K. S. Straley and S. C. Heilshorn, *Adv. Mater.*, 2009, **21**, 4148.
193. S. Moon, S. K. Hasan, Y. S. Song, F. Xu, H. O. Keles, F. Manzur, S. Mikkilineni, J. W. Hong, J. Nagatomi, E. Haeggstrom, A. Khademhosseini and U. Demirci, *Tissue Eng., Part C*, 2010, **16**, 157.
194. K. Pataky, T. Braschler, A. Negro, P. Renaud, M. P. Lutolf and J. Brugger, *Adv. Mater.*, 2012, **24**, 391.
195. S. Allazetta, S. Cosson and M. P. Lutolf, *Chem. Commun.*, 2011, **47**, 191.
196. E. F. Plow, T. A. Haas, L. Zhang, J. Loftus and J. W. Smith, *J. Biol. Chem.*, 2000, **275**, 21785.
197. E. Ruoslahti and M. D. Pierschbacher, *Cell*, 1986, **44**, 517.
198. E. Fong and D. A. Tirrell, *Adv. Mater.*, 2010, **22**, 5271.
199. E. L. Rexeisen, W. Fan, T. O. Pangburn, R. R. Taribagil, F. S. Bates, T. P. Lodge, M. Tsapatsis and E. Kokkoli, *Langmuir*, 2010, **26**, 1953.
200. K. Shroff, E. L. Rexeisen, M. A. Arunagirinathan and E. Kokkoli, *Soft Matter*, 2010, **6**, 5064.

CHAPTER 6

Chemomechanical Hydrogels: Selective Response towards External Effector Molecules

HANS-JÖRG SCHNEIDER

FR Organische Chemie der Universität des Saarlandes,
D-66041 Saarbrücken, Germany
Email: ch12hs@rz.uni-sb.de

6.1 Introduction

Chemoresponsive materials exhibit size changes by interactions with substances in the surrounding medium, representing a special kind of artificial muscle. Hydrogels bearing basic or acidic groups in the polymer backbone undergo volume changes by pH variation. Such gels can be used to detect, for example, glucose after incorporation of glucose oxidase.[1] The application of such enzymes, however, suffers from their tendency to degrade over time; in addition the *in vivo* presence of physiologic buffers attenuates the pH response. Until recently, chemoresponsive hydrogels were mostly triggered by unspecific interactions, *e.g.* by salts or solvents.[2] Glucose-sensitive hydrogels were already developed in 1992,[3] on the basis of covalent interactions with boronic acids (see Section 6.4). In contrast, the chemoresponsive hydrogels mostly dealt with in the present chapter bear binding sites which enable selective non-covalent associations. A related principle is based on hydrogelators, which can exhibit a macroscopic gel–sol transition by external stimuli.[4] Furthermore, gelation itself can be stimulated or disrupted by competitive guests, including cavitands as additives.[5]

Monographs in Supramolecular Chemistry No. 11
Polymeric and Self Assembled Hydrogels: From Fundamental Understanding to Applications
Edited by Xian Jun Loh and Oren A. Scherman
© The Royal Society of Chemistry 2013
Published by the Royal Society of Chemistry, www.rsc.org

Host groups which are covalently attached to polymer chains in hydrogels can, by non-covalent interactions with guest molecules, lead to volume changes by up to several 100%.[2] In the past few decades, supramolecular chemistry has opened the way to analyze and to use the corresponding molecular recognition mechanisms in much detail.[6] The present chapter intends to show what can be achieved with these principles, highlighting also the significant differences between host–guest complexation in solution and within hydrogels.

6.2 Basic Mechanisms: Water Content Changes

The fundamental mechanism leading to volume changes of hydrogels will be discussed on the basis of experiments with selected polymer units P1–P5 shown in Scheme 6.1. Figure 6.1 illustrates how the association of a guest molecule with two or more host groups at opposing sites of the polymer backbones can lead to non-covalent cross-linking, and subsequent contraction. The opposite expansion is observed with import of guest molecules associated primarily with host groups at one side of the polymer chain; the effector guest molecules need to carry with them water molecules for solvation, in particular as the used effector substrates invariably are bearing charges which also require the import of solvated gegenions. It is known that, for example, carboxymethylcellulose

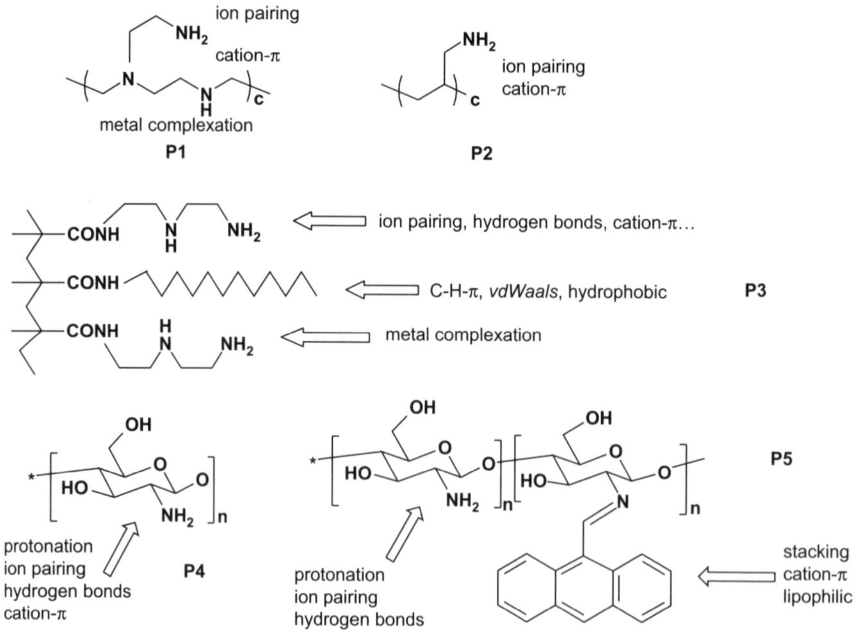

Scheme 6.1   Some structural elements in the chemomechanical polymers and a possible interaction mechanism. P1, polyethylenimine; P2, polyallylamine; P3, poly(methyl methacrylic) derivative (also contains other units, to a minor degree; see Schneider *et al.*[2b]); P4, chitosan; P5, chitosan-anthryl derivative.

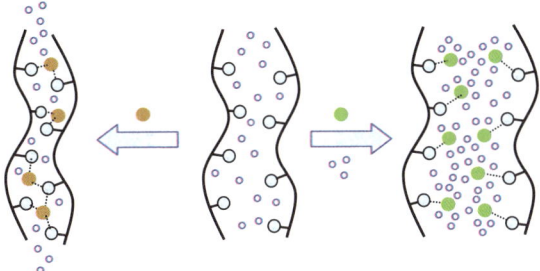

Figure 6.1 Hydrogel contraction by non-covalent cross-linking and water release (*left*) or expansion by binding of guest molecules with water uptake for solvation (*right*).

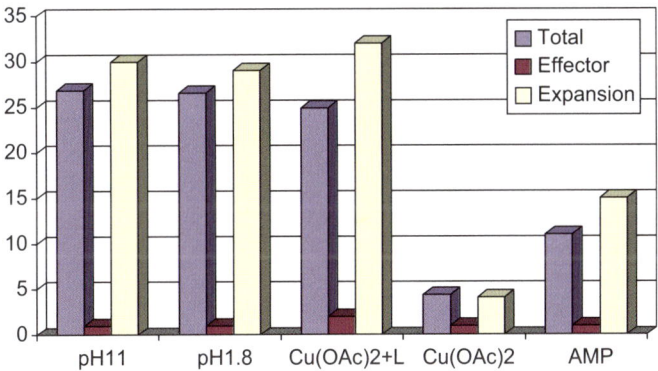

Figure 6.2 Weight increase (scaled per mg) compared to expansion (measurements with polymer P3, in vol%).
(Reproduced from Schneider *et al.*[2b] by permission of Wiley-VCH.)

gels exhibit water uptake as a function of the cross-linkers, which was studied by FT-IR spectroscopy.[7] For cross-linked chitosan/polyether hydrogels the amount of bound water is a function of the pH.[8] Hydrogels derived from fluorenyl-Ala-Ala dipeptides release 40% of their water upon addition of the strong binding ligand vancomycin.[9]

That water uptake for expansion and release for contraction is by far the major contribution to the observed size changes has been proven by gravimetric measurements. These show that the water content of such hydrogels increases, for instance, from 77% up to 99%; this also applies to pH-triggered changes. It should be noted that pH-controlled size changes depend strongly on the presence of neutral salts such as buffers.[2b,10] Obviously, the volume changes correspond to weight changes, with only small weight contributions from the effector substances (see Figure 6.2); all weight changes are essentially due to water uptake (Table 6.1).

Noticeably, there is a monotonous volume increase as a function of the effector concentration (Figure 6.3), with no indication of phase transitions

Table 6.1 Weight increase factor $f1$ upon interaction with different compared to volume increase factor $f2$, and water content before and after volume change.[a]

Effector	Water content (%)	f1 weight increase	f2 volume increase
1.0 mM NaOH; pH 11.0	99	26.85	30.0
10.0 mM HCl; pH 1.8	99	26.6	29.0
10.0 mM AMP; pH 11.0	98	11.5	15.0
1.0 mM Cu(OAc)$_2$ + Cuprizon; pH 4.8	99	25.0	32.0
2.0 mM Cu(OAc)$_2$; pH 6.0	95	4.4	4.1
Without effector	77	0.0	0.0

[a]Gravimetric measurements with polymer film P3; double measurements show deviations of f by ±10% on average; weight increase due to effector alone negligible (calculated from effector content in gel, determined by UV spectroscopy.[2b])

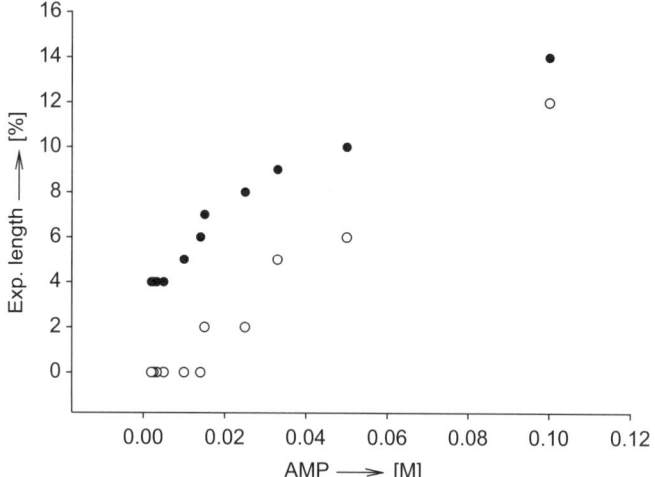

Figure 6.3 Expansion (length) as function of AMP concentration; polymer P3, lower trace (○), in the absence of buffer; upper trace (●), in the presence of 0.02 M NaH$_2$PO$_4$ buffer.
(Reproduced from Schneider *et al.*[2b] by permission of Wiley-VCH.)

which are normally characterized by some discontinuity. At the beginning there is a lag period due to the first necessary occupation of the polymer particle surface; in contrast, one observes by UV spectroscopy an effector concentration decrease immediately after adding the effector, without any time lag.

6.3 Speed and Sensitivity of Response

As expected, the kinetics of the volume changes are determined by the diffusion rates of the effector into the inside of the particles. Therefore one observes a distinct dependence on the particle surface/volume ratio (Figure 6.4);[10] with thin

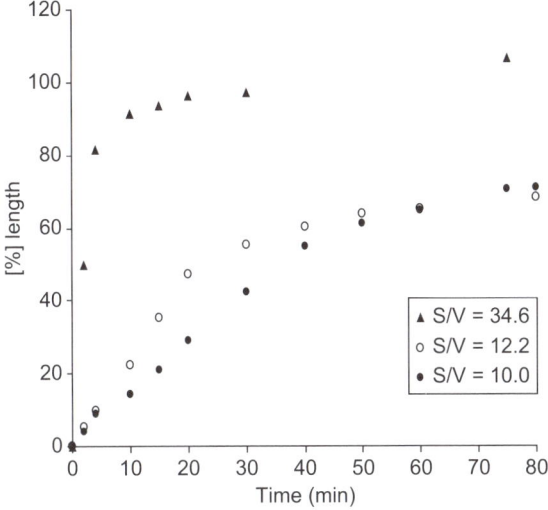

Figure 6.4   Rate dependence of chitosan gel particle P4 elongation induced by 50 mM L-histidine on the surface to volume ratio S/V. Approximate "half-life" time $t_{1/2}$ for 50% of the maximum expansion: $t_{1/2} = 42$, 32 and 3 min for $S/V = 10.0$, 12.2 and 34.6, respectively.
(Reproduced from Kato and Schneider[10] by permission of Wiley-VCH.)

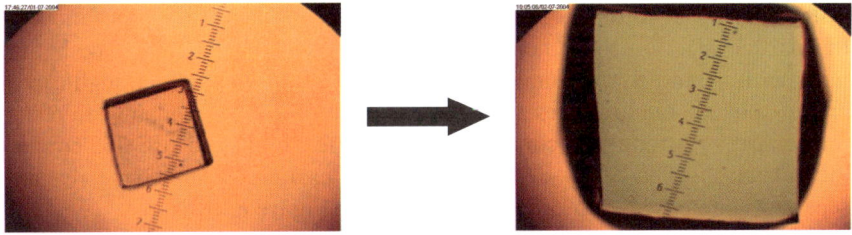

Figure 6.5   Size increase of a water-saturated chemomechanical hydrogel film piece ($1.1 \times 0.9 \times 0.4$ mm) induced by a 0.25 mM solution of a dipeptide (Gly-Gly) in the presence of Cu(II) ions, pH 4.5.
(Reproduced from Lomadze and Schneider[31] by permission of Elsevier.)

films the rates can be enhanced by orders of magnitude. The kinetics are essentially controlled by the diffusion rates of the substrates in the gel, after a short initial phase of loading the surface of the gel particles (for details, see Schneiderr *et al.*[2b]). With large effector molecules such as proteins the diffusion becomes slower; with cytochrome *c*, for instance, it may take weeks with larger gel particles. The experiments shown in Figures 6.2–6.4 were usually carried out with gel pieces of $1.4 \times 1.4 \times 0.5$ mm, in order to determine accurately the size with the help of a measuring microscope, before and after adding the effector. Figure 6.5 illustrates the size increase measured this way, here with a dipeptide as stimulus.

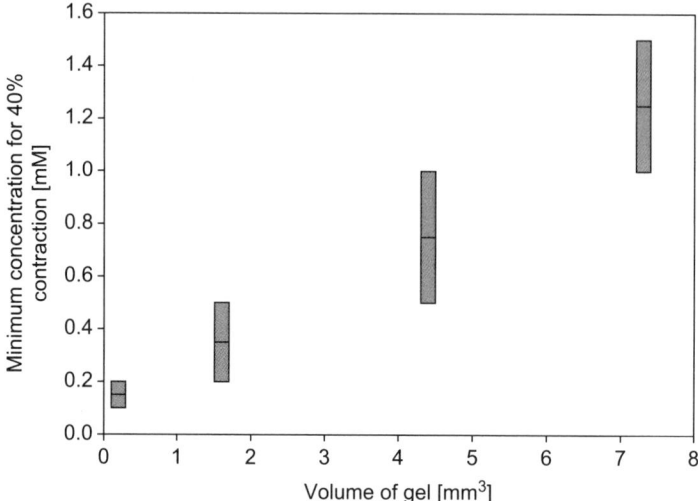

Figure 6.6 Sensitivity increase with decreasing particle size of polymer P4. Sensitivity
of 40% contraction (in one dimension) induced by D-*O*-dibenzoyl tartrate
as a function of the gel particle size.
(Reproduced from Schneider and Kato[28] by permission of Wiley-VCH.)

Besides the speed of response, the sensitivity is a critical issue for the practical
use of chemomechanical materials. The sensitivity of chemomechanical materials is analogous to that of all sensor particles:[11] on the one hand, a function of
the particle size, on the other hand it depends on the binding force between
polymer host and effector molecule. Only if the affinity is high enough can one
expect that all binding sites within one particle are occupied by the effector
molecules.[12] As long this condition is met, one can indeed use this compart-
mentalization effect;[12] by decreasing the gel particle volume, one reaches a
sensitivity increase by an order of magnitude (Figure 6.6).

6.4 Covalent Interactions: Glucose-Sensitive Hydrogels

Hydrogels containing phenylboronic acid side chains in the backbone alter in
the presence of glucose swelling properties, either by ionization or by formation
of glucose-mediated reversible cross-links; such systems allow the design of
closed-loop insulin delivery systems.[3,13] The anionic tetrahedral boronate
undergoes bidentate condensation with diols such as glucose; glucose has long
been known for to react better than other saccharides with boron esters[14] and
can thus be detected selectively. The binding of glucose then stabilizes the
charged form of boronic acid, which under normal conditions is present in
neutral form (Scheme 6.2). The anion formation promotes the development of
Donnan osmotic swelling forces. In biological fluids, however, high ionic
strength may render Donnan potential effects insignificant. In this case, water
import by the presence of the charged boronate with the corresponding

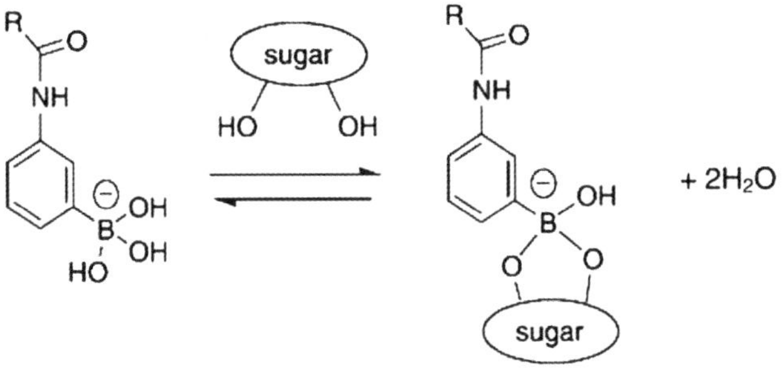

Scheme 6.2 Formation of the anionic boronate ester with sugars.

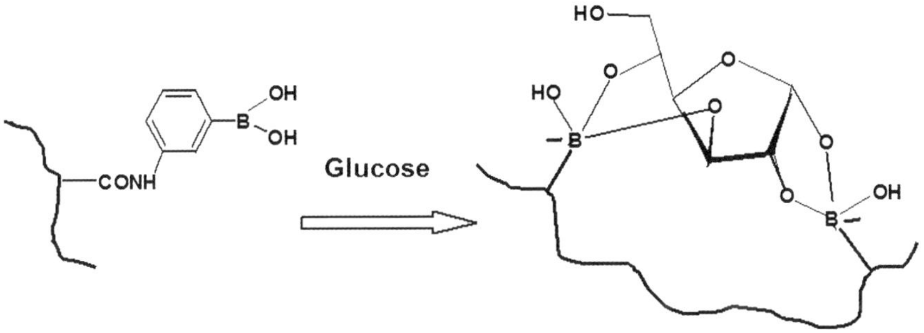

Scheme 6.3 Covalent cross-linking of polymer-bound boronic acid units with glucose.

gegenion will be the most import contribution. Related to this, crystalline colloidal arrays of spherical polystyrene colloids polymerized with thin hydrogel films containing boronic esters were used for optical glucose detection.[15] Boronic ester containing gels also exhibit a sol–gel transition as a function of sugar concentration,[16] which may be used in single-walled carbon nanotubes.[16b]

Shrinking may also occur due to ionic bridging by multivalent counterions, but in particular by the covalent cross-linking shown in Scheme 6.3. A hydrogel obtained from poly(methyl methacrylate) with arylboronic acid moieties exhibited reversible contractions without interference from common blood sugars other than glucose.[17] It is known that glucose reacts with boronic acid in the furanose form.[18] Incorporation of tertiary amines in another boronic acid hydrogel also leads to enhanced glucose-induced cross-linking.[19]

6.5 Inorganic Ions as Effectors

Inorganic anions with a polyallylamine-derived gel (P2) stimulate either expansion or contraction, depending on the presence of neutral salts (Figure 6.7).[20] Related effects of different ionic strength have been reported before,[21]

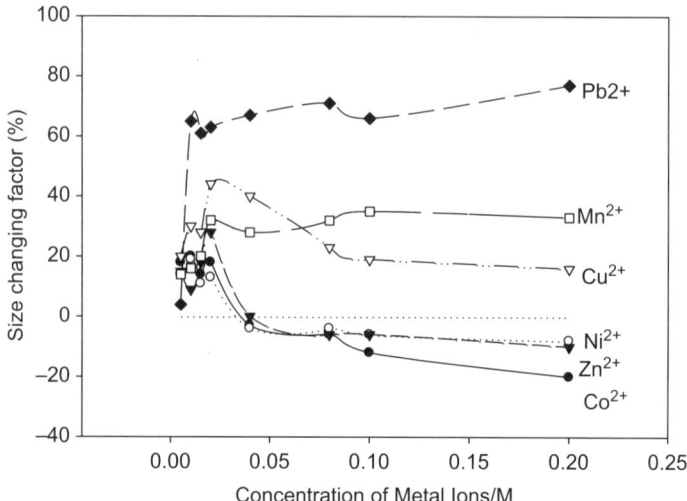

Figure 6.7   Volume changes (%) induced by different anions on the PAA-derived
                  polymer I; [NaX] = 0.10 M; *left bars* (always expansion): in pure water, pH
                  7.3; *right bars* (partial contraction), in the presence of 0.02 M phosphate
                  buffer, pH 7.0.
                  (Reproduced from Schneider *et al.*[2a] by permission of Wiley-VCH.)

also with β-hairpin peptides in hydrogel nanostructures consisting of semi-
flexible fibrillar assemblies. Circular dichroism spectra indicated that in the
absence of salt the peptides are unfolded. When the ionic strength is increased,
the electrostatic interactions between the charged amino acids within the peptide
are screened, which promotes the subsequent β-hairpin formation.[22] A hydrogel
containing benzo-18-crown-6-acrylamide exhibits expansion in the presence of
Ba^{2+} ions, and does not respond to K^+ or other ions which are known to bind
less efficiently; the expansion was ascribed to the repulsion among the charged
Ba^{2+} complex groups and the osmotic pressure within the hydrogel.[23]

   The ethylenediamine units in polymer P3 provide binding sites for heavy
metal ions and respond selectively to the their presence.[24,25] With smaller
cations, cross-linking effects seem to dominate, particularly at higher concentrations; only large ions such as Pb^{2+} lead invariably to expansion (Figure
6.8). Copper and zinc ions can coordinate with more than one ethylenediamine
unit of P1, leading to contraction by coordinative cross-linking.[26]

6.6   Aromatic Effector Molecules: from Arene Acids
to Nucleotides, Peptides, and Amino Acids

Organic effector molecules bearing anionic functions lead with polymer P3 to
expansion, which increases with the size of the residues, but also depends on the
applied pH (Scheme 6.4).[27] Noticeably, saturated frameworks are ineffective,
which points to cation–π interactions with the protonated ethylenediamine

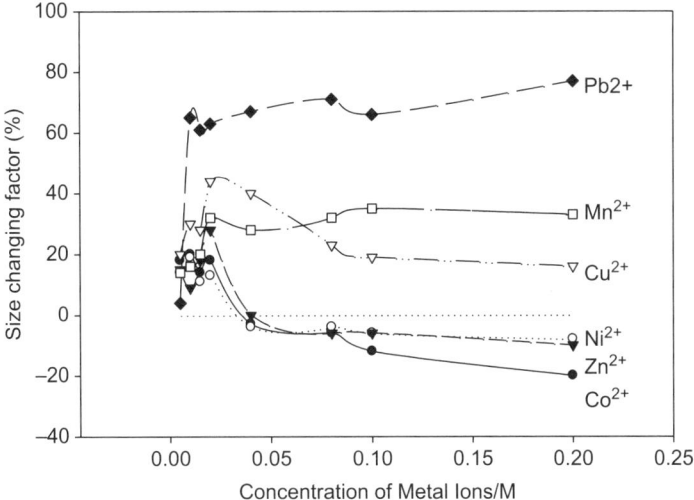

Figure 6.8 Size changes of the polymer film P3 with different metal ions. Changes given as length increase or decrease (50% length change, for example, corresponds to 125% square size, or 237% volume change).[24]

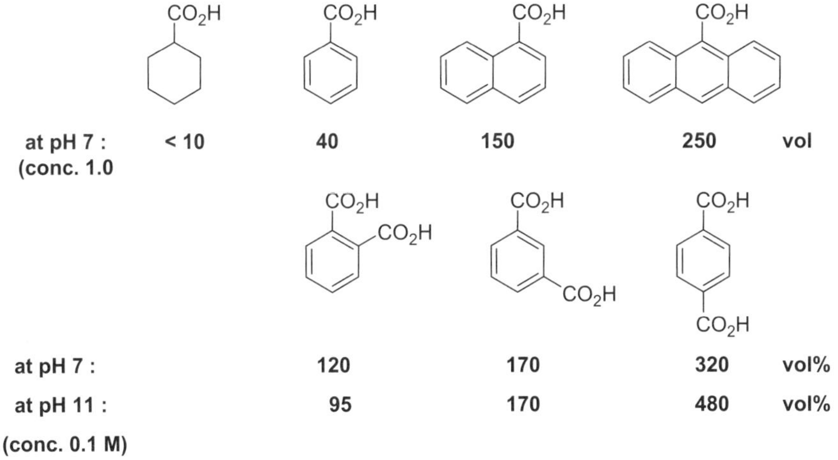

	CO₂H cyclohexane	CO₂H benzene	CO₂H naphthalene	CO₂H anthracene	
at pH 7 : (conc. 1.0	< 10	40	150	250	vol

	CO₂H / CO₂H (ortho)	CO₂H / CO₂H (meta)	CO₂H / CO₂H (para)	
at pH 7 :	120	170	320	vol%
at pH 11 :	95	170	480	vol%
(conc. 0.1 M)				

Scheme 6.4 Selectivity between different organic effector compounds [0.02–0.05 M phosphate buffer; the effects at pH 11 are corrected for differences between pH 7 and pH 11 alone (390 vol%)].
(Reproduced from Schneider *et al.*[27] by permission of Wiley-VCH.)

units as a major contribution. Similar effects are seen with nucleotides, again significantly depending on the applied pH (Scheme 6.5).

Non-covalent interactions with aromatic moieties are also responsible for the chiral recognition with the inherently chiral chitosan gel P4; this allowed for

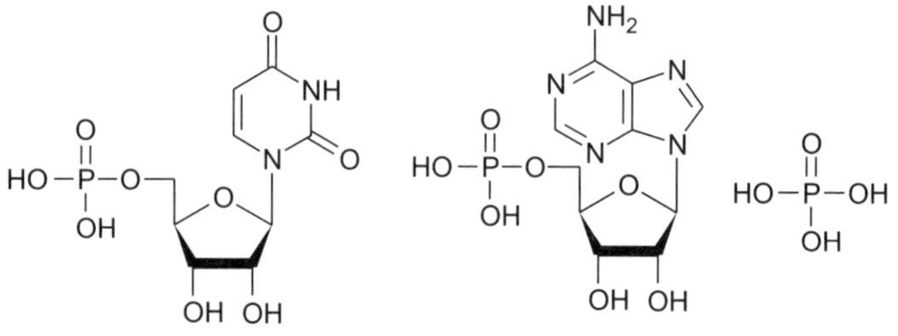

| | at pH 7: | 45 | 5 | <10 |
| | at pH 11: | 28 | 60 | <10 |

Scheme 6.5 Expansion with polymer P3 (in vol%) triggered by nucleotides at different pH values; the values at pH 11 are corrected for the effect of pH alone (about 70%).[2]

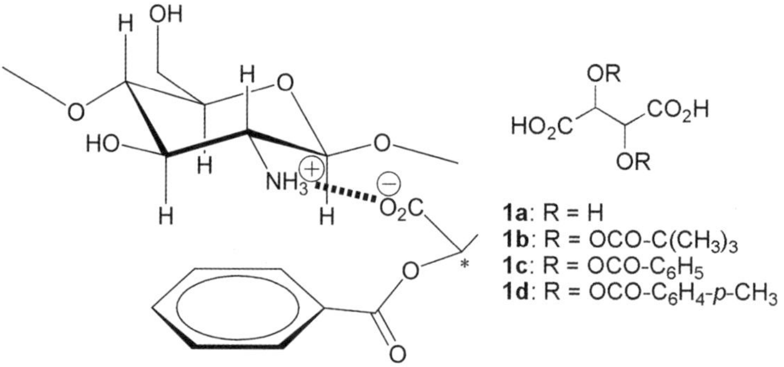

1a: R = H
1b: R = OCO-C(CH_3)_3
1c: R = OCO-C_6H_5
1d: R = OCO-C_6H_4-p-CH_3

Scheme 6.6 Chiral recognition of tartaric acids by cation–π interactions with the chitosan backbone P4; only the aromatic derivatives **1c** and **1d** show discrimination between D- and L-forms.

the first time the translation of chirality directly into macroscopic mechanical motion. The gel exhibits distinctly different volume changes upon exposure to enantiomeric tartaric acid derivatives (Scheme 6.6).[28] However, only with the corresponding phenyl esters does an enantiomer discrimination takes place; NMR spectroscopy has shown structures which are suitable for a cation–π interaction, possibly accompanied by weaker C–H–π hydrogen bonds.

Arenes bearing two anionic residues in opposite positions exhibit volume contractions with polyethylenimine (P1) with a maximum of –97%, which was ascribed to both cation–π interactions of the aryl moieties with the positively charged nitrogen centers in the polymer backbone, as well as to stacking interactions between the effector aryl groups.[26] Stacking interactions between aromatic residues in a hydrogel derived from P5 lead to selective gel size expansions with amino acids bearing different side chains (Scheme 6.7).[29] Stacking in addition to dispersive forces are visible in contractions of poly-allylamine-derived gels (P2) by dicarboxylic acids (Scheme 6.8);[20] nitro

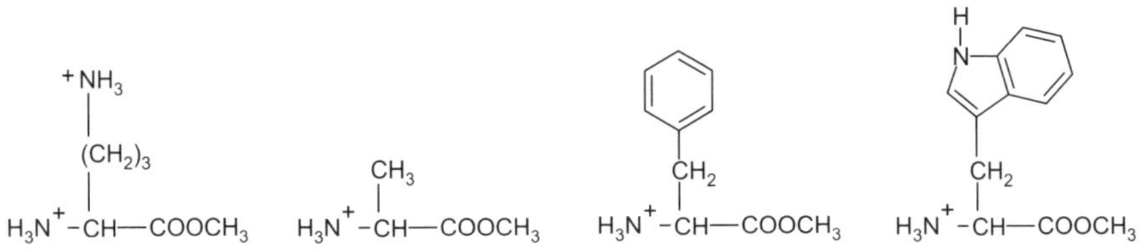

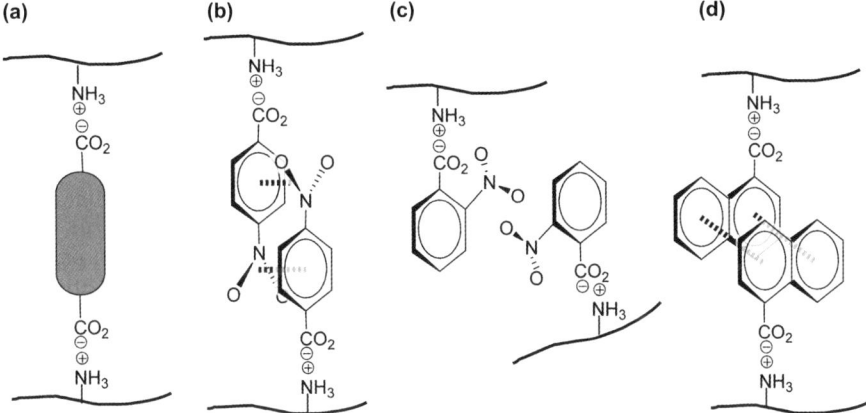

Scheme 6.7 Volume expansions (%) on chitosan-anthryl polymer P5 with different amino acid esters; pH and salt effects are deducted.[29]

Scheme 6.8 Stacking and dispersion forces between aromatic dicarboxylic acids with non-covalent cross-linking in a P2-derived hydrogel by (a) benzenedicarboxylic acid; (b) *p*-nitrobenzoic acid; (c) *o*-nitrobenzoic acid (no interaction with aryl groups, see text); (d) stacking of naphthyl groups.[20]

substituents allow dispersive interactions with opposing arenes only in the absence of steric distortions (in the *ortho* position there is no planarity).

6.7 Cooperative Effects and Logical Gate Functions: How to Bring Silent Effectors Alive

Chemoresponsive hydrogels exhibit cooperativity between different effectors more often than the corresponding supramolecular systems in solution. Thus, the size changes exerted by different anionic effectors such as nucleotides depend not only on the applied pH (Schemes 6.4 and 6.5), but also on the presence of various salts, performing this way as an AND logical gate.[30] The most practical application is the triggering of gel size changes by peptides, which occurs only in the simultaneous presence of metal ions such as Cu^{2+} or Zn^{2+}.[31] These metal ions bind to the diamine units of P3, and use their open

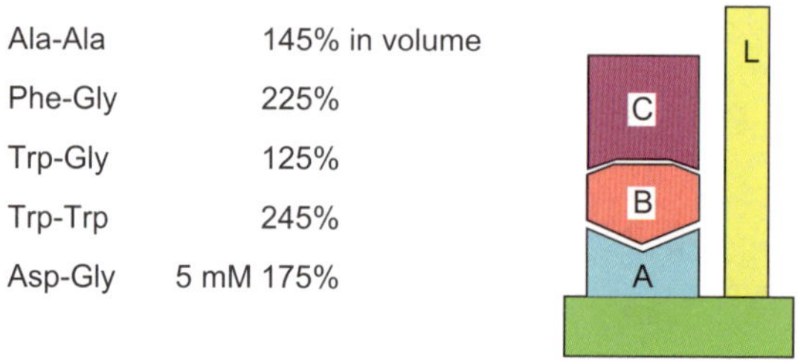

Ala-Ala	145% in volume
Phe-Gly	225%
Trp-Gly	125%
Trp-Trp	245%
Asp-Gly	5 mM 175%

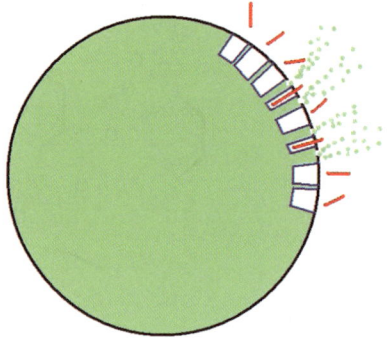

Scheme 6.9   Cooperativity between metal and peptide effects on the expansion of a hydrogel based on P3 loaded with Cu^{2+}. A, diamine unit; B, metal ion; C, peptide; L, lipophilic side chain in P3. In the absence of Cu^{2+} the peptides have no effect ($<5\%$).
(Reproduced from Lomadze and Schneider[31] by permission of Elsevier.)

Figure 6.9   Schematic representation of a capsule equipped with binding sites at the periphery; upon uptake of an effector molecule (*red stick*) the receptors (*blue*) contract and the drug (*green*) is released through opened spaces.

coordination sites to also bind peptides, which otherwise have no effect on the gel (Scheme 6.9). Interactions with the lipophilic side chains also present in P3 lead at the same time to a moderate selectivity with respect to the peptide structure.

6.8   Outlook

Chemomechanical gels are of particular interest for biomedical applications, particularly with biocompatible polymers. pH-sensitive hydrogels can be effective for tissue-specific targeting in view of different pH values in tissue and organs.[32] The gel networks discussed in this chapter, which incorporate selective binding sites in the polymer backbone, hold much promise for drug release by external stimuli (Figure 6.9) or for cell surface recognition, for example; detoxification by uptake of harmful agents is also feasible. Besides gels for smaller effector substances such as metal ions, nucleotides, peptides, carbohydrates, *etc.*, biopolymer-sensitive materials have already been reported, including enzyme-,[33] antigen-,[34] and DNA-responsive systems.[35] Thin film techniques as well as miniaturization down to micro- and nanoscales will make the response of such materials faster and more sensitive. The selectivity of such materials can be varied by implementation of the many specific recognition sites known from supramolecular chemistry in solution; cooperativity will allow selectivity as a function of several effectors, such as tumor markers at varying pH values.

References

1. S. R. Marek, C. A. Conn and N. A. Peppas, *Polymer*, 2010, **51**, 1237; see also T. E. Benavidez and A. M. Baruzzi, *Polymer*, 2012, **53**, 438.
2. (a) H.-J. Schneider, K. Kato and R. M. Strongin, *Sensors*, 2007, **7**, 1578; (b) H.-J. Schneider, T. J. Liu and N. Lomadze, *Eur. J. Org. Chem.*, 2006, 677; (c) H.-J. Schneider and K. Kato, *J. Mater. Chem.*, 2099, **19**, 569.
3. Y. Kitano, K. Koyama, O. Kataoka, T. Kazunori, Y. Okano and Y. Sakurai, *J. Controlled Release*, 1992, **19**, 161.
4. H. Komatsu, S. Matsumoto, S. Tamaru, K. Kaneko, M. Ikeda and I. Hamachi, *J. Am. Chem. Soc.*, 2009, **131**, 5580.
5. A. Foster and J. W. Steed, *Angew. Chem. Int. Ed.*, 2010, **49**, 6718.
6. H.-J. Schneider, *Angew. Chem. Int. Ed.*, 2009, **48**, 3924; H.-J. Schneider and R. M. Strongin, *Acc. Chem. Res.*, 2009, **42**, 1489.
7. R. Barbucci, A. Magnani and M. Consumi, *Macromolecules*, 2000, **33**, 7475.
8. Y.-L. Guan, L. Shao, J. Liu and K. DeYao, *J. Appl. Polym. Sci.*, 1996, **62**, 1253.
9. Y. Zhang, H. W. Gu, Z. M. Yang and B. Xu, *J. Am. Chem. Soc.*, 2003, **125**, 13680.
10. K. Kato and H.-J. Schneider, *Eur. J. Org. Chem.*, 2009, 1042.
11. H. A. Clark, R. Kopelman, R. Tjalkens and M. A. Philbert, *Anal. Chem.*, 1999, **71**, 4837.
12. H.-J. Schneider, L. Tianjun and N. Lomadze, *Chem. Commun.*, 2004, 2436.
13. R. A. Siegel, Y. D. Gu, M. Lei, A. Baldi, E. E. Nuxoll and B. Ziaie, *J. Controlled Release*, 2010, **141**, 303; A. Matsumoto, K. Yamamoto, R. Yoshida, K. Kataoka, T. Aoyagi and Y. Miyahara, *Chem. Commun.*, 2010, 2203.
14. J. Boeseken, *Adv. Carbohydr. Chem.*, 1949, **4**, 189.
15. V. L. Alexeev, S. Das, D. N. Finegold and S. A. Asher, *Clin. Chem.*, 2004, **50**, 2353.
16. (a) J. H. Jung and S. Shinkai, *Top. Curr. Chem.*, 2004, **248**, 223; (b) S. Tamesue, M. Numata and S. Shinkai, *Chem. Lett.*, 2011, **40**, 1303.
17. G. K. Samoei, W. H. Wang, J. O. Escobedo, X. Y. Xu, H.-J. Schneider, R. L. Cook and R. M. Strongin, *Angew. Chem. Int. Ed.*, 2006, **45**, 5319.
18. J. C. Norrild and H. Eggert, *J. Am. Chem. Soc.*, 1995, **117**, 1479.

19. K. E. S. Dean, A. M. Horgan, A. J. Marshall, S. Kabilan and J. Pritchard, *Chem. Commun.*, 2006, 3507.
20. K. Kato and H.-J. Schneider, *Eur. J. Org. Chem.*, 2008, 1378.
21. K. W. Seo, D. J. Kim and K. N. Park, *J. Ind. Eng. Chem.*, 2004, **10**, 794; T. Watanabe, K. Ito, C. Alvarez-Lorenzo, A. Y. Grosberg and T. Tanaka, *J. Chem. Phys.*, 2001, **115**, 1596.
22. B. Ozbas, J. Kretsinger, K. Rajagopal, J. P. Schneider and D. J. Pochan, *Macromolecules*, 2004, **37**, 7331.
23. X. J. Ju, L. Y. Chu, L. Liu, P. Mi and Y. M. Lee, *J. Phys. Chem. B*, 2008, **112**, 1112.
24. H.-J. Schneider, L. Tianjun and N. Lomadze, *Chem. Commun.*, 2004, 2436.
25. J. Ricka and T. Tanaka, *Macromolecules*, 1985, **18**, 83.
26. K. Kato and H.-J. Schneider, *Langmuir*, 2007, **23**, 10741.
27. H.-J. Schneider, T. J. Liu and N. Lomadze, *Angew. Chem. Int. Ed.*, 2003, **42**, 3544.
28. H.-J. Schneider and K. Kato, *Angew. Chem. Int. Ed.*, 2007, **46**, 2694.
29. N. Lomadze and H.-J. Schneider, *Tetrahedron*, 2005, **61**, 8694.
30. H.-J. Schneider, T. J. Liu, N. Lomadze and B. Palm, *Adv. Mater.*, 2004, **16**, 613.
31. N. Lomadze and H.-J. Schneider, *Tetrahedron Lett.*, 2005, **46**, 751.
32. H. J. Yoon and W. D. Jang, *J. Mater. Chem.*, 2010, **20**, 211; L. Du, S. J. Liao, H. A. Khatib, J. F. Stoddart and J. I. Zink, *J. Am. Chem. Soc.*, 2009, **131**, 15136.
33. A. R. Hirst, B. Escuder, J. F. Miravet and D. K. Smith, *Angew. Chem. Int. Ed.*, 2008, **47**, 8002; X. W. Du, J. F. Li, Y. Gao, Y. Kuang and B. Xu, *Chem. Commun.*, 2012, 2098.
34. T. Miyata, A. Jikihara, K. Nakamae and A. S. Hoffman, *J. Biomater. Sci., Polym. Ed.*, 2004, **15**, 1085.
35. Y. Murakami and M. Maeda, *Biomacromolecules*, 2005, **6**, 2927.

CHAPTER 7

Injectable Temperature- and pH/Temperature-Sensitive Block Copolymer Hydrogels

CONG TRUC HUYNH[†] AND DOO SUNG LEE*

Theranostic Macromolecules Research Center, Department of Polymer Science and Engineering, Sungkyunkwan University, Suwon, Gyeonggi-do 440-746, South Korea
*Email: dslee@skku.edu

7.1 Introduction

Hydrogels are three-dimensional hydrophilic polymeric networks that can absorb a considerable amount of water or biological fluid while maintaining their semi-solid morphology.[1–3] Injectable polymeric hydrogels are the most attractive materials for biomedical and pharmaceutical applications, especially for drug/protein delivery and tissue engineering.[1–16] Such injectable hydrogel systems are of particular interest because of the advantages they offer: (i) they require no organic solvent in processing; (ii) they require no surgical or implantation procedures; (iii) their precursor solutions can be easily mixed with bioactive molecules, such as drugs, proteins, DNA or cells prior to gelation or injection; (iv) the formed gels become drug carriers for localized delivery or act as scaffolds for tissue regeneration; (v) the formed gels have the ability to

[†]Current address: Australian Institute for Bioengineering and Nanotechnology, The University of Queensland, St Lucia, Brisbane, QLD 4072, Australia

Monographs in Supramolecular Chemistry No. 11
Polymeric and Self Assembled Hydrogels: From Fundamental Understanding to Applications
Edited by Xian Jun Loh and Oren A. Scherman
© The Royal Society of Chemistry 2013
Published by the Royal Society of Chemistry, www.rsc.org

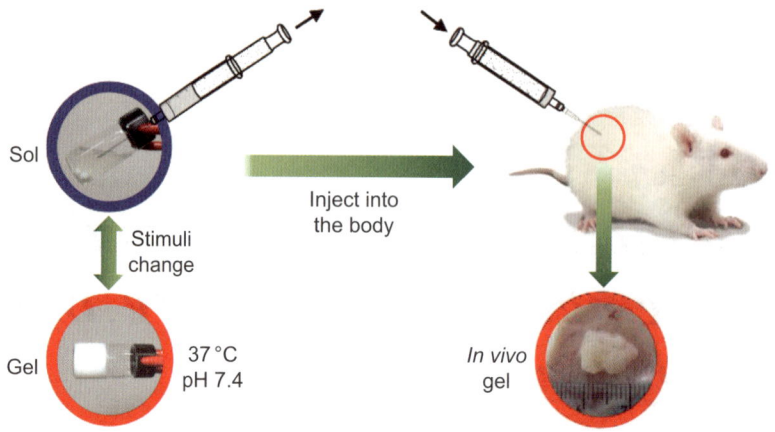

Figure 7.1   General sol–gel phase transitions of injectable stimuli-sensitive hydrogels. (Adapted from Huynh *et al.*[1] by permission of American Chemical Society.)

release both hydrophobic and hydrophilic drug and ionic bioactive molecules, with sustained release behaviour. The aqueous solutions of these polymeric hydrogels exhibit a sol–gel phase transition in response to external stimuli, such as pH, temperature, glucose, electric fields, magnetic fields, ionic strength or a combination of them. Bioactive molecules, such as drugs, proteins, DNA or cells, can be easily mixed with the hydrogel precursor solutions, prior to *in vitro* gelation or injection into the body, as shown in Figure 7.1. The formed gels become carriers of therapeutic agents for localized delivery, or scaffolds for tissue regeneration.[1–6]

Injectable hydrogels can be classified into two basic types, depending on their cross-linking method: chemical or physical. Chemically cross-linked hydrogels can be generated *via* a Schiff base,[17] enzymes,[18–20] photopolymerization[21–23] or Michael addition or other chemical reactions;[24–28] they usually exhibit a noticeable volume change. Although chemically cross-linked hydrogels possess strong mechanical properties, the requirement of cross-linking agents, enzymes, photoinitiators and/or organic solvents may damage cells as well as denature incorporated bioactive molecules which might limit their applications. In contrast, physically cross-linked hydrogels, formed *via* the self-assembly of amphiphilic block/graft copolymers in response to external stimuli, such as temperature or pH/temperature, offer a mild method of preparing hydrogels. Moreover, such preparations exhibit a sol–gel phase transition without any significant volume change.[1–6]

In this chapter, a short view of temperature- and pH/temperature-sensitive block copolymer physical hydrogels is presented. In addition, typical examples of temperature-sensitive hydrogels using poly(lactide-*co*-glycolide)-poly(ethylene glycol)-poly(lactide-*co*-glycolide) (PLGA-PEG-PLGA) triblock copolymers, and pH/temperature-sensitive hydrogels using cationic PEG-poly(amino ester urethane) [PEG-PAEU]$_x$ block copolymers, are reported in detail.

7.2 Injectable Temperature-Sensitive Block Copolymer Hydrogels

Temperature is among the easiest parameters to experimentally manipulate; thus, not surprisingly, temperature-sensitive hydrogels have been widely studied for both *in vitro* and *in vivo* applications. Temperature-sensitive hydrogels exist as a solution at certain temperatures, and undergo a sol–gel phase transition at physiological temperature.[1–6] Amphiphilic block copolymers, consisting of hydrophilic poly(ethylene glycol) (PEG or PEO) and hydrophobic poly(propylene oxide) (PPO) or aliphatic polyesters, such as PLGA, poly(ε-caprolactone) (PCL), poly(CL-*co*-lactide)-PEG-poly(CL-*co*-lactide) (PCLA-PEG-PCLA), poly[(*R*)-3-hydroxybutyrate] (PHB) or polypeptides or polyphosphazene, have been widely studied as temperature-sensitive hydrogels for drug delivery and tissue engineering.

The triblock copolymer PEO-PPO-PEO, known as Pluronic or Poloxamer hydrogels, can self-assemble in water to form micelles, and micellar association upon heating of the polymer solution results in a sol-to-gel phase transition, with the existence of a gel at physiological temperature (37 °C).[29–31] However, their weak mechanical strength, non-biodegradability, short residence time and high permeability may limit the applicability of Pluronic hydrogels.[1–3] Several efforts have been made to improve the biodegradability and/or mechanical strength of Pluronic hydrogels.[32–34] A biodegradable triblock copolymer composed of PEG-poly(L-lactic acid)-PEG (PEG-PLLA-PEG), which shows a gel-to-sol phase transition in water upon heating, was reported as an alternative to Pluronic hydrogel.[35] However, the encapsulation of bioactive molecules and injection must be carried out at high temperatures in the sol state, which may damage labile molecules, or be uncomfortable for patients. Subsequently, hydrogel systems using PLGA-based copolymers, such as PEG-PLGA-PEG or PLGA-PEG-PLGA triblock copolymers, PLGA-grafted to PEG or PEG-grafted to PLGA, have been developed.[36–46] These copolymers offer a mild method for incorporating bioactive molecules, as these solutions flow freely at room temperature but change to a gel at body temperature. Aqueous solutions of PCL-based[47–56] or PCLA-based[57–63] copolymers, including diblock methylene-PEG-PCL (mPEG-PCL), triblock PEG-PCL-PEG or PCL-PEG-PCL, triblock PCLA-PEG-PCLA, with or without being modified by Lys-Arg-Gly-Asp-Lys-Lys (KRGDKK) or Arg-Gly-Asp (RGD), were reported to undergo a sol-to-gel phase transition with an increase in temperature. Multi-block poly(ether ester urethane)s consisting of PHB, PEG and PPO (PHB/PEG/PPO) in aqueous solutions were found to undergo a reversible sol–gel phase transition upon temperature change at very low copolymer concentrations.[64–66]

The mechanism for the gelation of amphiphilic block copolymers is proposed as the formation of micelles and/or bridged micelles at low temperatures, and the association of micelles or bridged micelles at higher temperatures, caused by the increase in hydrophobicity of copolymers, which results in the gelation. Figure 7.2 shows a schematic diagram for the gelation mechanism of PLGA-PEG-PLGA as a representative of a PEG-contained amphiphilic B-PEG-B

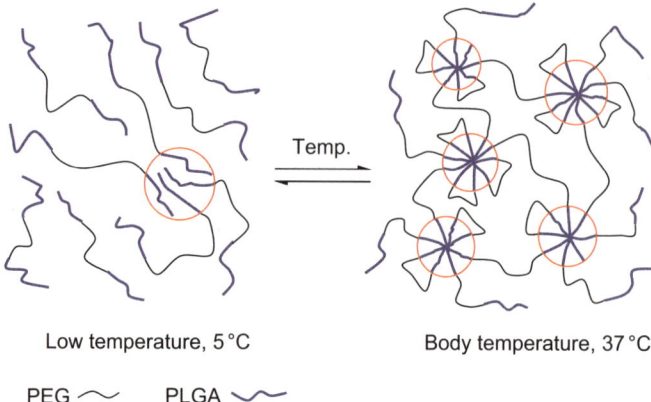

Low temperature, 5 °C   Body temperature, 37 °C

PEG ⌒   PLGA ⌒⌒

Figure 7.2   Schematic diagram showing the sol–gel phase transition of PLGA-PEG-PLGA hydrogel, as a representative of B-PEG-B triblock copolymer hydrogels (B is an amphiphilic block, such as PLGA, PCL or PCLA), with increasing temperature.

triblock copolymer hydrogel (B is an amphiphilic block, such as PLGA, PCL or PCLA). At low temperatures (*e.g.* 5 °C), the individual micelles and bridged micelles coexist in the sol state. With increasing temperature (to 37 °C), the fraction of bridged micelles increases, due to the increase in the hydrophobicity of the copolymer. Finally, ordered packing of the bridged micelles occurs, resulting in their gelation.[1,2]

Degradation products of PEG/polyester block copolymers are acidic molecules that may damage the encapsulated molecules, or promote inflammatory responses by the host tissue. Recently developed polypeptide-based hydrogels, which show a low degradation rate with zwitterionic amino acid degradation products, have the potential to overcome this limitation. Aqueous solutions of copolymers comprising PEG or Poloxamer (PLX) with polypeptides, such as polyalanine, poly(alanine-*co*-phenylalanine) (PAF-PLX-PAF), poly(alanine-*co*-leucine) (PAL), Gly-Phe-Gly-Asp (GFGD) or zwitterionic Gly-Arg-Gly-Leu (GRGL) or Gly-Arg-Gly-Asp (GRGD),[67–77] undergo a sol-to-gel phase transition with increasing temperature. Polyphosphazene and its derivatives are also promising materials, because their degradation product is phosphate.[78–80]

Besides these, several other block copolymers in aqueous solutions can also form a gel by changing the temperature, suggesting potential applications in drug delivery and tissue engineering. These copolymers include those based on poly(ethylene/butylene) (PEB),[81,82] poly(propylene fumarate) (PPF),[83] poly-(trimethylene carbonate) (PTMC),[84,85] poly(caprolactone-*co*-trimethylene carbonate) (PCTMC),[86] polyorthoester,[87] poly(δ-valerolactone) (PVL),[88] poly(ethylene glycol)–sebacic acid polyester (PEG-SA),[89] poly(ethyl 2-cyanoacrylate) (PEC)[90] and poly(*N*-vinylpyrrolidone-*b*-alanine),[91] amongst others.

The potential applications of temperature-sensitive hydrogels in drug/protein delivery and tissue engineering have been widely studied. They were employed for delivery of ketoprofen and spironolactone,[37] doxorubicin (DOX),[48,57,78] honokiol,[53,54] lidocaine,[55] bovin serum albumin (BSA),[53,65,76,82] insulin,[74] human growth hormone (hGH),[75,79,80] FITC–dextran,[89] basic fibroblast growth factors (bFGFs),[82] vascular endothelial growth factor (VEGF),[88] bone marrow-derived rat mesenchymal stem cells (rMSCs)[32] and for three-dimensional (3D) cell culture and tissue engineering.[69,70,86]

Although temperature-sensitive block copolymer hydrogels have been widely studied and show potential applicability, these hydrogels suffer from some limitations that restrict the applications in which they may be utilized. First, when temperature-sensitive hydrogels are injected into the body *via* a syringe, they tend to form a gel, as the needle is warmed by body temperature, making injection difficult, especially with a long guidance catheter. Second, their neutral properties limit their application in the release of ionic drugs/proteins/peptides. Third, biodegradable polymers could be degraded during storage and circulation for commercial use. Therefore, the reconstitution problem of the polymer solution is of concern. Fourth, the degradation of polyester-based copolymers sometimes generates acidic products that change the local pH. The resulting low pH damages incorporated proteins or cells. Thus, it is important to maintain neutral pH during the degradation.

7.3 Injectable pH/Temperature-Sensitive Block Copolymer Hydrogels

"Smart" polymers that are capable of responding to multiple stimuli, especially pH and temperature, have drawn considerable research interest because of their advantages in terms of chemical structure and ability to form an ionic complex with therapeutic agents. This dual response can prevent gelation within the injection needle.

Nonbiodegradable pH/temperature-sensitive hydrogels are of interest for therapeutics delivery, where bioactive agents, especially proteins, are protected from denaturation. A solution of the triblock copolymer PDPAEMA-PMPC-PDPAEMA {PDPAEMA = poly[2-(diisopropylamino)ethyl methacrylate], PMPC = poly(2-methacryloyloxyethyl phosphorylcholine)} or the pentablock copolymer PDEAEMA-PLX-PDEAEMA {PDEAEMA = poly(2-(diethylamino)-ethyl methacrylate)} shows a reversible sol-to-gel phase transition in the pH range 2–9 or 7.7–8.3, respectively.[92,93] Aqueous solutions of PLX end-capped by amine groups exhibit a closed-loop, sol–gel–sol phase transition, as a function of temperature and pH.[94] Recently, pH/temperature-sensitive hydrogels, using block copolymers composed of PEG and poly(amino urethane) (PAU) or poly-(amido amine) (PAA) or poly(amino urea urethane) (PAUU), were developed by Lee's group.[95–98] These copolymer aqueous solutions exist in the sol state at low pH (*e.g.* pH 6.5) in the whole range of experimental temperature (0–70 °C), but exhibited a sol-to-gel-to-sol phase transition with increasing temperature at pH ≥ 7.0, due to hydrophobic interactions and hydrogen bonding between the PAU or PAA or PAUU blocks.

Biodegradation of biomaterials is indispensable for the diffusion of therapeutic agents and/or formation of new tissues. It also eliminates the need for
surgery to remove the material introduced into the body. Cationic poly-
(β-amino ester) (PAE), which is non-cytotoxic and forms a good ionic
complex with negatively charged proteins or pDNA at physiological pH, was
introduced as a biodegradable pH-sensitive polymer.[99,100] Aqueous solutions
of PAE-PCL-PEG-PCL-PAE, or PAE-PCLA-PEG-PCLA-PAE pentablock
copolymers or a graft copolymer of (PAE-*g*-PCL)-PEG-(PAE-*g*-PCL) undergo
a sol-to-gel-to-sol phase transition as a function of both pH and temperature.[101–105] On the other hand, aqueous solutions of PAE-PEG-PAE or PEG-
(-PAE)$_4$ exhibit a sol-to-gel and gel-to-sol phase transition with increasing pH
and temperature, respectively.[106,107] These two polymers are easily dissolved in
water at a relatively low pH, because of free hydrophobic blocks. The multi-
block copolymer [PCL-PEG-PCL-PAU]$_x$ was reported to show a sol-to-gel-to-
sol phase transition in aqueous solution.[108] However, copolymers containing
PCL(A)-PEG-PCL(A) are difficult to dissolve in water. Efforts to prepare
biodegradable PAU-based copolymers without hydrophobic blocks led to the
recent development of poly(β-amino ester urethane)-based copolymers
(PAEU).[109–113] The [PEG-PAEU]$_x$ multiblock and PAEU-PEG-PAEU triblock copolymers are easily dissolved in water at low pH, and exhibit a sol-to-
gel phase transition with change of pH and temperature to physiological
conditions. A schematic diagram for the sol–gel phase transition mechanism of
the [PEG-PAEU]$_x$ hydrogel, as a representative of cationic copolymer hydrogels, is depicted in Figure 7.3. A solution exists at low pH and low temperatures
(such as pH 6.6, 5 °C), due to the hydrophilic character of the ionized PAEU
segments. However, a gel forms under physiological conditions (pH 7.4, 37 °C),

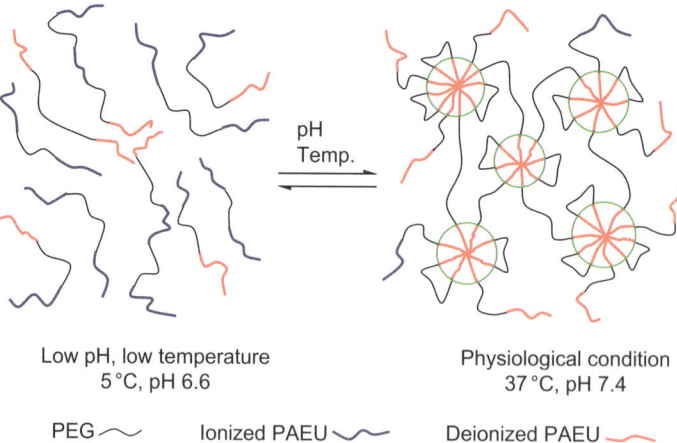

Low pH, low temperature　　　　　　　Physiological condition
　　　5 °C, pH 6.6　　　　　　　　　　　　37 °C, pH 7.4

PEG ⁓　　　Ionized PAEU ⌇　　　Deionized PAEU ⌐

Figure 7.3　Schematic showing the sol–gel phase transition of [PEG-PAEU]$_x$
hydrogel, as a representative of cationic block copolymer hydrogels, with
changing pH and temperature.

due to the strongly hydrophobic interaction and hydrogen bonding between the deionized PAEU segments.[110–113]

In contrast to cationic pH-sensitive polymers, anionic polymers can dissolve in water at basic pH (*e.g.* pH 8.0) and form ionic complexes with growth factors, mostly positively charged, but become insoluble at physiological pH. Typical examples of anionic biodegradable pH/temperature-sensitive copolymer hydrogels are oligomer sulfamethazine-based (OSM) copolymers. Aqueous solutions of the pentablock copolymers OSM-PCLA-PEG-PCLA-OSM or OSM-PLGA-PEG-PLGA-OSM exhibit a reversible sol-to-gel-to-sol phase transition as a function of both pH and temperature.[114–120] The existence of a gel at physiological conditions lets it serve as a drug carrier, especially as a growth factor depot for cell culture.

pH/Temperature-sensitive block copolymer hydrogels possess complicated structures, and their preparation requires several steps. In contrast, low molecular weight gelators, such as cationic oligo(β-amino ester urethane) (OAEU), oligo(amido amine)s (OAAs) and oligo(amido amine amino ester)s (OAAAEs), have simple structures, and can be prepared *via* a simple one-step synthesis.[121–123] Their aqueous solutions exist in the sol state at low pH (*e.g.* pH 6.5), but turn into the gel state under physiological conditions, and thus serve as depot carriers for bioactive molecules.

The potential applications of pH/temperature-sensitive hydrogels in drug delivery/tissue engineering have been widely studied. The use of these hydrogel systems for the delivery of chlorambucin,[95,96] lidocaine,[106] paclitaxel,[108,117] DOX,[110,111,121] FITC-BSA,[98] insulin,[101,102,123] hGH,[112] human mesenchymal stem cells (hMSCs) and recombinant human bone morphogenetic protein-2 (rhBMP-2)[119] has been reported.

7.4 Typical Examples of Injectable Temperature- and pH/Temperature-Sensitive Hydrogels

In this chapter, two representative examples of temperature- and pH/temperature-sensitive block copolymer hydrogels are reported in detail. The first is a typical temperature-sensitive hydrogel using triblock PLGA-PEG-PLGA copolymers, which is used to produce many different commercial products, such as ReGel and OncoGel.[124–129] The PLGA-PEG-PLGA triblock copolymers are synthesized by the ring-opening polymerization of D,L-lactide (DLLA) and glycolide (GA), initiated by hydroxyl groups at the ends of PEG, in the presence of tin(II) octoate as a catalyst, as shown in Scheme 7.1. Details of the synthesis and characterization of the copolymers are in previous reports.[39,40] Table 7.1 lists the detailed characteristics of the triblock PCLA-PEG-PCLA copolymers. The copolymer aqueous solutions exhibit sol-to-gel-to-sol phase transitions with increasing temperature. Their gelation mechanism, controlled factors and potential application have been discussed in detail.[39,40]

The second is a typical cationic pH/temperature-sensitive hydrogel using [PEG-PAEU]$_x$ block copolymers. The [PEG-PAEU]$_x$ block copolymers are

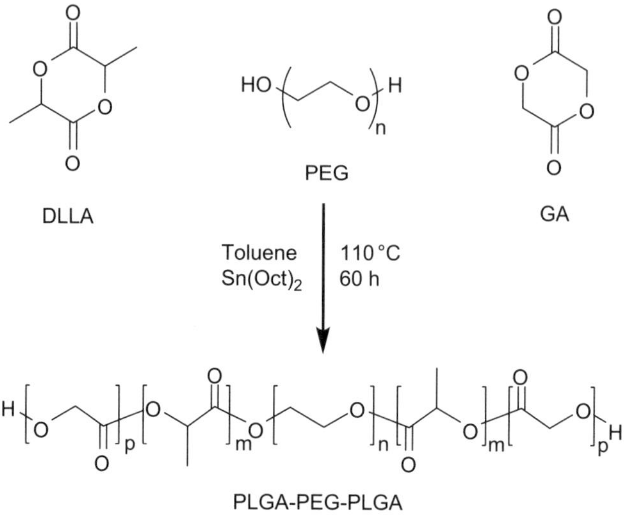

Scheme 7.1   Synthesis of PLGA-PEG-PLGA triblock copolymers.

Table 7.1   Detailed characteristics of the synthesized PLGA-PEG-PLGA triblock copolymers.

No.	M_n^a (PEG)	PLGA-PEG-PLGA^b	PLGA/PEG (wt%)^b	DLLA/GA (mol/mol)^b	PDI^c	T_g (°C)^d
A-1	1000	1148/309-1000-309/1148	2.9/1	3.0	1.28	−8
A-2	1000	1046/293-1000-293/1046	2.7/1	2.9	1.17	−10
A-3	1000	969/256-1000-256/969	2.5/1	3.0	1.24	−13
A-4	1000	809/205-1000-205/809	2.0/1	3.1	1.24	−18
B-2	1000	1051/317-1000-317/1051	2.7/1	2.6	1.21	−10
C-2	1000	922/396-1000-396/922	2.6/1	1.9	1.28	−11
D-2	1000	902/468-1000-468/902	2.7/1	1.5	1.28	−11
E-3	1500	1507/369-1500-369/1507	2.5/1	3.2	1.15	−

^{a}Provided by Sigma-Aldrich.
^{b}Calculated from ^{1}H MNR.
^{c}Measured by GPC (PS standard).
^{d}Measured by DSC.

synthesized by the polyaddition reaction between the isocyanate groups of
hexamethylene diisocyanate and the hydroxyl groups at the ends of PEG and
an amino ester dihydroxyl monomer in chloroform, in the presence of dibutyltin dilaurate as a catalyst, as shown in Scheme 7.2. The tertiary amine groups
are introduced into the copolymer as pH-sensitive moieties. Details of the
synthesis and characterization of the copolymers are in previous reports.[110,111]
Table 7.2 lists the detailed characteristics of the [PEG-PAEU]$_x$ block copolymers. The aqueous copolymer solutions show sol-to-gel and gel-to-sol phase
transitions with increasing pH and temperature, respectively. The controlled

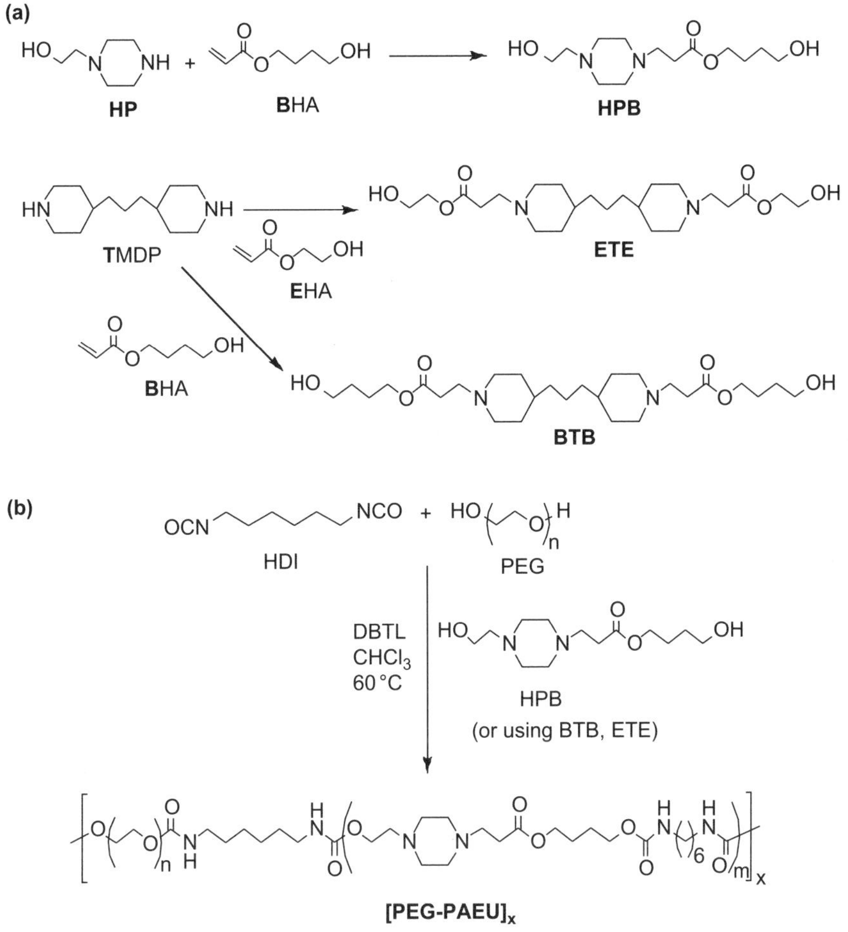

(a)

(b)

Scheme 7.2 Synthesis of (a) amino ester dihydroxyl monomers and (b) [PEG-PAEU]$_x$ block copolymers.

factors, degradation behaviour, cytotoxicity and potential application of the [PEG-PAEU]$_x$ hydrogel are also discussed.

7.4.1 Experiments to Examine the Copolymer and Hydrogel Properties

7.4.1.1 *Differential Scanning Calorimetry Measurements*

Differential scanning calorimetry (DSC) is a thermoanalytical technique in which the difference in the amount of heat required to increase the temperature of a sample and reference is measured as a function of temperature. This is an

Table 7.2   Characteristics of the synthesized [PEG-PAEU]$_x$ block copolymers.

No.	PEG^a	Monomer	$[PEG^a\text{-}PAEU^b]_x{}^c$ or $[EG_n{}^a\text{-}EU_m{}^b]_x{}^c$	$PAEU^b$ (wt%)	$M_n{}^c$	PDI^c
P-HPB-15-1	1500	HPB	$[EG_{34}\text{-HPB-EU}_6]_{2.2}$	62.5	8700	1.89
P-HPB-20-1	2000	HPB	$[EG_{45}\text{-HPB-EU}_6]_{1.9}$	55.5	8500	1.65
P-HPB-20-2	2000	HPB	$[EG_{45}\text{-HPB-EU}_{12}]_{1.2}$	71.4	8700	1.52
P-HPB-46-1	4600	HPB	$[EG_{104}\text{-HPB-EU}_6]_{1.2}$	35.2	8800	1.56
P-BTB-20-1	2000	BTB	$[EG_{45}\text{-BTB-EU}_6]_{2.0}$	66.6	11 900	1.78
P-BTB-20-2	2000	BTB	$[EG_{45}\text{-BTB-EU}_9]_{1.6}$	75.0	12 800	1.76
P-BTB-46-2	4600	BTB	$[EG_{104}\text{-BTB-EU}_9]_{1.4}$	56.6	14 300	1.69
P-ETE-20-2	2000	ETE	$[EG_{45}\text{-ETE-EU}_{10}]_{1.5}$	75.3	12 400	1.77

^{a}Provided by Sigma-Aldrich; n = number of repeating units of PEG.
^{b}Calculated from ^{1}H NMR; m = number of repeating units of PAEU segment.
^{c}Measured and calculated from GPC.

important technique that is usually used to measure the glass transition temperature (T_g) of copolymers.[39,40]

7.4.1.2   Sol–Gel Phase Transition Measurements

The sol (flow)–gel (non-flow) phase transition temperature of the copolymer in aqueous solution is an important factor to determine the pH and temperature sensitivity of the hydrogel. It can be determined using the tube inverting method. Briefly, the copolymers with designed concentrations are dissolved with stirring in distilled water or phosphate buffered saline (PBS) solution to make homogeneous solutions. The solutions of the pH/temperature-sensitive copolymers are adjusted to the designed pH values with 5 M NaOH or 5 M HCl to prevent dilution. The sample vials are placed in a water-bath and heated slowly from 0 to ∼70 °C within 10 min equilibrium at temperature intervals of 2 °C. The sol–gel transition is determined by inverting the vial. The gel state is defined to be non-flow in 30 s.[110–113]

7.4.1.3   Rheological Measurements

The variation of the mechanical properties of a copolymer solution, determined by a mechanical analyser (*e.g.* Bohlin Rotational Rheometer), is an important factor to examined gelation points as well as the gel strength. Briefly the copolymer aqueous solution in PBS is placed between two round-plates with a determined gap. An oscillation mode with a certain frequency, stress and heating rate is performed (*e.g.* at frequency of 1 rad s^{-1}, stress controlled at 0.4 Pa, heating rate 1 °C min^{-1}).[110–113]

7.4.1.4   In Vitro *Degradation Measurements*

The degradation of the injectable hydrogels is very important in some applications, because the excretion of the degraded products can eliminate the surgery to remove the implanted materials. The *in vitro* degradation of the copolymer hydrogels is the first step that should be performed. Briefly, 4 mL vials containing 0.5 mL of the copolymer solution at pH 7.4 are incubated at 37 °C for 15 min to form gels. Subsequently, 1 mL of PBS solution at pH 7.4 is added and the sample vials are incubated at 37 °C. The PBS solution in the sample vials is replaced every day by fresh solution. At a predetermined time, the samples are collected and freeze-dried. The remaining weight (ratio of the lyophilized degraded gels to the initial gels) is measured to determine the degradation rate of the copolymer hydrogel.[111]

7.4.1.5   In Vitro *Cytotoxicity Examination*

For biomedical applications, the materials are usually expected to elicit minimum cytotoxicity. The first step in examining the possible cytotoxicity of the polymer in the fabrication of hydrogels is the *in vitro* cytotoxicity test. The general description of the direct contact method to explore the *in vitro* cytotoxicity of the polymer is as follows. Cells are placed in a 96-well plate at a seeding density of 5000 cells per well in 200 µL of growth medium and maintained at 37 °C for 24 h. The growth medium is removed and fresh medium containing the desired amount of polymer is added. The cells are incubated for 48 h, after which their metabolic activity is measured using the methylthiazolydiphenyltetrazolium bromide (MTT) assay. Briefly, 100 µL of fresh growth medium containing 50 µg of MTT is added to each well and the cells are incubated at 37 °C for 4 h. The absorbance at 570 nm (SpectraMax M5 Microplate Reader, Molecular Devices) is directly proportional to the number of living cells. The survival percentage relative to the mock-treated cells (100% survival) is then calculated.[53,100,115]

7.4.1.6   In Vivo *Experiments*

Male Sprague-Dawley (SD) rats are usually used for the *in vivo* experiments. The rats (5–6 weeks old, average body weight 200 g) are handled in accordance with the National Institutes of Health (NIH) guidelines for the care and use of laboratory animals (NIH publication 85-23, revised 1985). To examine its injectability and *in vivo* gelation, the copolymer solution (~200 µL) is injected subcutaneously into the back of male SD rats using a syringe and needle. At a predetermined time point the rats are sacrificed and the gel morphology can be observed.[110–113] To examine the *in vivo* degradation, a known amount (~400 µL) of the copolymer solution is injected into the SD rat. At a predetermined time point the rats are sacrificed and the gels are collected and freeze-dried to obtain the residual weight. The remaining weight, which is calculated from the weight ratio of the lyophilized degraded gels to the initial gels, shows the degradation rate of the hydrogel.[110]

7.4.1.7   In Vitro *Release of Drug from the Copolymer Hydrogels*

The *in vitro* release of drug from the hydrogels is the first step to estimate the possible applicability of the materials as drug carriers. Briefly, a homogeneous solution of the copolymer and drug is prepared, adjusted to pH 7.4 and stabilized. Known amounts of drug-loaded copolymer solution are placed into 4 mL vials and incubated at 37 °C for 30 min to allow the gel formation. Subsequently, 3 mL of fresh release medium (PBS solution, 37 °C and pH 7.4) is added to each vial, which is then incubated at 37 °C and shaken at 20 rpm. At a predetermined time, 1.5 mL of the release medium is sampled and 1.5 mL of fresh release medium is added to the vials to maintain a constant volume of release medium. The drug concentrations in the release media and standard samples are analysed and calculated.[121]

7.4.2 Temperature-Sensitive PLGA-PEG-PLGA Triblock Copolymer Hydrogels

7.4.2.1 *Controlling the Sol–Gel Phase Transition Diagram*

Sol–gel phase transition diagrams of the PLGA-PEG-PLGA (ABA-type) triblock copolymer hydrogels are determined by the tube inversion method. These reveal that there are three different factors that affect the sol–gel phase transition, namely the PEG/PLGA ratios, the DLLA/GA ratios and the molecular weights of the PEG, which are shown in Figure 7.4. In general, the phase

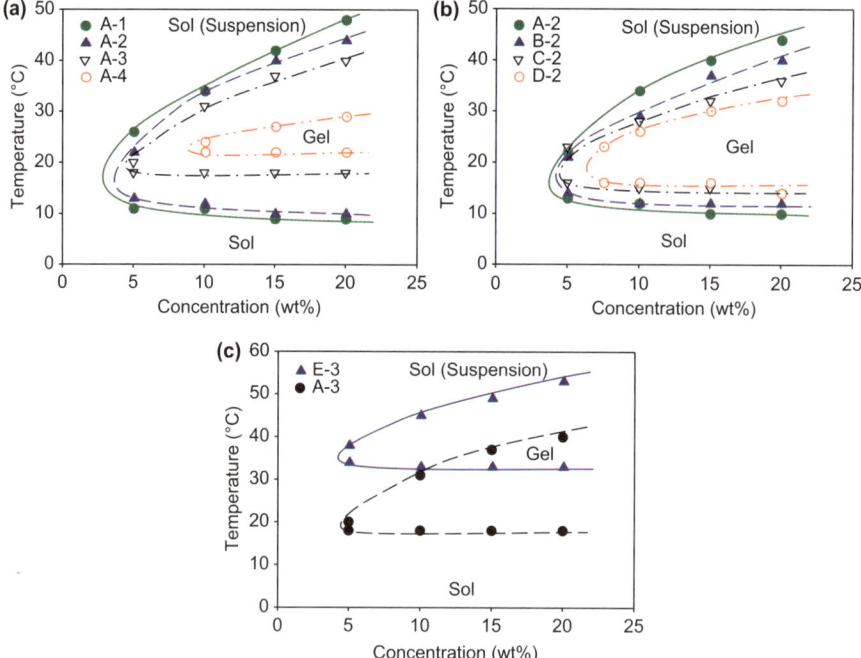

Figure 7.4 Sol–gel phase transition diagram of PLGA-PEG-PLGA hydrogels: (a) different PLGA/PEG ratios; (b) different DLLA/GA ratios; (c) different molecular weights of PEG.
(Adapted from Shim *et al.*[40] by permission of Wiley.)

diagrams show a critical gel concentration (CGC) and two critical gelation temperature (CGT) curves. The lower CGT (sol-to-gel) at low temperatures is less dependent on the polymer concentration than the higher CGT (gel-to-sol) at high temperatures, which is consistent with the BAB-type hydrogel systems.[36,39,40] It is noted that the CGC of the ABA-type triblock copolymers is much lower than that of the BAB-type, which is typically higher than 20 wt%.[36–38] This can be attributed to the structure of ABA-type copolymer, in which two hydrophobic PLGA blocks can easily form bridged micelles and facilitate the ordered packing of the bridged micelles, resulting in their gelation, as shown in Figure 7.2.

The effects of the PLGA/PEG ratios, the DLLA/GA ratios and molecular weights of PEG on the sol–gel phase transition of PLGA-PEG-PLGA hydrogels are shown in Figure 7.4. The gel regions become broader and shift to lower concentration with an increase in the hydrophobicity of the copolymer, by increasing the PLGA/PEG ratio (A-4 to A-1, Figure 7.4a) or DLLA/GA ratio (D-2 to A-2, Figure 7.4b). These increased hydrophobicities of the copolymers facilitate the forming of micelles, bridged micelles and ordered packing bridged micelles.[39,40] However, the effect of the molecular weight of the PLGA on the gel–sol transition diagram is stronger and more significant than that of the DLLA/GA ratio in PLGA. This is confirmed by the glass transition temperatures (T_g) of the copolymers, which are determined by DSC (see Table 7.1). The T_g increases from –18 to –8 °C with the increase of PLGA block length from 1000 to 1500, but is nearly independent of the DLLA/GA ratio. In addition, the gel regions can be modulated by changing the molecular weight of the PEG. The gel region shifts to higher temperatures without changing the CGC, by increasing the molecular weight of the PEG (A3 to E3, Figure 7.4c). This is attributed to the change of the total molecular weight of the copolymer but constancy of the copolymer composition (PLGA/PEG ratio).[39,40] In other words, the shape of the gel region is maintained with constant copolymer composition, but shows a strong vertical shift by changing the molecular weight of the PEG (or total molecular weight). Such interesting vertical shifts of the gel region are also observed when different additives are added to the copolymer solutions, such as PEG and NaSCN.[39,40] Adding a small amount of NaSCN raises the gel regions to higher temperatures, while the addition of PEG shifts the gel regions to lower temperatures, without influencing the CGC.

In summary, by increasing the hydrophobicity of the copolymer, the gel regions of the PCLA-PEG-PCLA copolymer hydrogel can be broadened, and can be shifted to lower CGC. The gel regions can be vertically shifted by changing the molecular weight of the PEG, without changing copolymer composition or addition of some additive, such as PEG or NaSCN.

7.4.2.2 In Vitro *and* In Vivo *Gelation of PLGA-PEG-PLGA Hydrogels*

The PLGA-PEG-PLGA copolymer aqueous solution exists in a sol state at low temperatures (*e.g.* at 5 °C, as shown in Figure 7.5a1), which facilitates

Figure 7.5   Photographs of (a) *in vitro* and (b) *in vivo* gelation of 10 wt% PLGA-PEG-
                PLGA solution (E-3): (a1) a solution at 5 °C; (a2) a gel at physiological
                temperature (37 °C); (b1) *in vivo* gel at 5 min after injection of 200 μL
                copolymer solution into SD rat; (b2) the gel after tissue removal.

formulation with bioactive molecules and injection into the body. Moreover, it
changes to a gel at physiological temperature (37 °C, Figure 7.5a2), or after
being injected into the SD rat body (Figure 7.5b), and can thus serve as a depot
carrier of bioactive molecules for sustained release. This result confirms the
injectability and *in situ* gel forming ability of the PLGA-PEG-PLGA temperature-sensitive hydrogel.

7.4.2.3   *Potential Application of PLGA-PEG-PLGA Hydrogels as Drug Delivery Systems*

Potential applications of PLGA-PEG-PLGA triblock copolymers as hydrogel
carriers for bioactive molecules or proteins have been widely investigated.
ReGel, a commercial PLGA-PEG-PLGA product (1500-1000-1500) formulated with paclitaxel (PTX) and manufactured under the trade name
OncoGel, has been extensively examined both *in vitro* and *in vivo* for its
potential to support the sustained release of PTX.[124] Injection of OncoGel into
the pancreas of a pig provided high and sustained localized concentrations of
PTX, which showed a potential minimally invasive local treatment option for
unresectable pancreatic tumors.[125,126] OncoGel can show a sustained PTX
delivery over approximately 6 weeks from a single administration. When
OncoGel was administered directly into the patients' solid tumors (0.06–2.0 mg

PTX per cm^3 tumor volume), the tumor volume was well tolerated, and the PTX remained localized at the injection site, which minimized systemic exposure.[127–129] The influence of the drug concentration, solvent composition and composition of the copolymer on the release of testosterone from PLGA-PEG-PLGA systems over the course of three months has been reported.[130] Hydrophilic 5-fluorouracil and hydrophobic indomethacin released from the PLGA-PEG-PLGA hydrogel over periods of five days and one month, respectively, with copolymer composition-dependent release rates.[131] The PLGA-PEG-PLGA hydrogel was used to enhance the equilibrium active fraction of topotecan (a derivative of camptothecin), a water-soluble drug with a reversible structural active–inactive conversion.[132] The release of human insulin from ReGel proceeded at a constant rate over the course of two weeks, both *in vitro* and *in vivo*, without an initial burst.[133,134] The zinc complex of incretin hormone glucagon-like peptide-1 (Zn-GLP-1) showed zero-order release from ReGel over two weeks, without an initial burst. After a single injection of Zn-GLP-1/ReGel into Zucker diabetic fatty (ZDF) rats, the plasma GLP-1 concentration was maintained at significantly higher levels than that of the control group over the course of two weeks, and the levels of plasma insulin increased.[135] Insulin and bee venom peptide (BVP) were released from the PLGA-PEG-PLGA copolymer hydrogel over the course of 15 and 40 days, respectively.[136] The application of PLGA-PEG-PLGA hydrogel as a potential bandage for corneal wound repair, due to its suitable gelling profile and biocompatibility properties, was also reported.[137]

7.4.3   pH/Temperature-Sensitive [PEG-PAEU]$_x$ Block Copolymer Hydrogels

7.4.3.1   Controlling the Sol–Gel Phase Transition Diagram

The tube inversion method was used to determine the sol–gel phase transition diagram of the copolymer solution. Figure 7.6 shows the sol–gel phase transition diagram of the [PEG-PAEU]$_x$ block copolymer hydrogels. The aqueous copolymer solutions show sol-to-gel and gel-to-sol phase transitions with increasing pH and temperature, respectively, with the gel regions covering physiological conditions (pH 7.4, 37 °C). The sol–gel phase transitions of the [PEG-PAEU]$_x$ block copolymer and PAE-PEG-PAE triblock copolymer are similar, and both of them differ from that of the [PEG-PAU]$_x$ block copolymer, which shows a sol-to-gel-to-sol phase transition with increasing temperature, due to the presence of the ester groups in the PAEU and PAE segments.[95,106] The copolymer aqueous solution exists as a sol in the entire range of experimental temperatures at low pH (such as pH 6.6), because of the ionized tertiary amine groups and hydrophilic properties of the PAEU segments. Figure 7.7a shows a photograph of the sol state of 20 wt% P-BTB-20-2 solution (at pH 6.6, 5 °C) in the form of a transparent solution. However, at basic pH, the tertiary amine groups are deionized completely, as the strong hydrophobic interactions

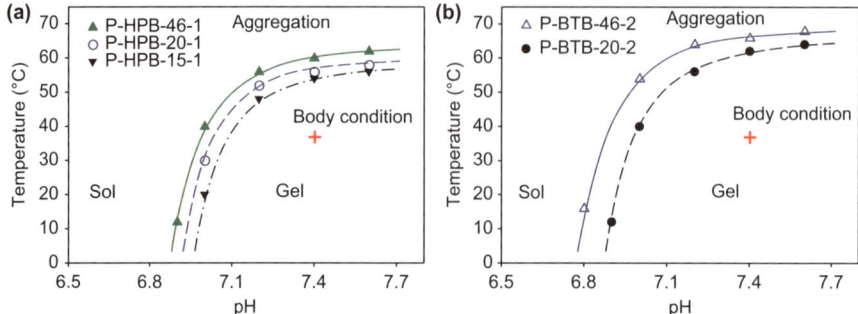

Figure 7.6   Sol–gel phase transition diagram of [PEG-PAEU]$_x$ hydrogels (20 wt%) with different molecular weights of PEG using monomer (a) HPB and (b) BTB.[111]
(Adapted in part from Huynh *et al.*[110] by permission of Taylor & Francis.)

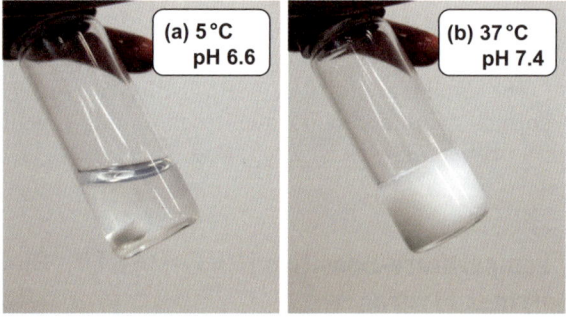

Figure 7.7   Photographs of the *in vitro* sol–gel phase transition of the 20 wt% copolymer P-BTB-20-2 upon heating and with increasing pH: (a) at 5 °C and pH 6.6 to (b) 37 °C and pH 7.4.

and hydrogen bonds between the deionized PAEU segments lead to the formation of a gel, as shown in Figures 7.3 and 7.7b (pH 7.4, 37 °C).[106,110–113] With further increasing temperature at basic pH (pH 7.4, 70 °C), a gel-to-sol transition occurs, due to the breaking of networks caused by the breaking of hydrogen bonds between the urethane groups in PAEU, and the partial dehydration of PEG at high temperatures.[106,110–113]

For controlling the gel region, many different factors of influence have been studied, including the molecular weight of PEG, the PAEU fraction, the monomer structure and the copolymer concentration. With an increase in the molecular weight of PEG, the gel regions become broader and shift to a lower pH (Figure 7.6), due to the larger size of the micelles.[110–113,138] It is found that the gel regions can be shifted to lower pH and higher temperature by: (i) increasing the hydrophobicity of the monomer (P-ETE-20-2 to P-BTB-20-2;

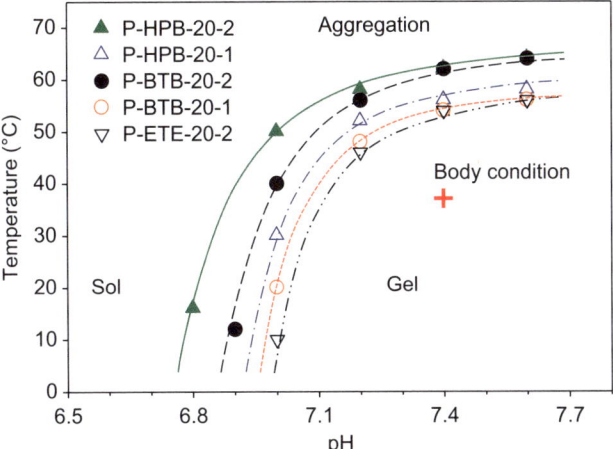

Figure 7.8 Sol–gel phase transition diagram of the [PEG-PAEU]$_x$ hydrogels with different influencing factors: monomer employed (P-HPB-20-2, P-BTB-20-2 and P-ETE-20-2) and PAEU fraction (P-HPB-20-2 and P-HPB-20-1 or P-BTB-20-2 and P-BTB-20-1).[111]
(Adapted in part from Huynh *et al.*[110] by permission of Taylor & Francis.)

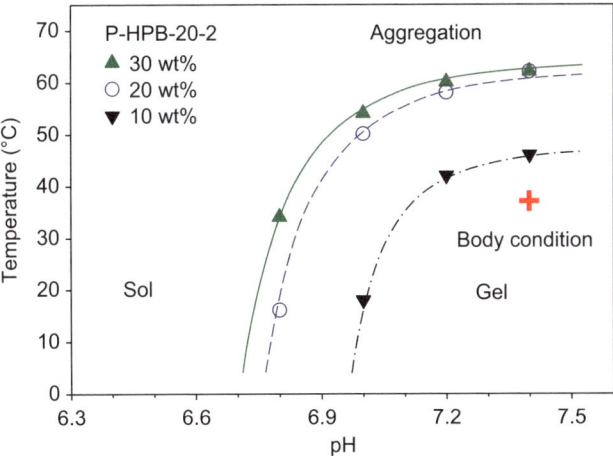

Figure 7.9 The influence of copolymer concentration on the sol–gel phase transition diagram of P-HPB-20-20 hydrogel (at pH 7.4).[111]

Figure 7.8); (ii) increasing the fraction of PAEU (P-HPB-20-1 to P-HPB-20-2, or P-BTB-20-1 to P-BTB-20-2; Figure 7.8); or (iii) increasing the copolymer concentration (Figure 7.9). This can be explained by the increase in the density of hydrogen bonding, and hydrophobic interactions between the deionized PAEU segments in the hydrogels.[110–113]

In summary, aqueous solutions of the [PEG-PAEU]$_x$ copolymer exhibit sol-to-gel and gel-to-sol phase transitions with increasing pH and temperature, respectively. The gel regions can be broadened and shifted to lower pH by increasing: (i) the molecular weight of the PEG; (ii) the hydrophobicity of the monomer; (iii) the fraction of PAEU; or (iv) the copolymer concentration.

7.4.3.2   The Properties of Hydrogels

Variation in the viscosity of the copolymer solution can be used to determine the sol–gel phase transitions and mechanical properties of the gel. A solution of 20 wt% P-HPB-20-2 exists in the sol state with low viscosity (< 10 Pa s) at pH 6.6 over the range of temperatures examined (Figure 7.10). However, the high viscosity ($\sim 10^4$ Pa s) of the sample at pH 7.4 and low temperatures demonstrates the presence of a gel. With increase of temperature, the viscosity decreases rapidly, indicating a gel-to-sol phase transition.[106,110–113] A similar rule can be observed for the copolymer solutions at pH > 6.8. The [PEG-PAEU]$_x$ hydrogel has a stronger mechanical property in comparison with PAE-PEG-PAE triblock copolymer hydrogels, due to the presence of abundant hydrogen bonds in this hydrogel.[106] A higher viscosity of the hydrogel can be obtained by increasing the density of functional deionized tertiary amine groups in the hydrogel (24 in P-HPB-20-2 compared with 18 in P-BTB-20-2, calculated for 1 molecule of PEG). The (red) dashed line in Figure 7.10

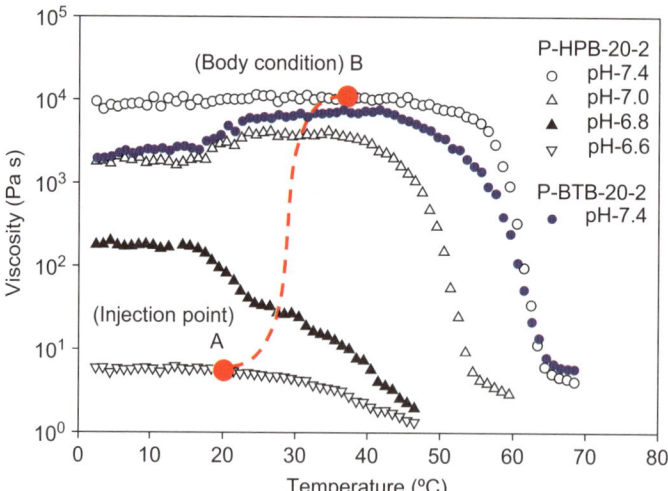

Figure 7.10   Change in viscosity of 20 wt% P-HPB-20-2 copolymer solutions as a function of temperature at different pH values and 20 wt% P-BTB-20-2 copolymer solution at pH 7.4. The (*red*) dashed line estimates the viscosity change in the injection process.[111]
(Adapted in part from Huynh *et al.*[110] by permission of Taylor & Francis.)

estimates the change of viscosity in the injection process. A copolymer solution with low viscosity at pH 6.6 (from A, 8 Pa s) was injected into the body and the viscosity increased under physiological conditions (to B, $\sim 10^4$ Pa s).

For biomedical applications, materials are usually expected to elicit minimum cytotoxicity. The [PEG-PAEU]$_x$ hydrogel system does not elicit serious cytotoxicity up to high copolymer concentrations (Figure 7.11). The cell viability is approximately 95% at a copolymer concentration of 500 μg mL^{-1}, and around 70% at a copolymer concentration of 1000 μg mL^{-1}. The injectability of the hydrogel system is also an important factor for potential applicability. The [PEG-PAEU]$_x$ copolymer solution exists in the sol state at low pH (*e.g.* pH 6.6, 5 °C; Figure 7.7a), which facilitates formulation with bioactive molecules and injection into the body. After being injected into the body, a gel forms, due to the changes in pH and temperature under physiological conditions (Figure 7.12), which offers a gel depot for the release of loaded bioactive molecules. In addition, the [PEG-PAEU]$_x$ hydrogel is a biodegradable system, which may eliminate the need for surgery to remove the injected hydrogel. The [PEG-PAEU]$_x$ hydrogels show rather fast degradation rates after the first 3 days, with a decrease in weight of around 30%, both *in vitro* and *in vivo* (Figure 7.13). However, the remaining weight *in vitro* of the hydrogels is > 50% after 2 weeks, which shows an improvement in comparison with the copolymer hydrogels using lactic acid as a modifier, which degrade completely in only one day.[68] The degradation rates of [PEG-PAEU]$_x$ hydrogels can be controlled by changing the structure of the monomer or copolymer. A copolymer with more ester groups in the structure results in a much faster degradation rate (P-BTB-20-2

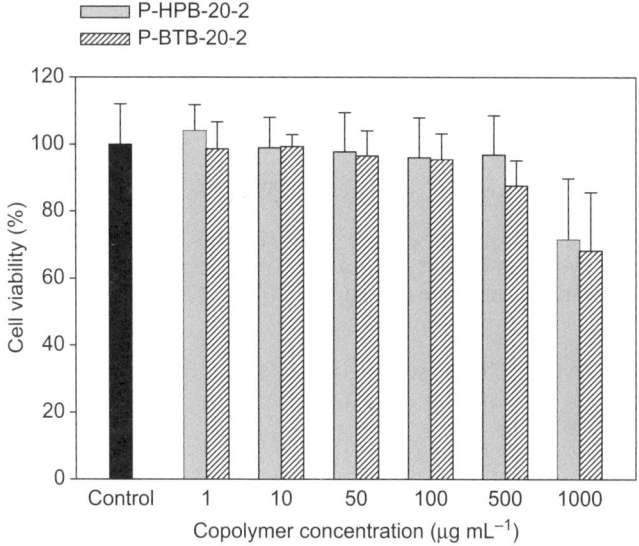

Figure 7.11 *In vitro* cytotoxicity of the [PEG-PAEU]$_x$ copolymers using L929 cells (± SD, $n = 3$).

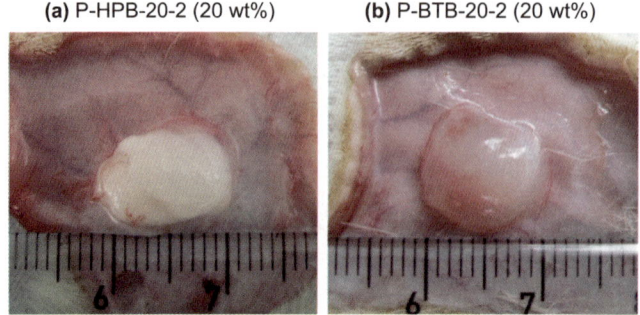

(a) P-HPB-20-2 (20 wt%)   **(b)** P-BTB-20-2 (20 wt%)

Figure 7.12   Photographs of *in vivo* gel at 10 min after injection of 200 μL copolymer
solution (20 wt%, pH 6.6) into SD rat: (a) P-HPB-20-2 and (b) P-BTB-
20-2.[111]
(Adapted in part from Huynh *et al.*[110] by permission of Taylor &
Francis.)

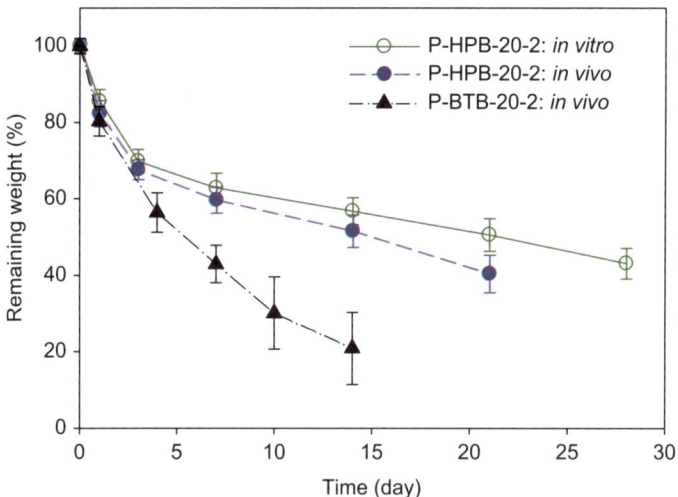

Figure 7.13   The remaining weights of P-HPB-20-2 and P-BTB-20-2 copolymer
hydrogels (20 wt%) checked by mass loss method ($\pm$ SD, $n = 3$). *In vitro*
experiments were carried out at pH 7.4 and 37 °C.[111]

compared to P-HPB-20-2), which is consistent with the PAEU-PEG-PAEU
triblock copolymer hydrogel system.[110–113]

7.4.3.3   Potential Application of [PEG-PAEU]$_x$ Hydrogels as Drug Delivery Systems

The potential application of the [PEG-PAEU]$_x$ hydrogels as drug carriers is
confirmed by the controlled release of doxorubicin (DOX). Figure 7.14 shows

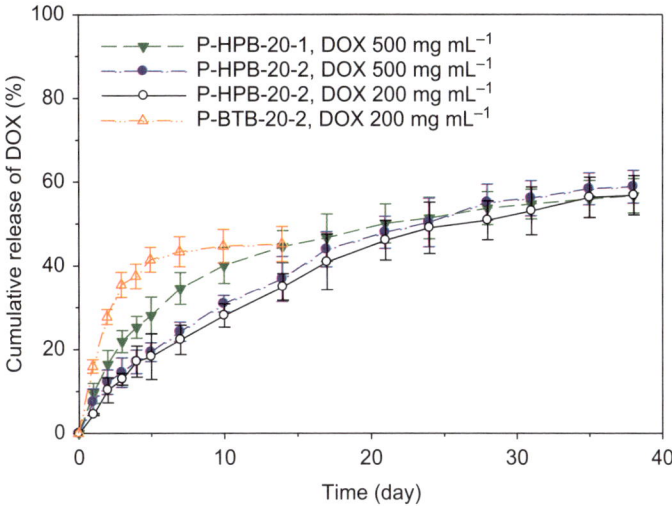

Figure 7.14   *In vitro* release profiles of DOX from [PEG-PAEU]$_x$ hydrogels ($\pm$ SD, $n = 3$).[111]
(Adapted in part from Huynh *et al.*[110] by permission of Taylor & Francis.)

the sustained released of DOX from [PEG-PAEU]$_x$ hydrogels for more than 5 weeks.[110,111] The release of DOX from the P-BTB-20-2 hydrogel is much faster than that from P-HPB-20-2 (1 week compared with 5 weeks), due to the faster degradation of the former. In the case of the P-HPB-20-2 hydrogel, the release profiles from different concentrations of the DOX-loaded hydrogels (500 and 200 μg mL^{-1}) show similar behaviour. This indicates that the release of DOX can be controlled by adjusting the degradation behaviour of [PEG-PAEU]$_x$ hydrogels, because the release profiles were not affected by the DOX-loaded concentration in the hydrogels. At the same DOX-loaded concentration (500 μg mL^{-1}), the hydrogel with a lesser fraction of PAEU (P-HPB-20-1) exhibited a faster release rate (compared to P-HPB-20-2), because of the fewer interactions (hydrophobic and/or hydrogen bonds) between the PAEU chains and DOX molecules, and lower stability of the lesser fraction of the PAEU hydrogel.

7.5   Conclusions

A short review about physically cross-linked injectable block copolymer hydrogels has been presented with two detailed typical examples, a temperature-sensitive hydrogel using the triblock copolymer PLGA-PEG-PLGA, and a pH/temperature-sensitive hydrogel using the cationic [PEG-PAEU]$_x$ block copolymer. Aqueous solutions of the PLGA-PEG-PLGA triblock copolymer undergo sol-to-gel-to-sol phase transitions with controllable gel regions when the temperature is increased. The formation of micelles and bridged micelles in the sol state at low temperatures, and the association of bridged micelles in the gel state at high temperatures, are attributed to the sol–gel phase transition mechanism. The gel regions can be controlled by adjusting the hydrophobicity of the copolymer, the molecular weight of the PEG or by adding some additives (*e.g.* PEG, NaSCN). On the other hand, aqueous solutions of the cationic [PEG-PAEU]$_x$ block copolymer exhibit sol-to-gel and gel-to-sol phase transitions with increasing pH and temperature, respectively. The hydrophilic character of ionized PAEU segments at low pH and temperatures in solution, and the formation of the gel at physiological conditions due to strong interactions (hydrophobic and hydrogen bond) between deionized PAEU segments, are proposed in the sol–gel mechanism. The gel regions can be adjusted by varying the PAEU fraction, the copolymer solution concentration, the molecular weight of the PEG or the monomer structure. These two hydrogel systems can easily be injected into the SD rat body and the *in situ* gels form rapidly, due to changing of pH and temperature in the body. The potential applicability as drug carriers for sustained release is also confirmed.

Acknowledgments

This study was supported by the Basic Science Research Program through a National Research Foundation of Korea (NRF) grant funded by the Korean Government (MEST) (2010-0027955) and the Converging Research Center Program funded by the Ministry of Education, Science and Technology (2011K000817).

References

1. C. T. Huynh, M. K. Nguyen and D. S. Lee, *Macromolecules*, 2011, **44**, 6629–6636.
2. M. K. Nguyen and D. S. Lee, *Macromol. Biosci.*, 2010, **10**, 563–579.
3. C. He, S. W. Kim and D. S. Lee, *J. Controlled Release*, 2008, **127**, 189–207.
4. M. H. Park, M. K. Joo, B. G. Choi and B. Jeong, *Acc. Chem. Res.*, 2012, **45**, 424–433.
5. X. J. Loh and J. Li, *Expert Opin. Ther. Pat.*, 2007, **17**, 965–977.
6. M. K. Joo, M. H. Park, B. G. Choi and B. Jeong, *J. Mater. Chem.*, 2009, **19**, 5891–5905.
7. L. E. Bromberg and E. S. Ron, *Adv. Drug Delivery Rev.*, 1998, **31**, 197–221.
8. L. Yu and J. Ding, *Chem. Soc. Rev.*, 2008, **37**, 1473–1481.
9. C. Tsitsilianis, *Soft Matter*, 2010, **6**, 2372–2388.
10. J. Kopeček, *Biomaterials*, 2007, **28**, 5185–5192.
11. J. D. Kretlow, L. Klouda and A. G. Mikos, *Adv. Drug Delivery Rev.*, 2007, **59**, 263–273.

12. N. S. Satarkar, D. Biswal and J. Z. Hilt, *Soft Matter*, 2010, **6**, 2364–2371.
13. H. J. Chung and T. G. Park, *Nano Today*, 2009, **4**, 429–437.
14. M.-T. Popescu, S. Mourtas, G. Pampalakis, S. G. Antimisiaris and C. Tsitsilianis, *Biomacromolecules*, 2011, **12**, 3023–3030.
15. K. P. Koutroumanis, K. Avgoustakis and D. Bikiaris, *Carbohydr. Polym.*, 2010, **82**, 181–188.
16. M. K. Joo, M. H. Park, B. G. Choi and B. J. Jeong, *Mater. Chem.*, 2009, **19**, 5891–5905.
17. Y. Zhang, L. Tao, S. Li and Y. Wei, *Biomacromolecules*, 2011, **12**, 2894–2901.
18. S. H. Lee, Y. Lee, S.-W. Lee, H.-Y. Ji, J.-H. Lee, D. S. Lee and T. G. Park, *Acta Biomater.*, 2011, **7**, 1468–1476.
19. N. Q. Tran, Y. K. Joung, E. Lih and K. D. Park, *Biomacromolecules*, 2011, **12**, 2872–2880.
20. C. W. Yung, W. E. Bentley and T. A. Barbari, *J. Biomed. Mater. Res.*, 2010, **95**, 25–32.
21. D. Wang, P. M. Fredericks, A. Haddad, D. J. T. Hill, F. Rasoul and A. K. Whittaker, *Polym. Degrad. Stab.*, 2011, **96**, 123–130.
22. J. B. Lee, J. J. Yoon, D. S. Lee and T. G. Park, *J. Biomater. Sci., Polym. Edn*, 2004, **15**, 1571–1583.
23. B. L. Dargaville, C. Vaquette, H. Peng, F. Rasoul, Y. Q. Chau, J. J. Cooper-White, J. H. Campbell and A. K. Whittaker, *Biomacromolecules*, 2011, **12**, 3856–3869.
24. R. Jin, L. S. M. Teixeira, A. Krouwels, P. J. Dijkstra, C. A. van Blitterswijk, M. Karperien and J. Feijen, *Acta Biomater.*, 2010, **6**, 1968–1977.
25. K. S. Kim, S. J. Park, J.-A. Yang, J.-H. Jeon, S. H. Bhang, B.-S. Kim and S. K. Hahn, *Acta Biomater.*, 2011, **7**, 666–674.
26. C. D. Pritchard, T. M. O'Shea, D. J. Siegwart, E. Calo, D. G. Anderson, F. M. Reynolds, J. A. Thomas, J. R. Slotkin, E. J. Woodard and R. Langer, *Biomaterials*, 2011, **32**, 587–597.
27. D. K. Wang, F. Rasoul, D. J. T. Hill, G. R. Hanson, C. J. Noble and A. K. Whittaker, *Soft Matter*, 2012, **8**, 435–445.
28. F. Lee, J. E. Chung and M. Kurisawa, *J. Controlled Release*, 2009, **134**, 186–193.
29. J. Rassing and D. Attwood, *Int. J. Pharm.*, 1983, **13**, 47–55.
30. M. J. Song, D. S. Lee, J. H. Ahn, D. J. Kim and S. C. Kim, *Polym. Bull.*, 2000, **43**, 497–504.
31. O. Glatter, G. Scherf, K. Schillen and W. Brown, *Macromolecules*, 1994, **27**, 6046–6054.
32. B. A. Borden, J. Yockman and S. W. Kim, *Mol. Pharm.*, 2010, **7**, 963–968.
33. M.-H. Cha, J. Choi, B. G. Choi, K. Park, I. H. Kim, B. Jeong and D. K. Han, *J. Colloid Interface Sci.*, 2011, **360**, 78–85.
34. A. Sosnik and D. Cohn, *Biomaterials*, 2005, **26**, 349–357.
35. B. Jeong, Y. H. Bae, D. S. Lee and S. W. Kim, *Nature*, 1997, **388**, 860–862.

36. B. Jeong, Y. H. Bae and S. W. Kim, *Macromolecules*, 1999, **32**, 7064–7069.
37. B. Jeong, Y. H. Bae and S. W. Kim, *J. Controlled Release*, 2000, **63**, 155–163.
38. Y. Duan, Y. Nie, T. Gong, Q. Wang and Z. J. Zhang, *J. Appl. Polym. Sci.*, 2006, **100**, 1019–1023.
39. D. S. Lee, M. S. Shim, S. W. Kim, H. Lee, I. Park and T. Chang, *Macromol. Rapid Commun.*, 2001, **22**, 587–592.
40. M. S. Shim, H. T. Lee, W. S. Shim, I. S. Park, H. J. Lee, T. H. Chang, S.W. Kim and D. S. Lee, *J. Biomed. Mater. Res.*, 2002, **61**, 188–196.
41. L. Yu, Z. Zhang and J. Ding, *Biomacromolecules*, 2011, **12**, 1290–1297.
42. L. Yu, G. Chang, H. Zhang and J. Ding, *J. Polym. Sci., Part A: Polym. Chem.*, 2007, **45**, 1122–1133.
43. L. Yu, H. Zhang and J. Ding, *Colloid Polym. Sci.*, 2010, **288**, 1151–1159.
44. J. Lee, Y. H. Bae, Y. S. Sohn and B. Jeong, *Biomacromolecules*, 2006, **7**, 1729–1734.
45. B. J. Tarasevich, A. Gutowska, X. S. Li and B. Jeong, *J. Biomed. Mater. Res., A*, 2009, **89**, 248–254.
46. B. Jeong, M. R. Kibbey, J. C. Birnbaum, Y. Won and A. Gutowska, *Macromolecules*, 2000, **33**, 8317–8322.
47. M. S. Kim, H. Hyun, G. Khang and H. B. Lee, *Macromolecules*, 2006, **39**, 3099–3102.
48. Y. M. Kang, G. H. Kim, J. I. Kim, D. Y. Kim, B. N. Lee, S. M. Yoon, J. H. Kim and M. S. Kim, *Biomaterials*, 2011, **32**, 4556–4564.
49. M. H. Kim, H. N. Hong, J. P. Hong, C. J. Park, S. W. Kwon, S. H. Kim, G. Kang and M.-J. Kim, *Biomaterials*, 2010, **31**, 1213–1218.
50. M. J. Hwang, M. K. Joo, B. G. Choi, M. H. Park, I. W. Hamley and B. Jeong, *Macromol. Rapid Commun.*, 2010, **31**, 2064–2069.
51. M. J. Hwang, J. M. Suh, Y. H. Bae, S. W. Kim and B. Jeong, *Biomacromolecules*, 2005, **6**, 885–890.
52. S. J. Bae, J. M. Suh, Y. S. Sohn, Y. H. Bae, S. W. Kim and B. Jeong, *Macromolecules*, 2005, **38**, 5260–5265.
53. C.-Y. Gong, S. Shia, P.-W. Dong, B. Kan, M.-L. Gou, X.-H. Wang, X.-Y. Li, F. Luo, X. Zhao, Y.-Q. Wei and Z.-Y. Qian, *Int. J. Pharm.*, 2009, **365**, 89–99.
54. M.-L. Gou, C.-Y. Gong, J. Zhang, X.-H. Wang, X.-H. Wang, Y.-C. Gu, G. Guo, L.-J. Chen, F. Luo, X. Zhao, Y.-Q. Wei and Z.-Y. Qian, *J. Biomed. Mater. Res.*, 2010, **93**, 219–226.
55. C.-Y. Gong, S. Shi, L. Wu, M.-L. Gou, Q.-Q. Yin, Q.-F. Guo, P.-W. Dong, F. Zhang, F. Luo, X. Zhao, Y.-Q. Wei and Z.-Y. Qian, *Acta Biomater.*, 2009, **5**, 3358–3370.
56. S. J. Bae, M. K. Joo, Y. Jeong, S. W. Kim, W.-K. Lee, Y. S. Sohn and B. Jeong, *Macromolecules*, 2006, **39**, 4873–4879.
57. W. Xun, D.-Q. Wu, Z.-Y. Li, H.-Y. Wang, F.-W. Huang, S.-X. Cheng, X.-Z. Zhang and R.-X Zhuo, *Macromol. Biosci.*, 2009, **9**, 1219–1226.

58. L. Yu, Z. Zhang, H. Zhang and J. Ding, *Biomacromolecules*, 2009, **10**, 1547–1553.
59. L. Yu, Z. Zhang, H. Zhang and J. Ding, *Biomacromolecules*, 2010, **11**, 2169–2178.
60. Z. Zhang, Y. Lai, L. Yu and J. Ding, *Biomaterials*, 2010, **31**, 7873–7882.
61. Z. Zhang, J. Ni, L. Chen, L. Yu, J. Xu and J. Ding, *Biomaterials*, 2011, **32**, 4725–4736.
62. J. Wu, F. Zeng, X.-P. Huang, J. C.-Y. Chung, F. Konecny, R. D. Weisel and R.-K. Li, *Biomaterials*, 2011, **32**, 579–586.
63. Y. M. Kang, S. H. Lee, J. Y. Lee, J. S. Son, B. S. Kim, B. Lee, H. J. Chun, B. H. Min, J. H. Kim and M. S. Kim, *Biomaterials*, 2010, **31**, 2453–2460.
64. X. J. Loh, S. H. Goh and J. Li, *Biomacromolecules*, 2007, **8**, 585–593.
65. X. J. Loh, S. H. Goh and J. Li, *Biomaterials*, 2007, **28**, 4113–4123.
66. X. J. Loh, S. H. Goh and J. Li, *J. Phys. Chem. B*, 2009, **113**, 11822–11830.
67. H. J. Oh, M. K. Joo, Y. S. Sohn and B. Jeong, *Macromolecules*, 2008, **41**, 8204–8209.
68. J. Y. Kim, M. H. Park, M. K. Joo, S. Y. Lee and B. Jeong, *Macromolecules*, 2009, **42**, 3147–3151.
69. B. G. Choi, M. H. Park, S.-H. Cho, M. K. Joo, H. J. Oh, E. H. Kim, K. Park, D. K. Han and B. Jeong, *Soft Matter*, 2011, **7**, 456–462.
70. B. G. Choi, M. H. Park, S.-H. Cho, M. K. Joo, H. J. Oh, E. H. Kim, K. Park, D. K. Han and B. Jeong, *Biomaterials*, 2010, **31**, 9266–9272.
71. S. H. Park, B. G. Choi, H. J. Moon, S.-H. Cho and B. Jeong, *Soft Matter*, 2011, **7**, 6515–6521.
72. Y. Y. Choi, J. H. Jang, M. H. Park, B. G. Choi, B. Chi and B. Jeong, *J. Mater. Chem.*, 2010, **20**, 3416–3421.
73. E. H. Kim, M. K. Joo, K. H. Bahk, M. H. Park, B. Chi, Y. M. Lee and B. Jeong, *Biomacromolecules*, 2009, **10**, 2476–2481.
74. Y. Jeong, M. K. Joo, K. H. Bahk, Y. Y. Choi, H.-T. Kim, W.-K. Kim, H. J. Lee, Y. S. Sohn and B. Jeong, *J. Controlled Release*, 2009, **137**, 25–30.
75. U. P. Shinde, M. K. Joo, H. J. Moon and B. Jeong, *J. Mater. Chem.*, 2012, **22**, 6072–6079.
76. H. J. Moon, B. G. Choi, M. H. Park, M. K. Joo and B. Jeong, *Biomacromolecules*, 2011, **12**, 1234–1242.
77. B. G. Choi, S.-H. Cho, H. Lee, M. H. Cha, K. Park, B. Jeong and D. K. Han, *Macromolecules*, 2011, **44**, 2269–2275.
78. G. D. Kang and S.-C. Song, *Int. J. Pharm.*, 2008, **349**, 188–195.
79. M.-R. Park, C.-J. Chun, S.-W. Ahn, M.-H. Ki, C.-S. Cho and S.-C. Song, *Biomaterials*, 2010, **31**, 1349–1359.
80. M.-R. Park, C.-J. Chun, S.-W. Ahn, M.-H. Ki, C.-S. Cho and S.-C. Song, *J. Controlled Release*, 2010, **147**, 359–367.
81. V. P. N. Nguyen, N. Kuo and X. J. Loh, *Soft Matter*, 2011, **7**, 2150–2159.
82. X. J. Loh, V. P. N. Nguyen, N. Kuo and J. Li, *J. Mater. Chem.*, 2011, **21**, 2246–2254.
83. R. Chapanian, M. Y. Tse, S. C. Pang and B. G. Amsden, *Biomaterials*, 2009, **30**, 295–306.

84. E. Behravest, A. K. Shung, S. Jo and A. G. Mikos, *Biomacromolecules*, 2002, **3**, 153–158.

85. S. Y. Kim, H. J. Kim, K. E. Lee, S. S. Han, Y. S. Sohn and B. Jeong, *Macromolecules*, 2007, **40**, 5519–5525.

86. S. H. Park, B. G. Choi, M. K. Joo, D. K. Han, Y. S. Sohn and B. Jeong, *Macromolecules*, 2008, **41**, 6486–6492.

87. E. Schacht, V. Toncheva, K. Vandertaelen and J. Heller, *J. Controlled Release*, 2006, **116**, 219–225.

88. J. Wu, F. Zeng, X.-P. Huang, J. C.-Y. Chung, F. Konecny, R. D. Weisel and R.-K. Li, *Biomaterials*, 2011, **32**, 579–586.

89. J. Lee, M. K. Joo, H. Oh, Y. S. Sohn and B. Jeong, *Polymer*, 2006, **47**, 3760–3766.

90. B. G. Choi, Y. S. Sohn and B. Jeong, *J. Phys. Chem. B*, 2007, **111**, 7715–7718.

91. J. O. Han, M. K. Joo, J. H. Jang, M. H. Park and B. Jeong, *Macromolecules*, 2009, **42**, 6710–6715.

92. Y. Ma, Y. Tang, N. C. Billingham and S. P. Armes, *Biomacromolecules*, 2003, **4**, 864–868.

93. M. D. Determan, J. P. Cox and S. K. Mallapragada, *J. Biomed. Mater. Res., A*, 2007, **81**, 326–333.

94. S. Y. Lee, Y. Lee, J. E. Kim, T. G. Park and C.-H. Ahn, *J. Mater. Chem.*, 2009, **19**, 8198–8201.

95. K. Dayananda, C. He, D. K. Park, T. G. Park and D. S. Lee, *Polymer*, 2008, **49**, 4968–4973.

96. C. T. Huynh, M. K. Nguyen, D. P. Huynh, S. W. Kim and D. S. Lee, *Polymer*, 2010, **51**, 3843–3850.

97. M. K. Nguyen, D. K. Park and D. S. Lee, *Biomacromolecules*, 2009, **10**, 728–731.

98. C. T. Huynh, Q. V. Nguyen, S. W. Kang and D. S. Lee, *Polymer*, 2012, **53**, 4069–4075.

99. D. M. Lynn and R. Langer, *J. Am. Chem. Soc.*, 2000, **122**, 10761–10768.

100. M. S. Lee, Y. L. Jang, D. P. Huynh, C. T. Huynh, Y. Lee, S. Y. Chae, S. H. Kim, T. G. Park, D. S. Lee and J. H. Jeong, *Macromol. Biosci.*, 2011, **11**, 789–796.

101. D. P. Huynh, M. K. Nguyen, B. S. Pi, M. S. Kim, S. Y. Chae, K. C. Lee, B. S. Kim, S. W. Kim and D. S. Lee, *Biomaterials*, 2008, **29**, 2527–2534.

102. D. P. Huynh, G. J. Im, S. Y. Chae, K. C. Lee and D. S. Lee, *J. Controlled Release*, 2009, **137**, 20–24.

103. D. P. Huynh, M. K. Nguyen, B. S. Kim and D. S. Lee, *Polymer*, 2009, **50**, 2565–2571.

104. D. P. Huynh, M. K. Nguyen and D. S. Lee, *Macromol. Res.*, 2010, **18**, 192–199.

105. Y. Zheng, M. K. Nguyen, C. He, C. T. Huynh and D. S. Lee, *Macromol. Res.*, 2010, **18**, 1096–1102.

106. M. K. Nguyen, C. T. Huynh and D. S. Lee, *Polymer*, 2009, **50**, 5205–5210.

107. C. T. Huynh, M. K. Nguyen, D. P. Huynh and D. S. Lee, *Colloid Polym. Sci.*, 2011, **289**, 301–308.
108. K. Dayananda, C. He and D. S. Lee, *Polymer*, 2008, **49**, 4620–4625.
109. Y. Zheng, C. He, C. T. Huynh and D. S. Lee, *Macromol. Res.*, 2010, **18**, 974–980.
110. C. T. Huynh, M. K. Nguyen, I. K. Jeong, S. W. Kim and D. S. Lee, *J. Biomater. Sci., Polym. Ed.*, 2012, **23**, 1091–1106.
111. C. T. Huynh, M. K. Nguyen, J. H. Kim, S. W. Kang, B. S. Kim, D. P. Huynh and D. S. Lee, *Soft Matter*, 2011, **7**, 4974–4982.
112. C. T. Huynh, S. W. Kang, Y. Li, B. S. Kim and D. S. Lee, *Soft Matter*, 2011, **7**, 8984–8990.
113. C. T. Huynh and D. S. Lee, *Colloid Polym. Sci.*, 2012, **290**, 1077–1086.
114. W. S. Shim, J. S. Yoo, Y. H. Bae and D. S. Lee, *Biomacromolecules*, 2005, **6**, 2930–2934.
115. W. S. Shim, J.-H. Kim, H. Park, K. Kim, I. C. Kwon and D. S. Lee, *Biomaterials*, 2006, **27**, 5178–5185.
116. W. S. Shim, S. W. Kim and D. S. Lee, *Biomacromolecules*, 2006, **7**, 1935–1941.
117. W. S. Shim, J.-H. Kim, K. Kim, Y.-S. Kim, R.-W. Park, I.-S. Kim, I. C. Kwon and D. S. Lee, *Int. J. Pharm.*, 2007, **331**, 11–18.
118. K. Dayananda, B. S. Pi, B. S. Kim, T. G. Park and D. S. Lee, *Polymer*, 2007, **48**, 758–762.
119. H. K. Kim, W. S. Shim, S. E. Kim, K.-H. Lee, E. Kang, J.-H. Kim, K. Kim, I. C. Kwon and D. S. Lee, *Tissue Eng., Part A*, 2009, **15**, 923–933.
120. D. P. Huynh, W. S. Shim, J. H. Kim and D. S. Lee, *Polymer*, 2006, **47**, 7918–7926.
121. C. T. Huynh, M. K. Nguyen and D. S. Lee, *Acta Biomater.*, 2011, **7**, 3123–3130.
122. M. K. Nguyen and D. S. Lee, *Chem. Commun.*, 2010, **46**, 3583–3585.
123. M. K. Nguyen, C. T. Huynh, G. H. Gao, J. H. Kim, D. P. Huynh, S. Y. Chae, K. C. Lee and D. S. Lee, *Soft Matter*, 2011, **7**, 2994–3001.
124. G. M. Zentner, R. Rathi, C. Shih, J. C. Mcrea, M. H. Seo, H. Oh, B. G. Rhee, J. Mestecky, Z. Moldoveanu, M. Morgan and S. Weitman, *J. Controlled Release*, 2001, **72**, 203–215.
125. K. Matthes, M. Mino-Kenudson, D. V. Sahani, N. Holalkere, K. D. Fowers, R. Rathi and W. R. Brugge, *Gastrointest. Endosc.*, 2007, **65**, 448–453.
126. C. Karaca, S. Cizginer, Y. Konuk, A. Kambadakone, B. G. Turner, M. Mino-Kenudson, D. V. Sahani, C. Macfarlane and W. Brugge, *Gastrointest. Endosc.*, 2011, **73**, 603–606.
127. S. J. Vukelja, S. P. Anthony, J. C. Arseneau, B. S. Berman, C. C. Cunningham, J. J. Nemunaitis, W. E. Samlowski and K. D. Fowers, *Anticancer Drugs*, 2007, **18**, 283–289.
128. J. B. Wolinsky, Y. L. Colson and M. W. Grinstaff, *J. Controlled Release*, 2012, **159**, 14–26.

129. N. L. Elstad and K. D. Fowers, *Adv. Drug Delivery Rev.*, 2009, **61**, 785–794.
130. S. Chen and J. Singh, *Int. J. Pharm.*, 2005, **295**, 183–190.
131. M. Qiao, D. Chen, X. Ma and Y. Liu, *Int. J. Pharm.*, 2005, **294**, 103–112.
132. G. Chang, T. Ci, L. Yu and J. Ding, *J. Controlled Release*, 2011, **156**, 21–27.
133. Y. J. Kim, S. Choi, J. J. Koh, M. Lee, K. S. Ko and S. W. Kim, *Pharm. Res.*, 2001, **18**, 548–550.
134. S. Choi and S. W. Kim, *Pharm. Res.*, 2003, **20**, 2008–2010.
135. S. Choi, M. Baudys and S. W. Kim, *Pharm. Res.*, 2004, **21**, 827–831.
136. M. Qiao, D. Chen, T. Hao, X. Zhao, H. Hu and X. Ma, *Int. J. Pharm.*, 2007, **345**, 116–124.
137. C. Pratoomsoot, H. Tanioka, K. Hori, S. Kawasaki, S. Kinoshita, P. J. Tighe, H. Dua, K. M. Shakesheff, F. Rosamari and A. J. Rose, *Biomaterials*, 2008, **29**, 272–281.
138. J. S. Yoo, M. S. Kim and D. S. Lee, *Macromol. Res.*, 2006, **14**, 117–120.

CHAPTER 8

Hydrogels for Biomedical Applications

XIAN JUN LOH,*[a] TUNG-CHUN LEE[b] AND YOSHIHIRO ITO[c]

[a] Melville Laboratory for Polymer Synthesis, Department of Chemistry, University of Cambridge, Lensfield Road, Cambridge CB2 1EW, UK; [b] Max Planck Institute for Intelligent Systems, Heisenbergstr. 3, 70569 Stuttgart, Germany; [c] Nano Medical Engineering Laboratory, RIKEN Advanced Science Institute, 2-1 Hirosawa, Wako-shi, Saitama 351-0198, Japan
*Email: xianjun_loh@scholars.a-star.edu.sg

8.1 Introduction to Hydrogels

Hydrogels are typically defined as chemically cross-linked three-dimensional macromolecular networks of hydrophilic copolymers swollen in water.[1–3] With technological progress, covalent cross-linking agents are no longer required for the formation of a hydrogel. There have now been many reports in the literature of physically cross-linked hydrogels. The reader is referred to Chapter 1 for a thorough introduction to hydrogels. Dynamic and reversible supramolecular linkages between the polymer chains can be utilized for the formation of hydrogels. These cross-links can be formed *via* hydrophobic interactions between polymer chains, host–guest complexation, metallo-assisted binding or layer-by-layer electrostatic approaches. Examples of supramolecular hydrogels include thermogelling polymers,[4–14] which contain no chemical cross-links. These polymers form hydrogels when the temperature is raised, leading to a switch in the physical state of the polymer to a hydrophobic state.

Monographs in Supramolecular Chemistry No. 11
Polymeric and Self Assembled Hydrogels: From Fundamental Understanding to Applications
Edited by Xian Jun Loh and Oren A. Scherman
© The Royal Society of Chemistry 2013
Published by the Royal Society of Chemistry, www.rsc.org

The hydrophobic interactions between the polymer chains result in the formation of non-covalent cross-links which hold the hydrogel in place. Host–guest interactions have also been utilized in the formation of hydrogels. Cyclodextrin-based pseudo-rotaxanes have been reported to form paste-like materials which could be used for drug delivery and tissue engineering applications.[15–17] These materials are highly attractive for applications as their physical properties are not rigidly defined and can be modified as required. The water retention capacity of the hydrogels can be adjusted by changing the hydrophilic content in the system, varying from 10% to over 99.9%.[18]

Hydrogels were first used for biomedical applications in 1960 when cross-linked poly(hydroxyethyl methacrylate) (PHEMA) hydrogels were fabricated for eye applications.[19] Since then, they have been used in a wide variety of applications such as wound dressings, soft contact lenses and soft tissue substitutes.[20–26] With new and innovative hydrogel architectures and designs, there has been an increase in the types of biomedical applications that require the use of hydrogels. This chapter summarizes the latest research in hydrogels used for biomedical applications, ranging from cardiac regeneration, glaucoma treatments, nerve repair, bone re-engineering and cartilage formation (Figure 8.1). Hydrogels are used in these applications as drug or cell carriers, tissue engineering scaffolds, solid supports for holding up collapsed organs or as interfacial agents to improve adhesion between separated parts of organs.

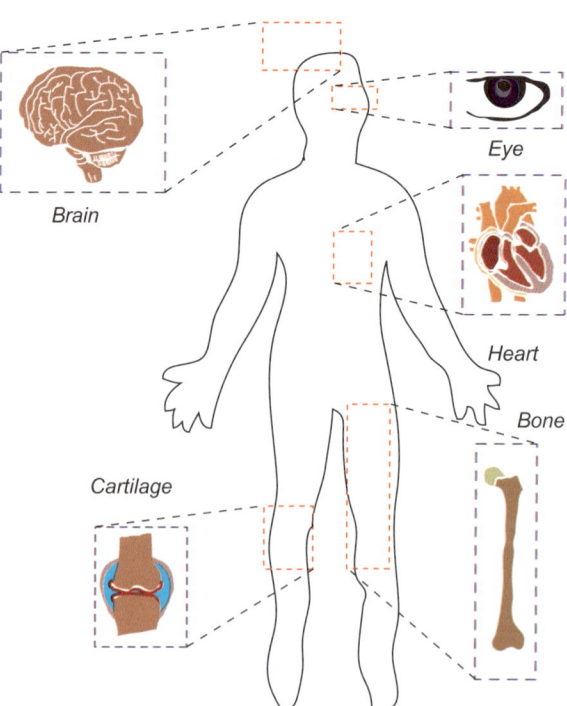

Figure 8.1   Parts of the body where hydrogels are used for biomedical applications.

8.2 Types of Hydrogels

There are different types of hydrogels which have been studied for biomedical applications. Both natural and synthetic polymers have been used, with different degrees of success in various applications. Hydrogels are promising materials from the clinical viewpoint. First, the high water content is attractive for biomedical applications due to its increased biocompatibility. Second, the mechanical properties of hydrogels can be tuned by a variety of approaches to obtain the desired physical properties. Third, chemical cues can be easily attached to the polymers to obtain hydrogels with superior interfacial interactions between the cells and the gel. Fourth, the interconnected 3D porous structure of hydrogels can potentially facilitate cell growth within their framework. Fifth, the easy functionalization of the polymers allows them to act as robust and versatile vehicles for drug and biomolecule delivery. Finally, hydrogels can be made injectable so that there will not be a need for complicated surgical intervention to insert the piece of gel into the body.

The use of hydrogels for a particular application is highly dependent on various factors, such as the site of application, the microenvironment of the implant site and the duration of the application, as well as specific requirements. The underlying principle governing the choice of the material is the compatibility of the hydrogel and the body. As a result, factors such as the mechanical properties of the gel should be tailored to match the properties of the organ as much as possible. The body is made up of organs with different moduli ranging from the hard (bone) to the soft (brain) and anything in between (muscles, cartilage). Certain hydrogels are also degradable, either hydrolytically or enzymatically. These factors play a crucial role in determining the lifetime of the hydrogels in the body. The mode of delivery of the hydrogel to the body is also a fundamental concern that clinicians have to consider. Gels which are pre-cross-linked have to be implanted *via* conventional surgical procedures, whereas injectable hydrogels which cross-link *in situ* or by the application of either a temperature or chemical stimulus could be attractive candidates for keyhole surgery, where the implantation of the gel simply involves the injection of the solution through a catheter to the target site. The following sections describe very briefly the major types of hydrogels that have found their place in the biomedical field for various applications.

8.2.1 Types of Biological (Natural) Hydrogels Used in Biomedical Applications

8.2.1.1 Protein-Based Hydrogels

8.2.1.1.1 Fibrin. Fibrin is a protein that is involved in the clotting of the blood. When fibrinogen and thrombin react together, fibrin is formed as a 3D mesh. The polymerization process takes place *via* mechanisms similar to those involved in normal blood clotting. This property has been exploited for the formation of a biopolymer scaffold for tissue engineering applications.[27–30] Fibrin can be applied in a high-viscosity "glue" form, as hydrogels or as microspheres. Fibrin glue has been commonly used as a delivery vehicle in injectable cardiac tissue engineering.[31,32]

8.2.1.1.2 Collagen. Collagen is a biological protein with a high glycine (Gly) content. It is found in the extracellular matrix (ECM) of many tissues, such as the skin, bone, cartilage, tendons, blood vessels and teeth. Its role is to provide structural and mechanical support.[33] Its biodegradability, cell-binding properties and low antigenicity make it a valuable material for tissue engineering applications.[33,34] However, collagen degrades rapidly, which leads to a rapid loss of mechanical properties.[35] Improvements have been made in attempting to extend the lifetime of collagen, either by adding mineral crystals or by combining collagen with either natural polymers, *e.g.* elastin or glycoaminoglycans (GAGs), or synthetic polymers, *e.g.* methacrylate derivatives, or by applying various cross-linking methods.[36–39]

8.2.1.1.3 Matrigel. Matrigel is a gelatinous protein mixture secreted by Engelbreth-Holm-Swarm (EHS) mouse sarcoma cells. This product is marketed by BD Biosciences and by Trevigen under the name Cultrex BME. Matrigel resembles the complex ECM environment found in many tissues and is commonly used by cell biologists as a substrate for cell culture.[40,41] The temperature-sensitive nature of this gel allows a cold Matrigel solution, kept at 4 °C, to exist in a fluid state. When incubated at 37 °C, the Matrigel proteins will self-assemble to produce a thin semi-solid film as a coating on the cell culture dish. Matrigel allows cells to exhibit complex behavior in a 3D environment, for example endothelial cells form intricate networks on Matrigel-coated surfaces but not on plastic surfaces.[42]

8.2.1.2 Polysaccharide-Based Hydrogels

8.2.1.2.1 Alginate. Alginate is a polysaccharide isolated from seaweed which has been used as an injectable cell delivery vehicle in cardiac tissue engineering. It forms hydrogels under gentle conditions in the presence of divalent ions such as calcium. Alginate is relatively biocompatible and is approved by the US Food and Drug Administration (FDA) for human use as wound dressing material.[43–45]

8.2.1.2.2 Hyaluronic Acid. Hyaluronic acid (HA) is a polysaccharide with alternating units of α-1,4-D-glucuronic acid and β-1,3-*N*-acetyl-D-glucosamine. HA is available in a wide range of molecular weights from 10^3 to 10^7 g mol^{-1}.[46] HA is the major component of the ECM in various connective tissues as a structural element, and can act as an interfacial agent with binding proteins, proteoglycans and other bioactive molecules. HA contributes to the regulation of water balance and behaves as a lubricant, protecting the articular cartilage surface, and acts as a scavenger molecule for free radicals. In addition, HA is recognized by specific cell receptors regulating cell behavior, inflammation, angiogenesis and healing processes, and acts as a selective and protective coating around the cell membrane.[46] Commercially available HA is obtained mainly by extraction from umbilical cord, synovial fluid, rooster comb and vitreous humor, but it can be also produced through large-scale microbial fermentation, without the risk of animal-derived pathogens.[47] Given its good biocompatibility and viscoelastic properties, HA has been extensively studied and employed in the biomedical field for delivery systems, cell encapsulation and tissue engineering.[48–53] It is an attractive biomaterial due to its non-immunogenic properties, widespread availability and ease of chain size manipulation. It also has direct effects on tissue organization *via* interactions with cell-surface receptors, promoting the migration of cells and facilitating ECM remodeling.[54] HA has been conjugated to alginate, chitosan and fibrin gel matrices to provide artificial ECM environments.[55] The ethyl and benzyl esters of hyaluronan, named HYAFF-7 and HYAFF-11, respectively, are two of the most characterized hyaluronan derivative polymers from both physicochemical and biological viewpoints, degrading at predictable rates through hydrolytic degradation of the ester bonds (around two months for complete hydrolysis).[54,56] Human chondrocytes grown onto HYAFF-11 3D scaffolds are able to re-express *in vitro* their differentiated phenotype and to reduce the expression and production of molecules involved in cartilage degenerative diseases.[57,58] Histological/morphological evaluation of repaired tissue by the HYAFF-11 scaffold, employed in chondrocyte transplantation *in vivo*, demonstrated a significant improvement in the quality of the healing in comparison to defects without grafted chondrocytes.[59]

8.2.1.2.3 Chitosan. Chitosan is a linear biocompatible and biodegradable cationic polysaccharide composed of randomly distributed β-(1-4)-linked D-glucosamine (deacetylated unit) and *N*-acetyl-D-glucosamine (acetylated unit). Its degradation products are oligomeric biocompatible chitosan fragments of variable length. Chitosan has been widely used in the tissue engineering of skin, bone, cartilage, liver, nerve and blood vessel in the past 25 years. Chitosan exists in different forms, with different applications in tissue engineering, such as porous structures, chitosan-based nanofibrous structures and chitosan hydrogels. Chitosan hydrogels are responsive to external stimuli such as light and temperature. Temperature-responsive chitosan hydrogels have a wide array of applications. The temperature-responsive chitosan-glycerol phosphate (GP) hydrogels are appealing because cytokines, genetic material and supportive cells relevant for the repair and regeneration of the tissue can be incorporated into the polymer solution. When injected into the body, the polymer solution forms a semi-solid gel *in situ*, entrapping the bioactive components within the injected area.

8.2.2 Types of Synthetic Hydrogels Used in Biomedical Applications

8.2.2.1 Stable Hydrogels

8.2.2.1.1 Poly(2-hydroxyethyl Methacrylate). Poly(2-hydroxyethyl methacrylate) (PHEMA) was one of the first synthetic hydrogels used as a biomaterial. In the late 1950s, Czechoslovakian chemist Otto Wichterle and his assistant Drahoslav Lim worked on producing PHEMA gels. This work was eventually published in 1960.[19] In 1961, using a home-made setup constructed from his son's erector set and parts from a bicycle, Wichterle created the prototype of a spin-casting machine and used it to produce the world's first soft contact lenses. PHEMA hydrogels have high swellability and have been shown to be biocompatible materials.[60,61] They are largely used in ophthalmic applications as material for contact lenses.[62]

8.2.2.1.2 Poly(Vinyl Alcohol). Poly(vinyl alcohol) (PVA) is a synthetic hydrogel that has been used in several biomedical applications, including contact lenses, ophthalmic materials, tendon repair and drug delivery. PVA hydrogels have been reported to be biocompatible[63] and have been shown to resist protein adsorption and cell adhesion.[64] Recent biomedical applications of PVA hydrogels have focused on physical cross-linking of the polymer chains brought about by chain entanglements.[65] The utility of PVA physical hydrogels is enhanced by submolecular, molecular and supramolecular engineering of these materials. Along with newly developed bioconjugation techniques, this transformation from nanoscale engineering to microscale precision and to macroscale materials design makes PVA a useful biomaterial for many different biomedical applications.

8.2.2.2 Biodegradable Hydrogels

8.2.2.2.1 Poly(Lactic Acid). Poly(lactic acid) or polylactide (PLA) is a thermoplastic aliphatic polyester which is derived from renewable resources, such as corn starch, tapioca products or sugar cane. Lactic acid is produced from bacterial fermentation of corn starch or cane sugar. Two lactic acid molecules undergo a single esterification and then catalytically cyclize to make a cyclic lactide ester. PLA of high molecular weight is produced from the dilactate ester by a ring-opening polymerization using a catalyst such as tin(II) octoate [tin(II) 2-ethylhexanoate]. The biodegradability of the PLA-based polymers have been studied under *in vitro* and *in vivo* conditions.[66] These PLA-containing hydrogels have been used for both drug delivery and tissue engineering applications.[67-71]

8.2.2.2.2 Poly(ε-caprolactone). Poly(ε-caprolactone) (PCL) is a FDA approved biodegradable polyester with a low melting point of around 60 °C and a glass transition temperature of about $-60\,^{\circ}\text{C}$. PCL is prepared by ring-opening polymerization of ε-caprolactone using a catalyst such as tin(II) octoate. PCL is an implantable biomaterial that can be degraded by hydrolysis of its ester linkages in physiological conditions. It degrades much more slowly that PLA and is interesting for the fabrication of long-term implantable devices. PCL-based hydrogels have been investigated for tissue engineering applications.[72–75]

8.2.2.2.3 Poly[(R)-3-hydroxybutyrate]. Poly[(R)-3-hydroxybutyrate] (PHB) is a natural biodegradable polyester which is highly crystalline and hydrophobic, showing greater hydrophobicity than either poly(lactic acid) or poly(ε-caprolactone).[76,77] PHB-based hydrogel films have been prepared for tissue engineering applications.[8,9,78–80] Additionally, cell growth on PHB-containing surfaces appears to be enhanced compared with the surfaces without PHB.[10,76,77,80]

8.2.2.3 Peptide Hydrogels

Peptides are self-assembling motifs that can form extended 3D structures due to their propensity for hydrogen bonding. Peptide folding is the physical process by which a peptide structure assumes its functional 3D shape from a random coil conformation. The folding of peptides can be exploited for the cross-linking of the chains for the formation of peptide hydrogels. Peptide hydrogels have been widely studied for tissue engineering applications due to the fact that the ECM is made up of proteins. Peptide hydrogels have been shown to promote the ECM production.[81] The biocompatibilities of these systems are also advantageous for many biomedical applications, such as tissue engineering and drug delivery.[82–87]

8.2.2.4 Stimuli-Responsive Hydrogels

8.2.2.4.1 Pluronics. Pluronics are a class of non-ionic triblock copolymers, composed of a central hydrophobic chain of poly(propylene glycol) (PPG) flanked by two hydrophilic chains of poly(ethylene glycol) (PEG), invented by Irving Schmolka in 1973. The most popular thermogelling form of Pluronics is Pluronics F127, which forms semi-solid gels when it is heated and solubilizes when the temperature is reduced. It has a molecular weight of $12\,700\ \text{g mol}^{-1}$ and has a PEG:PPG weight ratio of about 2:1. These thermogelling copolymers have been used for a variety of different applications, ranging from wound healing to ophthalmic applications.

8.2.2.4.2 Poly(N-isopropylacrylamide). Poly(N-isopropylacrylamide) (PNI-PAAm) is a temperature-sensitive polymer that has been incorporated into hydrogels to fabricate environmentally sensitive polymer systems for biomedical applications, *e.g.* from tissue engineering, wound healing to drug delivery.[88] PNIPAAm exhibits a temperature-responsive phase transition property similar to other acrylamide analogs. The common characteristic of temperature-sensitive polymers is the presence of hydrophobic groups, such as methyl, ethyl or propyl groups. PNIPAAm has a lower critical solution temperature (LCST) of about 32 °C, being water soluble at temperatures below the LCST and precipitating out of solution above the LCST. PNIPAAm hydrogels have been widely investigated for drug delivery[3,89] and tissue engineering cell sheet detachment technologies.[90-94]

8.3 Applications

8.3.1 Hydrogels for Bone Repair

8.3.1.1 *Bone Fractures, Degenerative Diseases and Conventional Treatments*

8.3.1.1.1 Fractures and Natural Regeneration. Fractures, inflammatory and degenerative problems of bones affect millions of people globally. According to a survey conducted by the Bone and Joint Decade's Musculoskeletal Portal in 2007 (http://www.boneandjointdecade.org), osteogenetic problems account for half of all chronic diseases in people over 50 years old in developed countries. It is predicted that the percentage of people suffering from bone diseases will double by 2020.

For bone fractures, natural healing either occurs *via* direct intramembranous bone formation, given the fractures are well apposed and stabilized, or *via* endochondral bone formation if non-stabilized or large defects need to be bridged.[95,96] The latter go through four temporally distinguishable stages (inflammation, soft callus formation, hard callus formation and remodeling), leading to the re-establishment of functional bone.[95,96] Notably, during bone tissue formation, homeostasis and regeneration, the immobilization and release of growth factors probably depends on the respective microenvironments. By understanding the detailed mechanism of bone tissue remodeling and regeneration, direct intervention in cell signaling events could be made so as to trigger and influence such processes.

8.3.1.1.2 Degenerative Diseases. Common degenerative bone diseases include osteoarthritis, osteoporosis and lumbar spinal stenosis (LSS). In fact, degenerative spinal changes can be observed in 95% of people by the age of 50,[97] and stenosis is the most common cause of serious back pain in adults $\geq$ 60 in age.[98] LSS is caused by the narrowing of the spinal canal and neural foramen, while stenosis leads to the painful condition of neurogenic intermittent claudication (NIC). The degree of stenosis is exacerbated in the extension of the spine, owing to the corresponding increase in neurologic compression.

8.3.1.1.3 Conventional and State-of-the-art Treatments. In cases of serious bone fracture and joint deterioration, bone replacement surgery is a common practice to remedy the situation. In this context, autografts refer to transplantation of bone (or segments of bone) from one part of the body to another (damaged) part within the same patient, while allografts refer to transplantation of bone from a donor to a recipient who is suffering from bone damage. For example, spinal fusion (spondylodesis or spondylosyndesis) and laminotomy are conventional methods to treat spinal injury or diseases such as LSS. Nevertheless, such operations present several limitations, namely donor site scarcity, rejection, disease transfer, harvesting costs and post-operative morbidity.[99–101] On the other hand, various bone fractures, osteoporosis, scoliosis, low back pain and other musculoskeletal problems can be remedied by means of permanent, temporary or biodegradable devices. In such cases, orthopedic biomaterials are implanted into the human body to perform certain biological functions by substituting or repairing different tissues such as bone, cartilage or ligaments and tendons. Recently, the emergence of tissue engineering has been boosted as an alternative potential solution to tissue transplantation and grafting. Research in tissue engineering and regenerative medicine is exploring how to repair and regenerate organs and tissues *via* direct intervention of the natural cell signaling pathways. Natural components such as stem cells, growth factors, gene and peptide sequences are often employed in conjunction with synthetic or biochemically modified natural scaffolds.[102] Although gene therapy and stem/progenitor cell transplantation have shown huge promise in small animal models, thorough and complex investigations in safety and efficacy are inevitable before such concepts become clinically available. For further information on these two approaches, the reader is referred to two specific reviews and references therein.[103,104]

8.3.1.2 Hydrogels as Controlled Delivery Media for Regenerative Medical Treatment

The regeneration of bone tissue over a large-scale defect caused by trauma or disease remains a significant clinical problem. Although autologous bone transplantation, a so-called autograft, remains a typical way to treat bone defects, osteoinductive growth factors have begun to enter clinics. In fact, the discovery of factors that induce bone growth, such as bone morphogenic proteins (BMPs), has generated tremendous opportunities in triggered bone tissue regeneration in defected regions.[105,106] Unfortunately, the promising results from studies in small laboratory animals[107] could only partially be translated to the human situation.[108–110] A major limitation of growth factor therapy is that the required quantities of protein therapeutics for an effective treatment of bone defects exceed the physiological doses by several orders of magnitude, resulting in excessive treatment costs and significant risks of adverse side effects, such as ectopic bone formation or osteolysis.[111] For example, the amount of recombinant BMP-7 (3.5 mg) used for the treatment of a bone defects corresponds to double the entire amount of BMP-7 found in a human being.[112] The cause of the low efficacy in growth factor [*e.g.* vascular endothelial growth factor (VEGF), fibroblast growth factor (FGF) and BMP] treatment is still not entirely clear. Inefficient delivery of growth factors is proposed to be a major cause as this can lead to fast inactivation and clearance of the growth factor.[113,114] Other contributing factors could also be the induction of growth factor antagonists or inhibition of agonists, and limited functionality or availability of responsive (bone progenitor) cells. These issues have motivated the research in novel design of biomaterials, such as hydrogels, for controlled biomolecule delivery from the solid phase as well as guided cell proliferation within the solid matrices (Figure 8.2).

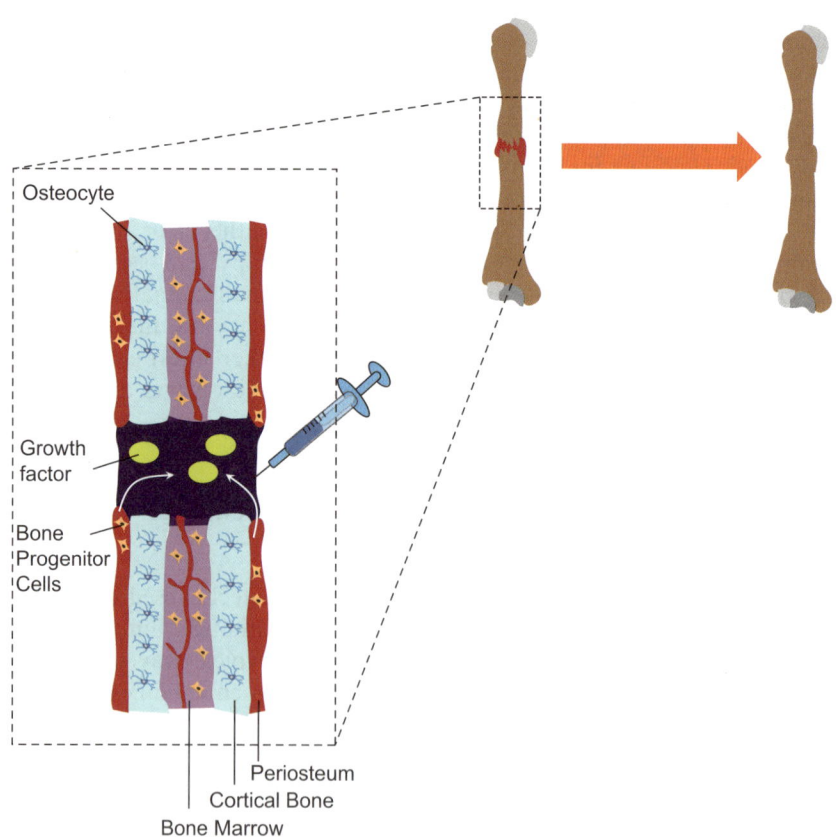

Figure 8.2   Utilizing an injectable hydrogel for bone healing applications.

8.3.1.3 ECM-Mimicking Hydrogel Matrices

Cell adhesion is mediated by protein–cell interactions, specifically biological recognition of protein sequences by transmembrane cell receptors, and can be enhanced by specific sequences [arginine–glycine–aspartic acid (RGD)]. In this context, the design of biomimetic scaffolds is inspired by the role of the natural ECM in guiding cell attachment and regulating tissue regeneration.[115,116] In particular, such scaffolds are considered to take advantage of the bodies' inherent capacity to heal, essentially using the body as a "bioreactor" to regenerate bone.

Common strategies based on this concept involve functionalization of biomaterial surfaces using proteins and peptides that mimic the ECM chemistry, for instance RGD sequences,[117,118] and addition of growth factors into a hydrogel scaffold for a controlled delivery to the surrounding cells. For the latter approach, various growth factors have already been used, including FGF, platelet-derived growth factor and insulin-like growth factor, transforming growth factor (TGF), BMPs, interleukins and interferons.[119,120]

Ideally, delivery of growth factors should be highly localized in the desired region of the body, physiologically present in relevant doses and its activity be preserved in a pre-programmed timeframe. In addition, the delivery scaffolds should provide a favorable biochemical environment to facilitate communication with the cells in the host body for active involvement in the tissue regeneration process. Synthetic hydrogels offer great promise as biomaterials to fulfill the above requirements.[121,122] With highly hydrated and swollen 3D networks of macromolecules, hydrogels can hold up and deliver growth factors (and other biomolecules) with physiologically relevant doses in a localized and sustainable manner. These kinds of hydrogels have been fabricated either from naturally derived biomolecules, such as alginate, collagen and fibrin, or from synthetic polymers, employing chemical or physical cross-linking reactions.[27]

Growth factor loading can be done either in the form of a conventional physical mixture or in the form of chemical/supramolecular tethering. In the former approach, the freely embedded growth factors are released by passive diffusion and material degradation, where release kinetics can be controlled by changing the growth factor quantity and/or by varying the material degradation rate.[114,123] The following subsections highlight the concepts of growth factor immobilization (and the corresponding release mechanism) *via* covalent grafting and supramolecular binding.

8.3.1.3.1 Tethering of Growth Factors to Hydrogels *via* Covalent Grafting. The covalent grafting of growth factors onto a hydrogel matrix requires complementary chemical functionalities on the growth factor, as well as on the hydrogel scaffold. In particular, growth factors have been chemically modified or genetically engineered to contain functional groups such as thiols, acrylates, azides and glutamine (Gln) tags. Initial examples involved tethering of growth factors to collagen matrices by employing homobifunctional PEG-based cross-linkers containing terminal and primary amine selective succinimidyl groups.[124,125] The PEG-based cross-linkers can (in a statistical manner) simultaneously cross-link the collagen matrices to yield the hydrogel and covalently tether growth factors onto the gel scaffold. Sustained delivery of tethered TGF-β2 and VEGF165 resulted in an enhanced and prolonged response *in vivo*, compared to the unmodified growth factors.

The above strategies have the advantage of wide applicability towards a broad spectrum of proteins. Nevertheless, the exact site and the stoichiometry of modification are often difficult to control and thus may adversely interfere with the bioactivity of the tethered growth factors.[126,127] In order to avoid such potential problems, novel strategies for site-specific modification of growth factors have been developed. For instance, introduction of an exogenous cysteine to the C-terminus of VEGF121 and VEGF165 by recombinant technology has been achieved.[128] While reduced (*i.e.* unpaired) cysteines are of relatively low abundance in natural proteins, the exogenous cysteine can serve as a well-defined anchoring point for covalent tethering to the PEG macromers *via* a Michael-type addition reaction with vinyl sulfone groups on the PEG scaffolds. Release of covalently bound VEGF was shown to be mediated by matrix metalloproteinases (MMP) or plasmin-mediated gel degradation.

Growth factor release can be further engineered by the introduction of a plasmin-sensitive linker domain between the growth factor and a Gln-containing peptide within a modified fibrin gel matrix.[129] The tethered growth factors can be released *via* the cleavage of the immobilization domain induced by the localized cell-mediated proteolytic activity of recruited cells. Such cell-demanded release is highly sustained and requires only low doses of growth factors. Through this approach, release of the tethered growth factor cargo from fibrin matrices is triggered by proteolytic activity of invading cells, and thus is synchronized with the tissue morphogenesis. The Gln sequence has been grafted onto a wide variety of growth factors by recombinant engineering, including beta-nerve growth factor (β-NGF),[130] VEGF121,[129,131–133] truncated angiopoietin-1 (ΔAng-1),[134] ephrin-B2,[135] the sixth Ig-like domain of L1 (L1Ig6),[136–138] parathyroid hormone (PTH),[139] IGF-1[140] and BMP.[141]

8.3.1.3.2 Tethering of Growth Factors to Hydrogels *via* Supramolecular Binding. The requirement of chemical or genetic modifications of growth factors for covalent immobilization renders it both time and labor intensive, and might even compromise the growth factor activity.[142] The development of hydrogels that immobilize growth factors by supramolecular interactions thus offers promising opportunities to overcome such inherent drawbacks of covalent tethering.[143] Chemical/biochemical functionalization is required only on the hydrogel scaffolds and direct chemical or genetic modification of the growth factors is unnecessary. Hence, this approach offers a higher degree of synthetic convenience and flexibility. Moreover, since proteolytic activity is not required for growth factor release, the delivery of growth factors can be orthogonally tailored *via* hydrogel engineering.[142] The above advantages motivate and direct recent research in functionalization of hydrogels by naturally derived growth factor binding components such as heparin, chondroitin sulfate, hyaluronic acid or fibronectin.

An early example based on this approach involves the covalent incorporation of 1-ethyl-3-(3-dimethylaminopropyl)carbodiimide/N-hydroxysuccinimide (EDC/NHS)-activated heparin into collagen matrices using NHS chemistry.[144] Heparin-modified collagen matrices were later used for trapping and delivery of VEGF,[145–147] FGF-2[148,149] and stromal cell-derived factor-1 α (SDF-1α).[150] However, the use of heparin is limited by its anticoagulant properties. It has been shown that heparin-modified collagen matrices reduces the thrombogenic activity of collagen and may therefore prevent platelet adhesion and blood coagulation.[149]

Purely synthetic hydrogels were functionalized with heparin *via* a similar strategy, where amine-functionalized star-PEG was modified with heparin by EDC/sulfo-NHS chemistry.[151] Notably, heparin was also employed both as the main component of the hydrogel matrix and as a growth factor binding site.[152] FGF was immobilized in this assembled hydrogel by heparin binding and released as a function of matrix erosion. In another report, hydrogel formation was mediated *via* cross-linking of heparin functionalized PEG by dimeric, heparin-binding growth factors, as shown for VEGF.[153] Interestingly, growth factor release from the resultant supramolecularly cross-linked hydrogel is mediated *via* receptor-induced gel erosion. Hence, the hydrogels can specifically target cells that express the receptor corresponding to the growth factor used for cross-linking. Nevertheless, simultaneous addition of growth factors and cells during cross-linking represent a potential disadvantage of strategies relying on chemical conjugation of heparin.

8.3.1.4 Hydrogels as Structural Implants

In addition to their artificial analogs, hydrogels are common in nature and constitute a major portion of joint spaces for viscoelasticity and shock absorption.[154] Naturally derived hydrogels, such as hyaluronic acid, can be obtained from bacterial fermentation processes or directly extracted from tissues.[155] Hyaluronic acid-based biopolymers can form gels with up to 99% water content and have molecular weights in the range of hundreds of kDa. Their chemical structures are very complex and often incompletely characterized. Given the fact that the major portion of their weight is in the form of water, a number of hydrogels do not provide high enough mechanical strength for practical structural implants. Nevertheless, a notable class of hydrogels does in fact offer reasonable strength in mechanical compression, tension and shear resistance and thus has been recently developed for medical applications, where mechanical strength and durability is a requirement.[156] An example of such material is hydrolyzed polyacrylonitrile (HPAN). A segmented block copolymer can be designed and synthesized to contain hydrophilic acrylic acid and amide groups (the polar, globular region) and a nitrile crystalline block that provides mechanical stiffness and strength. The pendant nitrile groups can then be hydrolyzed to yield HPAN. Interestingly, the resultant HPAN hydrogel can be shaped into a wide variety of 3D structures, while dehydration of the gel under mechanical forces yields another distinct structure of almost any desired configuration. Upon rehydration, the HPAN gel can attain its original hydrated shape and retains its "memory".[157]

Medical treatments for spinal stenosis range from non-surgical pain management, such as physiotherapy or steroid injections, to serious surgical operations. Decompression laminectomy and fusions are typical surgical means to LSS, yet inherent risks are involved in both kinds of surgery.[158,159] Interspinous spacer devices (ISDs) are a class of lumbar spinal devices which are becoming more popular for treating LSS.[160] An ISD distracts the two adjacent spinous processes at the afflicted level and prevents the pathological extension. However, ISD surgery has recently been associated with a higher rate of early post-operative spinous process fractures. While often concealed by the metallic wings of the implants, these fractures are difficult to discover.[161]

Hydrogels offer an appealing opportunity in ISD applications, on top of the conventional ISD based on metal or hard plastics such as poly(ether ether ketone) (PEEK). In particular, owing to their viscoelasticity, HPAN hydrogels can conform to the bony anatomy of the interspinous space and allow normal motion while selectively restricting painful extension. Notably, a HPAN hydrogel device can be surgically implanted in the dehydrated, collapsed state, facilitating minimally invasive surgery and re-form its original shape within eight hours after rehydration by body fluids. A clinical case has been documented for using a HPAN ISD to treat a 59-year-old patient experiencing significant back and leg pain and diagnosed with stenosis at L4/5 and L5/S1 and degenerative disk disease at L5/S1.[157] A marked decrease in back and leg pain was reported in a Visual Analogue Scale (VAS) and Oswestry Disability questionnaires.

8.3.1.5 Multi-Functional Hydrogel Composites

Osseous tissue, the primary tissue of bone, is a relatively hard and lightweight composite material, which is mainly composed of calcium phosphate in the form of calcium hydroxyapatite (HAp). Owing to its excellent bioactivity and biocompatibility, HAp has been widely used as bone implant in clinical settings. Although it provides a favorable environment for osteoconduction, protein adhesion and osteoblast proliferation,[162–164] the application as implants is limited by its brittleness. Over the last decade, effort has been made to overcome such problems with the aim of widening the scope of application of HAp-based materials. Dispersion of HAp within a polymer matrix to yield HAp–polymer composites was found to be generally effective. The resulted composite materials have improved mechanical properties with retention of bioactivity and capability for cell attachment and proliferation.[164]

Recent advances allow routine synthesis of nano-hydroxyapatite (n-HAp) *via* hydrothermal precipitation methods.[165–167] The n-HAp can promote the formation of bone-like apatite on its surface and has been widely used in various biomedical fields.[168] On the polymer front, research has been focused on hydrogels which can form *in situ* because they can be applied by non-invasive techniques such as injection to the desired region of the body and rapidly set to form a scaffold in any shape of defect. In particular, thermosensitive hydrogels have received much attention in orthopedic tissue engineering as they offer convenience in practical treatments, where the hydrogels can be injected *in vivo* as a liquid (at a different temperature) with *in situ* gelation at physiological temperature without the need of any additional chemical or photo triggers.[89] Here, hydrogels can serve as an injectable scaffold for the delivery of a wide variety inorganic materials and even living cells.

A novel composite material based on the biodegradable and thermosensitive triblock copolymer poly(ethylene glycol)-poly(ε-caprolactone)-poly(ethylene glycol) (PEG-PCL-PEG, PECE) and n-HAp has recently been reported by Qian *et al.*[169] The gelation temperature of the composite can be tailored by adjusting the PECE:n-HAp ratio. In addition, the 3D interconnected porous structures of n-HAp-PECE hydrogel nanocomposites suggested that they could be used as scaffold material for cell proliferation. On the other hand, an injectable cement bio-composite has been prepared by the introduction of undifferentiated bone marrow stromal cells into a hydrogel-containing CaP particles in suspension.[170] After implantation of this bio-composite material in a mouse model, *in vivo* results revealed a good vascularization and integration into the host tissue.

8.3.2 Hydrogels for Cardiac Regeneration

Modern unhealthy lifestyle and eating patterns have led to more people to suffer from heart-related ailments. Ischemic heart disease is the worldwide leading cause of death.[171] Myocardial infarction (MI), or heart attack, restricts the amount of blood to the heart. This causes cardiac tissue to die and eventually leads to heart failure. Recently, new strategies such as regenerative cell therapies for repairing cardiac tissues have emerged, opening up new and exciting avenues of research in this direction. The direct application of cells results in poor cardiac performance due to unsatisfactory cell retention. Hydrogel-based biomaterials can be used to improve the cell retention, survival and differentiation of cardiac cells or stem cells. Hydrogels present a tissue-like interior environment with high water content which is very similar to that of the ECM of cardiomyocytes. Furthermore, the mechanical properties of hydrogels are very similar to that of the heart tissue. Hydrogels made from hydrophilic and biocompatible polymers or peptides are excellent candidates as cell delivery agents because they can form gels *in situ*.[89,172,173]

Hydrogels are attractive candidates for this application because of their moldability and their easy chemical modification. For example, targeting ligands can be attached to hydrogels for controlled drug delivery.[174,175] Cell guiding surfaces can be generated by modifying the surface of the

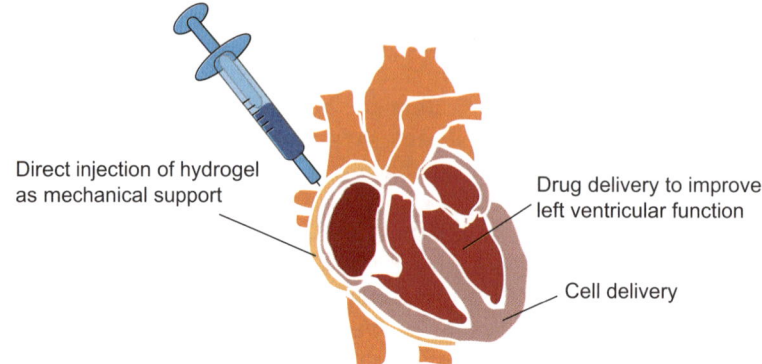

Direct injection of hydrogel
as mechanical support

Drug delivery to improve
left ventricular function

Cell delivery

Figure 8.3   Different approaches to using a hydrogel for heart tissue regeneration
therapies.

hydrogels with different chemical gradients. Stem cell differentiation can be
controlled by microfabrication of hydrogels, generating different shapes and
dimensions.[176–178] Furthermore, the use of injectable cardiac cell hydrogels
depots for the repair of cardiac tissues presents a less invasive approach
compared to that of an *in vitro* engineered epicardial patch implantation. This
section describes the major advances achieved in this area of cardiac tissue
regeneration with the aid of hydrogels (Figure 8.3).

8.3.2.1   Direct Injection of Acellular Hydrogel Materials

Hydrogels can be implanted to provide mechanical and structural support to
the heart after MI to reduce wall stress and to prevent ventricle remodeling.[179]
Collagen gel has been used in such an application with the result of improving
the left ventricular (LV) stroke volume and ejection fraction and preventing
paradoxical systolic bulging.[180] Alginate has been used in myocardial tissue
engineering, resulting in improved LV function compared to that achieved by
neonatal cardiomyocyte transplantation.[181] By chemically conjugating an
arginine–glycine–asparagine peptide (RGD) to the alginate hydrogel, angiogenesis induction and enhanced LV function was observed in a chronic rodent
model of ischemic cardiomyopathy.[182] A fibrin–alginate composite hydrogel
was developed and was found to prevent infarct expansion in a pig MI
model.[183]

Hydrogels derived from natural polymers promote cell recruitment, reduce
inflammatory response and promote the formation of new blood vessels. A
hyaluronic acid-based *in situ* forming hydrogel was developed for cardiac
regeneration.[184] Four weeks after the implantation of the hydrogel, the HA-
based hydrogel alone regenerated the myocardial structure. Fibrous tissue
formation was prevented and heart function was significantly recovered in a rat
MI model. A cell interactive RAD16-II peptide that could form a 3D hydrogel

structure was shown to induce the infiltration of progenitor cells and promote the vascular structure formation in a mouse MI model.[185] Fibrin glue has been injected into rat myocardium post-infarction. Wall thickness and cardiac function were preserved and neovasculature formation appeared to be observed within the infarct scar.[186,187]

8.3.2.2 Controlled Drug Delivery with Hydrogels

The modulation of the ischemic microenvironment of an infarcted heart can be achieved using growth factors. These cytokines induce cell homing, survival, growth and differentiation. The encapsulation of these biofactors in gel depots allows their controlled release as well as preserving their bioactivities. An important cytokine involved in angiogenesis is the basic fibroblast growth factor (bFGF). It has been shown that the bioactivity of bFGF can be enhanced when encapsulated in hydrogels.[13] The perivascular delivery of bFGF encapsulated in heparin–alginate hydrogel microspheres was performed in a pig model of chronic MI, leading to a significant improvement in myocardial function.[188] Heparin–alginate microcapsules containing bFGF have been implanted in ischemic and viable but ungraftable myocardial areas in a phase I clinical trial, proving to be safe and effective.[189] bFGF-loaded gelatin hydrogel microspheres improved the cardiac function in the infarcted myocardium of both rat and pig models following the intramyocardial injection (4 weeks post-infarction).[190,191] An alginate hydrogel was reported for the sequential intra-myocardial delivery of VEGF and platelet-derived growth factor (PDGF) for treating MI.[192] When these two factors were combined, better angiogenic induction and therapeutic efficacy was observed compared to either of the single-factor treatments. The fibrin glue injection of DNA encoding for the angiogenic protein pleitropin into ischemic rat myocardium promoted neovasculature formation.[193] The sustained release of SDF-1α and PEG/fibrin bi-component hydrogel matrix stimulated higher myocardial recruitment of c-Kit+ cells and improved LV function.[194] Erythropoietin was delivered with an injectable supramolecular α-cyclodextrin pseudo-rotaxane hydrogel system in a rat MI model and this system reduced cell apoptosis, increased neo-vasculature formation, reduced infarct size and improved cardiac function.[195]

8.3.2.3 Cell Delivery with Hydrogels

Injectable hydrogels can be used in cardiac regenerative therapy as cellular delivery systems. Stem cells can be differentiated into cardiogenic cells in the microenvironment of the hydrogel.[196] Fibrin patches embedded with porcine bone marrow-derived mesenchymal stem cells (MSCs) were surgically implanted on the scarred myocardial surface area in pigs.[197] The regeneration of the cardiac cells resulted in improved cardiac function. This cell delivery strategy can be adapted to a catheter-based delivery, opening inroads into even more clinical applications. Co-delivery of human umbilical cord blood mononuclear cells (HUCBCs) and bFGF was performed using a fibrin matrix, leading to enhanced neovascularization, a reduction of the infarct area and apoptosis in the ischemic myocardium.[198] In another study based on the rat MI model, human MSCs encapsulated in RGD-modified alginate hydrogel microspheres resulted in an increase in angiogenesis and cell survival.[199]

The heart's ECM comprises mostly collagens, making collagen an ideal material for cell encapsulation and delivery applications. When a collagen-based matrix patch seeded with HUCBCs was fixed onto the epicardium in combination with HUCBC intramyocardial injection, LV wall thinning was prevented and LV function was preserved.[200] Human bone marrow derived-MSCs entrapped in a rat tail type I collagen gel formed a patch that was subjected to a short-term *in vitro* culture.[201] The epicardial application of the cultured patch in an infarcted rat heart led to a 23% cell engraftment within a week, with significant cell migration into the myocardium. After four weeks, LV wall thickness and function were significantly improved. In a similar study, Matrigel generated a high degree of restoration with an engrafted-to-infarct area ratio of 45%.[202]

Synthetic PEG-based hydrogels are used for cell delivery and can be made cell-responsive by programming chemical cues into the material *via* chemical bioconjugation techniques. A two-component hybrid hydrogel of PEG and fibrinogen designed for cardiac tissue engineering rendered the hydrogel bio-material susceptible to protease degradation and subsequent cell-mediated remodeling.[203–205] Mouse bone marrow-derived MSCs were encapsulated in a hepatocyte growth factor (HGF)-conjugated PEGylated fibrin hydrogel.[206] This system resulted in an inhibition of cell apoptosis and tissue fibrosis in the infarct site and led to improved function of the infarcted heart. In another report, an α-cyclodextrin/PEG-PCL-PEG hydrogel was used for cell trans-plantation therapy for bone marrow derived-MSCs.[207] Four weeks after transplantation, the hydrogel was absorbed with an increase in cell retention and vessel density in the infarcted tissue.

8.3.2.4 In Vitro *Cardiac Tissue Engineering*

Functional myocardial tissue graft can be engineered *in vitro* and applied to replace damaged heart tissue. A patient's own stem cells can be used to dif-ferentiate into functional cardiomyocytes and used to repair the damaged cardiomyocytes. The cardiogenic differentiation of stem cells can be induced by biochemical, topographic and physical factors.[208,209] Stem cells cultured in a 3D hydrogel cell culture environment system can undergo *in vitro* cell expan-sion and cardiogenic differentiation. Skeletal muscle-derived stem cells (MDSCs) were encapsulated in collagen gel to strengthen the cell–cell inter-actions, and resulted in the increased expression of cardiac genes, such as connexin 43 and cardiac troponin-T. Contractile force and intracellular cal-cium ion transients similar to those of native cardiac cells in the same gel system were observed.[210] Chemically and physically controllable cardiac differentiation *in vitro* was demonstrated using cell adhesive peptide conjugated to PEG-based hydrogels. In addition, matrix elasticity and peptide concentration plays an important role in controlling the differentiation behavior of the encapsulated pluripotent P19 embryonal carcinoma cells.[211] The alignment of the myocytes and the matrix fibers in the native myocardium results in mechanical and electrical anisotropy.[212]

In order to mimic the design of an artificial myocardium for functional improvement, a fibrin gel-based culture system, with both aligned and isotropically cultured cardiac cells, was studied.[213] The twitch force associated with electrical pacing was significantly increased in the aligned culture, possibly due to the up-regulated gene expression of connexin 43. Additionally, the culture of different cell types is possible with the hydrogel system, promoting cell–cell interactions and cell organization within the 3D environment. Using a peptide hydrogel system, cardiomyocytes were co-cultured with endothelial cells. The formation of capillary-like networks by the endothelial cells promoted cardiomyocyte reorganization along the endothelial cells.[214] A self-assembling peptide-based hydrogel (Puramatrix™) was used for the co-culture of cardiomyocytes and cardiac fibroblasts and an MMP-dependent cell alignment was observed.[215] Alignment of cardiac side population (SP) cells was done by restraining cells within a 3D micropatterned gelatin hydrogel system.[216] A micropatterned HA hydrogel and fibronectin coating could be used for the organized growth of cardiac organoids of several millimeters in length.[217]

Mechanical stimulation of cardiomyocytes cultured in a collagen gel resulted in the formation of cardiac muscle bundles which resemble adult cardiac tissue.[218] In another approach, cyclic mechanical stretch was applied to neonatal rat cardiomyocytes in collagen/Matrigel hybrid gels during *in vitro* culture and generated large, contractile heart grafts. These grafts supported contractile function of infarcted hearts.[219] Large-scale engineering of myocardial cell sheets for cell grafting can also be achieved using the cell sheet technique.[220] A temperature-responsive polymer, PNIPAAm, was grafted onto tissue culture dishes. At 37 °C, the surfaces are hydrophobic and cell adhesive; when cooled below 32 °C, the surfaces becomes hydrophilic and the surface becomes non-adhesive. A cardiomyocyte cell sheet was prepared for the construction of a cardiac tissue by layering several cell sheets.[220] Cardiac patches fabricated by layering cell sheets can be directly applied for cell transplantation, showing good engraftment.[221]

8.3.3 Hydrogels for Glaucoma Treatment

Glaucoma is the second leading cause of irreversible blindness and about 67 million people worldwide are afflicted with the disease.[222–224] Glaucoma is caused by damage to the optic nerve, which could permanently damage vision in the affected eye and lead to blindness if left untreated. Typically, it is associated with increased intraocular pressure (IOP) in the aqueous humor. Current glaucoma therapy relies on drugs that lower the IOP, and there are several glaucoma medications which are effective at lowering the IOP. However, patient compliance is often poor, leading to an ineffective treatment program. Novel delivery systems have great potential to improve patient compliance by providing local, sustained delivery of the drug while reducing side effects. There are many drugs in existence for glaucoma, and the major challenge lies in the delivery of the drug. With the correct drug delivery system in place, significantly improved patient care and clinical outcomes can be achieved. When deciding on suitable drug delivery systems for glaucoma treatment, it is important to understand the chemical structure and mechanism of the specific drug to be delivered. Effective topical glaucoma medications that lower the IOP include prostaglandin analogs (latanoprost), β-blockers (timolol), α-adrenergics (brimonidine), carbonic anhydrase inhibitors (dorzolamide) and cholinergics (pilocarpine). Hydrogels are typically used in the treatment of glaucoma as drug carriers.

8.3.3.1   Drug Delivery from Hydrogels for Glaucoma Treatment

Contact lenses are interesting candidates for the delivery of glaucoma drugs because of their wide use among patients.[225] As soft contact lenses are made from cross-linked hydrophilic polymers, water-soluble drugs used in glaucoma treatment tend to elute very quickly from the highly hydrated polymer networks.[2] Soft contact lenses have been used to deliver pilocarpine for the treatment of glaucoma in certain patients.[226,227] However, there have been reports of patient discomfort, spontaneous discharge of the medical device and the lack of vision correction.

Sustained timolol release has been achieved from hydrogels derived from *N*,*N*-diethylacrylamide and methacrylic acid (approximately 24 h).[228] A pilot study of timolol-loaded contact lenses (on three patients) showed that these contact lenses can lower the IOP.[229] Cross-linked gelatin hydrogels were explored as a potential delivery carrier for glaucoma medications.[230] Components such as the hydrophilic methacrylic acid concentration had an effect on the hydrogel water content and the delivery of timolol maleate from the contact lenses. When the methacrylic acid content was higher, more drug could be loaded in the gel.[231] Molecular imprinting can be used to control the composition of the polymer, which allows greater control of uptake and release of timolol.[62,228] Timolol was delivered using silicone hydrogel contact lenses to beagle dogs suffering from spontaneous glaucoma. Vitamin E was entrapped in the gels to extend the release duration of the drug. The IOP reduction from the lenses without vitamin E was comparable with that by eye drops with similar drug dosing. The use of the contact lenses resulted in higher bioavailability compared to eye drops. However, inclusion of vitamin E into the lenses did not improve the IOP reduction.[232]

Clinical studies were carried out to investigate the uptake and release kinetics of two common glaucoma drugs, timolol maleate and brimonidine tartrate, delivered using hydrogel contact lenses in a limited number of volunteers.

Vasurfilcon A contact lenses (CIBA Vision) were used for the uptake and release experiments. Vasurfilcon A is a copolymer of methyl methacrylate, vinylpyrrolidone and other methacrylates, with the addition of a proprietary UV-absorbing monomer. The maximum uptake and release of both drugs were completed within 60 minutes. The use of the lenses maintained the IOP at levels equivalent to those obtained with previous treatments. Gelatin hydrogel (GH) containing a chymase inhibitor (CI) was applied in a canine model of glaucoma surgery and its effects on the IOP and conjunctival scarring were studied. Implanting GH alone maintained the IOP reduction, whereas GH containing a CI enhanced the IOP-reducing effect by suppressing cell proliferation. This drug delivery system might be useful for maintaining filtering blebs for a longer duration after glaucoma surgery.[233]

A biocompatible hydrogel self-assembled from a peptide that has a peptide backbone containing an Arg-Gly-Asp (RGD) sequence and a hydrophobic *N*-fluorenyl-9-methoxycarbonyl (FMOC) tail was loaded with 5-fluorouracil (5-Fu). The peptide hydrogel was administered in the filtering surgery of rabbit eyes. Administration of the hydrogel without any drug led to low levels of connective tissue growth factor (CTGF) mRNA as well as reduced IOP at 21 days after the operation. The sustained release of 5-Fu from the hydrogel inhibited the scleral flap fibrosis; pathology and immunohistochemistry showed that the filtration fistula is patent without postoperative scarring formation. The IOP of the rabbit eyes was relatively low at 28 days after the operation. The release of 5-Fu from the hydrogel reduced the toxicity of 5-Fu to the surrounding ocular tissues, suggesting a feasibility of this peptide hydrogel as a potential implanted drug delivery system for the inhibition of postoperative scarring formation.[234,235] Forskolin nanocrystals were encapsulated in Noveon AA-1 polycarbophil/poloxamer 407 hydrogels.[236] These hydrogels were pH-sensitive and thermoreversible with a pH_{gel}/T_{gel} close to eye pH/temperature. The hydrogels sustained the release of forskolin for over 5 h. Studies on dexamethasone-induced glaucomatous rabbits indicated that the intraocular pressure lowering efficacy for nanosuspension/hydrogel systems was 31% and lasted for 12 h, which is significantly better than the effect of traditional eye suspension (18%, 4–6 h).

8.3.4 Hydrogels for Nerve Repair Applications

Damage to the central nervous system (CNS) is a serious clinical problem, causing massive disruption of axonal pathways that control neurologic functions essential for life. The effects are often crippling and could leave the patient in a vegetative state. It has been suggested that about 52 000 spinal injuries occur every year globally.[237] Adult mammalian CNS cannot repair its own nervous structures. Even when the healing of the CNS occurs, the build-up of nonvital connective tissue forms scars that impede the growth of regenerating axons.[238–240] Varying degrees of neurologic deficits result, depending on the topographic level and extent and severity of the injury. CNS damage can result from trauma, degenerative processes or stroke. Surgery involving the physical disturbance and removal of neural tissue can also cause significant damage. For example, CNS damage can occur through focal injury or during elective oncologic surgery involving the removal of a rim of vital tissue around a tumor. The restoration of tissue structure/function and the enhancement of axonal recovery in damaged parts of the CNS can be achieved by neurotransplantation, where the tissue reconstruction of neural parenchyma is achieved by cell replacement.

Damaged parts of the nerve occasionally present a gap which prevents the electrical signals to be transmitted to the neurons (Figure 8.4a). In order to repair the "path", hydrogels have been utilized as nerve conduits to help to bridge the gap (Figure 8.4b). Recently, a new approach based on repair by tissue engineering has been demonstrated. This technique involves the implantation of a 3D polymer hydrogel into the site of injury. The organization of migrating wound-healing cells is induced by the physical properties of the hydrogel, and axons are regenerated within its 3D structure. A polymer hydrogel, NeuroGel™, which has a defined macromolecular structure that promotes tissue building, has been implanted into the brain and spinal cord.

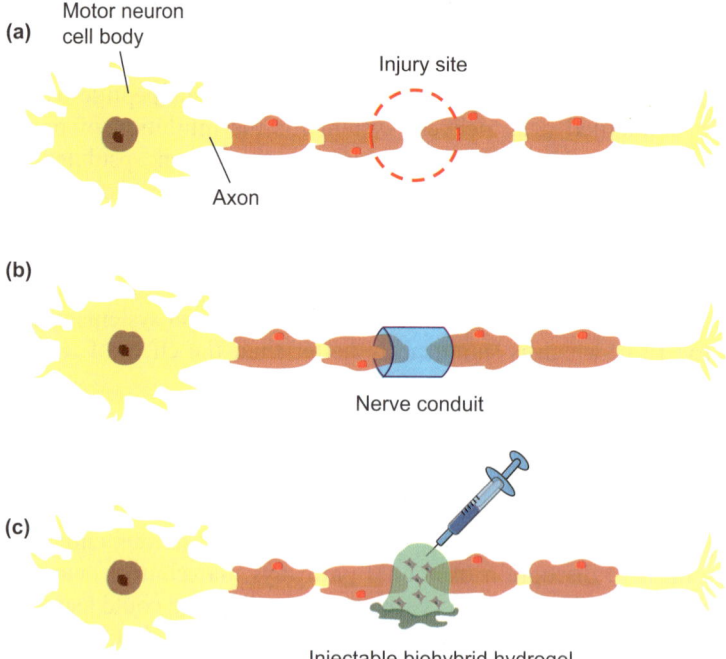

Figure 8.4   Different approaches to using a hydrogel for nerve regeneration therapies: (a) a CNS injury with a severed injury site; (b) using a 3D hydrogel as a nerve conduit to repair the nerve; (c) using an injectable biohybrid hydrogel for nerve regeneration.

The hydrogel derivatives with biologically active molecules further promote selective cell interactions. Biohybrid hydrogels with encapsulated neural tissue cells, embryonic carcinoma-derived neurons or neurotrophic factor secreting cells can also be used for nerve repair protocols (Figure 8.4c). Through these methods, bioartificial tissues with neural tissue specificity can be engineered.

8.3.4.1 Hydrogels as Nerve Regeneration Conduits

Polymer-based porous hydrogels have been explored as a tool in nerve regeneration protocols such as brain lesion repair.[241–244] NeuroGel, a biocompatible hydrogel of poly[*N*-2-(hydroxypropyl)methacrylamide] (PHPMA), was used as a soft superstructure for tissue repair of the CNS.[245] The interior structure of NeuroGel is a mix of pores of different sizes which allow small/large molecules, cells and blood vessels to move freely. The pore structure of NeuroGel forms a tortuous network of channels suitable for cell infiltration. The viscoelastic properties of NeuroGel match those of the host tissue. The recovery of spinal cord injuries (SCI) requires the anatomic restoration of damaged spinal axonal pathways across the lesion and the re-establishment of connectivity. In an acute model of SCI involving a double section of the spinal cord, the cut segment is replaced with a NeuroGel implant to bridge the gap between the two cord segments. This model was applied to cat spinal cord to study the feasibility of reconstructing a tissue defect made by transection at the T6–T7 junction, at two points 3 mm apart, using a NeuroGel implant. A high degree of integration of the polymer implant was observed more than 12 months after reparative surgery. The NeuroGel structure was not observable and the rostral and caudal polymer/spinal cord junctions were difficult to distinguish. Tolerance of the gel implant was excellent without any signs of necrosis, neoplastic growth or other pathologic conditions.

In another study, NeuroGel was implanted into the central cavity of the lesioned spinal cord three months after the compression injury.[246] A severe atrophy of the cord was observed at the traumatized area. The NeuroGel implant was placed into the cavity and the gel was wrapped in the white matter. In this model the gel was placed into an artificially created cavity. Seven months after reconstructive surgery, the gel was observed to have occupied the entire central cavity of the spinal cord and merged with its inner surface. The tissue matrix that formed within the polymer gel was made of astroglia cells that had sent out cytoplasmic expansions between the polymer microspheres and arrays of collagen fibres, thus forming cell bridges between the polymer particles. NeuroGel could be implanted in the developing spinal cord post-natally to restore the neurologic structure and function. This provides a method of repairing localized tissue defects. This could treat severe degenerative alterations associated with spina bifida aperta/cystica, including spine-related disorders such as paraplegia and incontinence.[247]

The midthoracic spinal cord of 2-day-old rats was lesioned and NeuroGel implants were used to bridge the tissue defect. After 4 weeks, excellent integration of NeuroGel into the spinal tissue was observed. Progressive recovery of reflex responses was observed in the animals, suggesting that the developing spinal cord has the capacity of spontaneous recovery of function following injury.[248] NeuroGel was tested for its ability to enhance tissue restoration and axonal regeneration in experimental lesions of the septohippocampal brain system. The regeneration potential of exogenous septal grafts in the presence of NeuroGel for re-innervating the lesioned hippocampus was studied.[249,250] NeuroGel was implanted into a lesion created between the septum and the hippocampus after removing the fimbria-fornix pathway, which includes the septohippocampal cholinergic pathway. Six months after reconstructive surgery, no signs of toxicity were observed on the brain surface. The hydrogel appeared tissue-like due to tissue ingrowth into the porous polymer network. The ingrowth completely filled the lesion cavity, providing a structural continuity across the defect. It was well-integrated with the host's hippocampus, cortex, dorsal striatum, dorsal thalamus and septum. The good integration of the NeuroGel implant and the host parenchyma promoted cell migration into the hydrogel. Next, NeuroGel was implanted into lesions created by aspiration of the right frontal cortex of 17-day-old rats.[242] Merging of the hydrogel implants with surrounding host cerebral cortical tissue was observed, leading to the restoration of the physical continuity across the defect. Seven weeks after reconstructive surgery, immunocytochemistry of the reconstructed brain region revealed a large number of infiltrating astrocytes and processes throughout the hydrogel implant, showing its potential use in repairing defects in the developing brain.

8.3.4.2   Hydrogels as Cell Carriers for Nerve Regeneration

Biohybrid hydrogel systems (BHS) are formed by the combination of polymer hydrogels with living cells and offer another means for structural tissue repair of the CNS. Neurons, astrocytes and Schwann cells can be incorporated in PHPMA polymer matrices to create BHS.[243] There are various methods for retaining cells within polymeric hydrogels: BHS can be created *in vitro* by entrapment at low temperature (cryopolymerization), by rehydrating the gel in a cell suspension, or by microinjection of the gel with a cell suspension.

Gel entrapment involves the simultaneous formation of the macromolecular network and cell immobilization within the forming hydrogel network.[251] The potential of BHS to replace specific tissue functions depends on the nature and functional status of the cells enclosed within the polymer network. As a model system, immature astrocytes and embryonic neurons were entrapped within PHPMA polymer matrices. The formation of BHS using gel entrapment relies on introducing developing cells into the reaction mixture prior to polymerization. This approach is limited by the possible loss of physiological or functional activity of the entrapped cells. However, this may be overcome by controlling the chemical properties of the pre-polymerization mixture. Immature astrocytes and neuroblasts were isolated from cerebral hemispheres of 17-day-old rat embryos. Immobilization of the cells within polymeric hydrogels was done by cryopolymerization of an oxygen-free solution of the monomer

HPMA diluted in Hank's balanced salt solution (HBSS) in the presence of a cross-linking agent and a redox initiating system. Polymerization at low temperatures allows cell inactivation to be minimized. Embryonic CNS cells were successfully entrapped within hydrogels and can survive and differentiate normally, as observed under monolayer culture conditions.

BHS can also be prepared by cell absorption, which was done with a P19 embryonal carcinoma cell clone treated with retinoic acid. This method for introducing cells into the pores of the polymer gel relies on capillary forces.[244] Glycidyl methacrylate and PHPMA gels were dehydrated and then incubated for 12 h in a cell suspension at a concentration of 10^6 cells mL^{-1}. On rehydration and swelling, the cells were incorporated into the pores of the gel network. Alternatively, cells can be microinjected into hydrogels.[251,252] Taking the study *in vivo*, BHS of PHPMA containing neonate Hoechst-labelled Schwann cells were prepared by gel entrapment and were implanted into lesion cavities of the frontal cortex of adult rats.[241] Ten months after implantation, there was evidence of axonal ingrowth from host tissue into the BHS implant, suggesting that Schwann cells promoted their regrowth within the implant. There was also no significant scarring that might restrict axonal regeneration.

Ciliary neurotrophic factor was introduced into preformed PHPMA hydrogels and the BHS were transplanted into surgically created cavities in the optic tracts of neonatal rats between the dorsal lateral geniculate nucleus and the superior colliculus.[253] Results showed that cells differentiated in the hydrogel implant and induced long-distance extension of regenerating axons in the optic tract. There was a massive increase of axonal growth in the gels acting as an *in vivo* "delivery system" of neurotrophic factors. Schwann cells have been entrapped within PHEMA hydrogels containing collagen and were used to bridge lesion cavities in the optic tracts of rats.[254] Cells survived for several weeks after implantation and promoted axonal growth within the polymer matrix and, eventually, myelination of the cylindraxes. In another study, nerve cells were immobilized within agarose gels made of polysaccharides.[255] However, agarose gels are thermoreversible polymers not stable enough for long-term *in vivo* implantation and they may degrade at body temperature.

8.3.5 Hydrogels for Cartilage Repair

One often hears of the elderly or sportsmen complaining of "pain in the joints". In fact, this condition is so widespread that over 15 million people per year are affected by this medical condition. The root cause of this is cartilage degeneration. However, this condition can either be caused by old age or by traumatic events. Articular cartilage degeneration is part of a clinical condition known as osteoarthritis (OA). OA causes chronic musculoskeletal pain and mobility disability. On the other hand, vigorous sporting activities can cause traumatic injuries that lead to either chondral or osteochondral cartilage focal lesions. Chondral lesions stay wholly within the cartilage and do not heal spontaneously. However, osteochondral lesions can be invaded by mesenchymal chondroprogenitor cells and form cartilage that penetrates through the vascularized subchondral bone, leading to a certain extent of spontaneous repair.[256]

Cartilage can be replaced either by autograft or with a full allograft, although synthetic prosthetic cartilages have also been considered. Autografts can be obtained from cartilage taken from the patella, the femoral condyle or the proximal fibula.[257,258] However, these are limited due to the small amount of cartilage available in the body for transplantation. Furthermore, tissue obtained from a site that bears less weight may not be able to withstand the forces imparted at joint surfaces.[259,260] While allograft transplantations have shown long-term efficacy, the induction of immune response remains an issue.[257,258] Cartilage degeneration can be treated by creating a full thickness defect through the subchondral bone. When the site is drilled, a clot forms over the bone surface, which can provide a scaffold for the migration of MSCs and allows the subsequent differentiation to chondrocytes and osteocytes.[260] However, the long-term effects of these protocols are still being debated.

The use of an artificial matrix for filling up small focal defects appears to be attractive for treatment of defects which, if left untreated, could lead to osteoarthritis. Hydrogels applied to the injury site directly could improve mechanical stability and facilitate tissue regeneration. Mechanically induced cartilage erosion could be counteracted and may prevent the development of osteoarthritis. Ideally, the hydrogel should also be an adhesive compatible material that can be applied in analogy to wound dressings.[261] The material should ideally be a fluid initially so as to be able to fill up the irregular defects (Figure 8.5). The hydrogel should be able to provide mechanical support as well

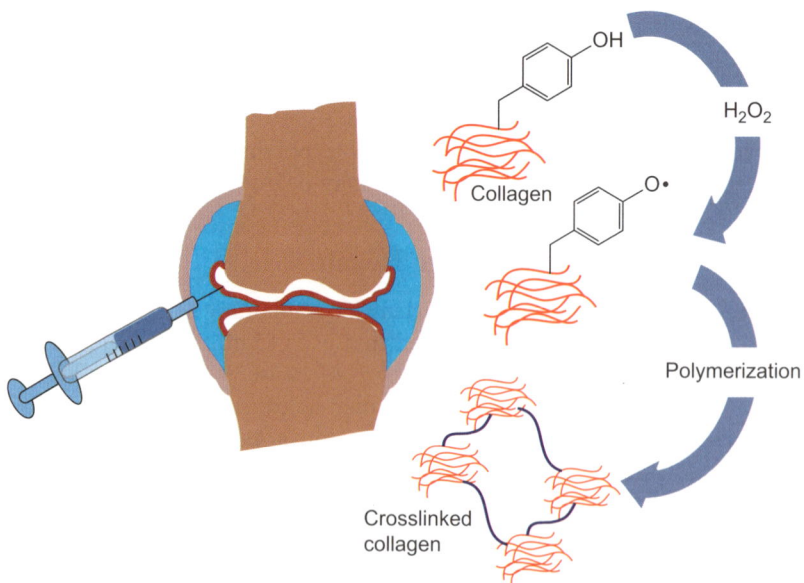

Figure 8.5   General strategy for using injectable hydrogels for cartilage repair.

as cartilaginous matrix properties. Typical approaches involve the injection of a hydrophilic polymer containing tyrosine groups. When an oxidizing agent such as hydrogen peroxide is injected, the tyrosine groups are oxidized to form tyrosine radicals. These radicals form on the polymer component as well as the collagen in the cartilage. The radicals react with each other in a cross-linking reaction, forming a gel-like network structure.

8.3.5.1 Conventional Hydrogels as Prosthetic Cartilage Biomaterials

PVA hydrogels are attractive replacement materials for articular cartilage. Articular cartilage is a natural fiber-reinforced hydrogel made of proteoglycans and type II collagen. PVA hydrogels may be prepared to have the same water content as articular cartilage and possess a low coefficient of friction, which is an important characteristic for lubrication of articular joints.[262] The biocompatibility of PVA hydrogels has been reported, with no inflammatory or degenerative changes in the articular cartilage or synovial membrane surrounding PVA hydrogel implants after 8–52 weeks.[63] The mechanical properties of PVA hydrogels still require enhancement to withstand the severe loading conditions imposed on articular joint surfaces.[63,263,264] Strategies such as the introduction of composite materials like rubber or glass, the use of cross-linking agents such as glutaraldehyde, and the use of freeze–thawing procedures to induce partial crystallinity have been attempted to improve the mechanical properties of the hydrogels.[262,265,266] A PVA hydrogel marketed under the name Salubria™ (Salumedica, Atlanta) has water content similar to natural cartilage tissue and is mechanically strong and biocompatible.[267] It has been proposed to be potentially strong enough to be used as an artificial vascular graft. Salubria is made of a freeze–thawed organic polymer containing the PVA. The mechanical properties of the Salubria biomaterial are influenced by the water and PVA weight ratio, the molecular weight of the PVA, the number of freeze–thaw cycles and the duration of the freeze cycle. The freeze–thaw cycle promotes a mesh entanglement between molecules of PVA to enhance its mechanical strength. This does not rely on traditional cross-linking methods that could potentially include toxic cross-linking agents and thereby decrease the biocompatibility of the resulting biomaterials. This hydrogel can be easily molded into customized shapes as it is in a soft gel state prior to the freeze–thaw processing. As PVA is nondegradable, it can serve as a permanent replacement for damaged articular cartilage or a temporary solution to improve joint function and delay total joint replacement.

8.3.5.2 Injectable Hydrogels for Cartilage Repair

Chitosan derivatives grafted with glycolic acid (GA) and phloretic acid (PA) (CH-GA/PA) form biodegradable injectable chitosan hydrogels through enzymatic cross-linking with horseradish peroxidase (HRP) and H_2O_2.

The *in vitro* culturing of chondrocytes in these hydrogels resulted in viable cells after two weeks.[268] HA grafted with a dextran-tyramine conjugate (Dex-TA) hydrogels were rapidly formed within 2 min *via* enzymatic cross-linking of the tyramine residues in the presence of HRP and H_2O_2.[269] Bovine chondrocytes incorporated in the hydrogels remained viable with enhanced chondrocyte proliferation, and matrix production was observed in the HA-*g*-Dex-TA hydrogels compared to Dex-TA hydrogels. Hydrogels with HA and PEG were also reported.[270] Dex-TA hydrogels could also be cross-linked with similar enzymatic cross-linking reactions for the encapsulation of chondrocytes.[271]

In another study, injectable hydrogels were prepared by similar enzymatic cross-linking of Dex-TA and heparin–tyramine (Hep-TA) conjugates and used as scaffolds for cartilage tissue engineering.[272] Chondrocytes were incorporated in these gels and these hydrogels induced an enhanced production of chondroitin sulfate and a more abundant presence of collagen compared to Dex-TA hydrogels. Small cartilage defects were filled with an acellular injectable hydrogel comprising Dex-TA and Hep-TA that can be applied during arthroscopic procedures.[273] Enhanced adhesion was observed between the cartilage and Dex-TA hydrogels and improved cell growth was observed by the incorporation of Hep-TA conjugates. Interfacial cartilage/hydrogel morphology and Raman spectroscopy revealed the reorganization of the collagen and showed the coupling of tyramine (TA) to tyrosine residues in collagen. Both the *in vitro* and *ex vivo* studies showed covalent bonding of TA-containing hydrogels to tyrosine residues in cartilaginous matrix proteins. The cell-attracting ability of these hydrogels could be exploited to guide tissue repair in focal cartilage defects, preventing or delaying the onset of osteoarthritis.

Peptide hydrogels have been studied for cartilage tissue engineering and regeneration. Primary chondrocytes were entrapped in a peptide solution which undergoes transition and co-acervation at $35\,^\circ$C.[274] The cells and their accumulated matrix could be recovered after 10 days of culture by gentle agitation at room temperature and plated in monolayer culture on Transwell membranes. This work showed that the peptide hydrogels could be used to promote chondrogenesis and be used for cartilage tissue engineering applications. The synthesis and accumulation of articular cartilage ECM for both primary chondrocytes[275] and adult stem cells (hADAS cells) has also been demonstrated with peptide hydrogels.[276] However, these systems possess shear moduli that are four orders of magnitude below that of articular cartilage.[275] Enzymatically cross-linked glutamine and lysine-containing peptide gels were developed to increase load-bearing capabilities.[277] The cross-linked hydrogels have the ability to promote the synthesis and retention of cartilage matrix by entrapped primary cells.[277]

In another work, a lysine-containing elastin-like polypeptide (ELP) with β-[tris(hydroxymethyl)phosphino]propionic acid (betaine) (THPP) was cross-linked.[278–280] The mechanical properties of the ELPs after a five-minute cross-linking reaction that enabled entrapment of cells was three orders of magnitude greater than uncross-linked ELPs.[278] These hydrogels were used as injectable scaffolds for cartilage regeneration in a goat model of an osteochondral defect, resulting in the synthesis of a new cartilage-like matrix.[281] The optimization of the ELPs is required for its use *in vivo*. A slight change in the properties of ELP can lead to very different biological and mechanical outcomes.[282] There is a significant dependence of mechanical properties, chondrocyte viability, metabolism and matrix synthesis on cross-link density (lysine frequency), followed by ELP solution concentration, with little dependence of biological or physical outcomes on molecular weight.[282]

Block copolymers of silk and elastin peptide sequences (SELPs) are produced recombinantly using similar methods to those used to produce ELPs, and contain an amino acid motif from *Bombyx mori* (silkworm) silk (GAGAGS) as well as ELP repeats.[283] These SELPs have been studied as an acellular therapy for cartilage repair in a rabbit model of an osteochondral defect. In this work, injectable SELPs were introduced into osteochondral defects on the femoral condyles of rabbit knees. The material remained in the defect for 12 weeks. This work demonstrates an ability to deliver SELPs in solution phase and trigger a cross-linking reaction and subsequent gel formation *in vivo*. This was further investigated in a goat model. SELPs have been reported to promote cartilage matrix synthesis by human mesenchymal stem cells (hMSCs).[284] hMSCs were encapsulated in the temperature-sensitive peptide gel at 37 °C and cultured for 28 days in the presence of transforming growth factor (TGF-β3). The peptide was found to maintain cell viability with or without TGF-β3, and promote increasing accumulation of glycosaminoglycan (GAG) and collagen over time in the presence of TGF-β3. ELPs created from exons of the human tropoelastin gene and cross-linked with genepin to form a turgid hydrogel have been used for cartilage tissue engineering applications in the knee and intervertebral disc (IVD).[285,286] The implant remained in the defect with no significant inflammation and limited local degradation. A greater amount of new hyaline cartilage deposition was observed compared to unfilled controls.[285] In other studies, a composite system combined EP4 [an ELP which is EP20-24-24-24-24, composed of five hydrophobic (exons 20 and 24 from human elastin) and four cross-linking (exons 21 and 23 from human elastin) domains] with a thiol-modified hyaluronan and a poly(ethylene glycol) diacrylate) (PEGDA) cross-linker for nucleus pulposus repair and/or treatment of early degenerative disc degeneration (DDD).[286] This composite material was evaluated for its ability to support viability and gene expression of NP-associated genes for pathologic human disc cells. These gels were shown to possess an aggregate modulus of 27.6 kPa and support cell viability and phenotype. However, there was limited effect in the injection of the gel into the defect.

8.4 Conclusions and Perspectives

This chapter has provided some insights into the applications of hydrogels in biomedicine and offers a glimpse of this exciting new frontier in the treatment of medical conditions. With its ability to mimic natural biological environments in a controlled way whilst being able to retain and release active pharmacological/bioactive agents, it is able to be part of novel treatment regimens that offer new hope for mankind. Much work has already been done in the areas of drug/cell delivery using hydrogels for cardiac and orthopedic conditions in the treatment of MI, old age degeneration of the cartilage and traumatic bone injuries. Whilst further continuing efforts are currently being applied in these areas, the biocompatibility and bioefficacy of the material with the cell remains a serious issue. Biocompatibility, though necessary, is not a sufficient condition for bioefficacy, which requires a clear understanding of the role of certain functional groups and moieties and how they interact with the biological system. For example, cytokines have been shown to affect the properties of stem cells and direct their differentiation.[287–290] Physical properties such as matrix stiffness are also known to direct cell behavior.[291–299]

Fundamental to all these approaches is the basic understanding of cell signaling and the complex interplay of cellular/biochemical factors which is currently being unveiled by biomedical researchers. Future work on hydrogels can be expected to evolve to incorporate a greater degree of control and tunability over the systems. For example, to create a more cell-friendly environment, biochemical cues can be conjugated to the hydrogels. Targeted drug delivery using hydrogels has been mentioned in many reports, but truly targeting systems have yet to be realized. It remains the grand challenge for oncology research to develop a system that is able to differentiate the cancerous cells from the non-cancerous ones. Smart and responsive hydrogels are also attractive for the potential that they can offer to biomedical applications. Controllable drug release rates, cell-responsive behavior, the ability to adapt to disease and biological signals are some of the capabilities that can be unlocked with further development of smart hydrogels. Above all, these initial steps have only revealed how much more ground needs to be covered, as new horizons emerge from the vantage points of summits already surmounted.

Acknowledgements

Xian Jun Loh is grateful for the invaluable assistance of J. G. Lim and M. J. Loh for proof-reading the chapter, providing critical insights to key concepts in the text and, above all, being there to provide all the emotional support during the preparation of the manuscript.

References

1. K. Y. Lee and D. J. Mooney, *Chem. Rev.*, 2001, **101**, 1869–1879.
2. N. A. Peppas, P. Bures, W. Leobandung and H. Ichikawa, *Eur. J. Pharm. Biopharm.*, 2000, **50**, 27–46.
3. Y. Qiu and K. Park, *Adv. Drug Delivery Rev.*, 2001, **53**, 321–339.
4. S. J. Bae, J. M. Suh, Y. S. Sohn, Y. H. Bae, S. W. Kim and B. Jeong, *Macromolecules*, 2005, **38**, 5260–5265.

5. L. E. Bromberg and E. S. Ron, *Adv. Drug Delivery Rev.*, 1998, **31**, 197–221.
6. M. J. Hwang, J. M. Suh, Y. H. Bae, S. W. Kim and B. Jeong, *Biomacromolecules*, 2005, **6**, 885–890.
7. M. K. Joo, M. H. Park, B. G. Choi and B. Jeong, *J. Mater. Chem.*, 2009, **19**, 5891–5905.
8. X. J. Loh, S. H. Goh and J. Li, *Biomacromolecules*, 2007, **8**, 585–593.
9. X. J. Loh, S. H. Goh and J. Li, *Biomaterials*, 2007, **28**, 4113–4123.
10. X. J. Loh, S. H. Goh and J. Li, *J. Phys. Chem. B*, 2009, **113**, 11822–11830.
11. X. J. Loh and J. Li, *Expert Opin. Ther. Pat.*, 2007, **17**, 965–977.
12. X. J. Loh, Y. X. Tan, Z. Y. Li, L. S. Teo, S. H. Goh and J. Li, *Biomaterials*, 2008, **29**, 2164–2172.
13. X. J. Loh, P. N. N. Vu, N. Y. Kuo and J. Li, *J. Mater. Chem.*, 2011, **21**, 2246–2254.
14. V. P. N. Nguyen, N. Y. Kuo and X. J. Loh, *Soft Matter*, 2011, **7**, 2150–2159.
15. J. Li, X. Li, X. P. Ni, X. Wang, H. Z. Li and K. W. Leong, *Biomaterials*, 2006, **27**, 4132–4140.
16. J. Li, X. P. Ni and K. W. Leong, *J. Biomed. Mater. Res., A*, 2003, **65**, 196–202.
17. X. P. Ni, A. Cheng and J. Li, *J. Biomed. Mater. Res., A*, 2009, **88**, 1031–1036.
18. A. S. Hoffman, *Adv. Drug Delivery Rev.*, 2002, **54**, 3–12.
19. O. Wichterle and D. Lim, *Nature*, 1960, **185**, 117–118.
20. J. Berger, M. Reist, J. M. Mayer, O. Felt and R. Gurny, *Eur. J. Pharm. Biopharm.*, 2004, **57**, 35–52.
21. S. H. Hyon, W. I. Cha, Y. Ikada, M. Kita, Y. Ogura and Y. Honda, *J. Biomater. Sci., Polym. Ed.*, 1994, **5**, 397–406.
22. R. Landers, A. Pfister, U. Hubner, H. John, R. Schmelzeisen and R. Mulhaupt, *J. Mater. Sci.*, 2002, **37**, 3107–3116.
23. J. B. Leach, K. A. Bivens, C. W. Patrick and C. E. Schmidt, *Biotechnol. Bioeng.*, 2003, **82**, 578–589.
24. P. C. Nicolson and J. Vogt, *Biomaterials*, 2001, **22**, 3273–3283.
25. K. H. Schmedlen, K. S. Masters and J. L. West, *Biomaterials*, 2002, **23**, 4325–4332.
26. H. Shin, P. Q. Ruhe, A. G. Mikos and J. A. Jansen, *Biomaterials*, 2003, **24**, 3201–3211.
27. J. A. Hubbell, *Curr. Opin. Biotechnol.*, 2003, **14**, 551–558.
28. S. E. Sakiyama, J. C. Schense and J. A. Hubbell, *FASEB J.*, 1999, **13**, 2214–2224.
29. J. C. Schense, J. Bloch, P. Aebischer and J. A. Hubbell, *Nat. Biotechnol.*, 2000, **18**, 415–419.
30. S. J. Taylor, J. W. McDonald and S. E. Sakiyama-Elbert, *J. Controlled Release*, 2004, **98**, 281–294.
31. E. C. Goldsmith, J. A. Stewart, M. O. Morales and W. E. Carver, *FASEB J.*, 2007, **21**, A972–A973.

32. V. Nehls, R. Herrmann, M. Huhnken and A. Palmetshofer, *Cell Tissue Res.*, 1998, **293**, 479–488.
33. C. H. Lee, A. Singla and Y. Lee, *Int. J. Pharm.*, 2001, **221**, 1–22.
34. C. B. Weinberg and E. Bell, *Science*, 1986, **231**, 397–400.
35. P. Angele, J. Abke, R. Kujat, H. Faltermeier, D. Schumann, M. Nerlich, B. Kinner, C. Englert, Z. Ruszczak, R. Mehrl and R. Mueller, *Biomaterials*, 2004, **25**, 2831–2841.
36. W. F. Daamen, H. T. B. van Moerkerk, T. Hafmans, L. Buttafoco, A. A. Poot, J. H. Veerkamp and T. H. van Kuppevelt, *Biomaterials*, 2003, **24**, 4001–4009.
37. J. Glowacki and S. Mizuno, *Biopolymers*, 2008, **89**, 338–344.
38. J. K. Rao, D. V. Ramesh and K. P. Rao, *Biomaterials*, 1994, **15**, 383–389.
39. S. Woerly, R. Marchand and C. Lavallee, *Biomaterials*, 1991, **12**, 197–203.
40. C. S. Hughes, L. M. Postovit and G. A. Lajoie, *Proteomics*, 2010, **10**, 1886–1890.
41. G. Benton, J. George, H. K. Kleinman and I. P. Arnaoutova, *J. Cell. Physiol.*, 2009, **221**, 18–25.
42. I. Arnaoutova, J. George, H. K. Kleinman and G. Benton, *Angiogenesis*, 2009, **12**, 267–274.
43. S. Schneider, P. Feilen, H. Cramer, M. Hillgartner, F. Brunnenmeier, H. Zimmermann, M. M. Weber and U. Zimmermann, *J. Microencapsul.*, 2003, **20**, 627–636.
44. Y. S. Choi, S. R. Hong, Y. M. Lee, K. W. Song, M. H. Park and Y. S. Nam, *Biomaterials*, 1999, **20**, 409–417.
45. G. S. Schultz, R. G. Sibbald, V. Falanga, E. A. Ayello, C. Dowsett, K. Harding, M. Romanelli, M. C. Stacey, L. Teot and W. Vanscheidt, *Wound Repair Regen.*, 2003, **11**, S1–S28.
46. T. C. Laurent, U. B. G. Laurent and J. R. E. Fraser, *Ann. Rheum. Dis.*, 1995, **54**, 429–432.
47. P. B. Malafaya, G. A. Silva and R. L. Reis, *Adv. Drug Delivery Rev.*, 2007, **59**, 207–233.
48. G. D. Prestwich, D. M. Marecak, J. F. Marecek, K. P. Vercruysse and M. R. Ziebell, *J. Controlled Release*, 1998, **53**, 93–103.
49. E. M. Ehlers, P. Behrens, L. Wunsch, W. Kuhnel and M. Russlies, *Ann. Anat.*, 2001, **183**, 13–17.
50. Y. Ji, K. Ghosh, X. Z. Shu, B. Q. Li, J. C. Sokolov, G. D. Prestwich, R. A. F. Clark and M. H. Rafailovich, *Biomaterials*, 2006, **27**, 3782–3792.
51. J. A. Wieland, T. L. Houchin-Ray and L. D. Shea, *J. Controlled Release*, 2007, **120**, 233–241.
52. F. B. Stillaert, C. Di Bartolo, J. A. Hunt, N. P. Rhodes, E. Tognana, S. Monstrey and P. N. Blondeel, *Biomaterials*, 2008, **29**, 3953–3959.
53. J. Y. Kang, C. W. Chung, J. H. Sung, B. S. Park, J. Y. Choi, S. J. Lee, B. C. Choi, C. K. Shim, S. J. Chung and D. D. Kim, *Int. J. Pharm.*, 2009, **369**, 114–120.
54. D. D. Allison and K. J. Grande-Allen, *Tissue Eng.*, 2006, **12**, 2131–2140.

55. K. Lindenhayn, C. Perka, R. S. Spitzer, H. H. Heilmann, K. Pommer-ening, J. Mennicke and M. Sittinger, *J. Biomed. Mater. Res.*, 1999, **44**, 149–155.
56. D. Campoccia, P. Doherty, M. Radice, P. Brun, G. Abatangelo and D. F. Williams, *Biomaterials*, 1998, **19**, 2101–2127.
57. B. Grigolo, G. Lisignoli, A. Piacentini, M. Fiorini, P. Gobbi, G. Mazzotti, M. Duca, A. Pavesio and A. Facchini, *Biomaterials*, 2002, **23**, 1187–1195.
58. B. Grigolo, L. De Franceschi, L. Roseti, L. Cattini and A. Facchini, *Biomaterials*, 2005, **26**, 5668–5676.
59. B. Grigolo, L. Roseti, M. Fiorini, M. Fini, G. Giavaresi, N. N. Aldini, R. Giardino and A. Facchini, *Biomaterials*, 2001, **22**, 2417–2424.
60. T. V. Chirila, I. J. Constable, G. J. Crawford, S. Vijayasekaran, D. E. Thompson, Y. C. Chen, W. A. Fletcher and B. J. Griffin, *Biomaterials*, 1993, **14**, 26–38.
61. S. H. Gehrke, D. Biren and J. J. Hopkins, *J. Biomater. Sci., Polym. Ed.*, 1994, **6**, 375–390.
62. H. Hiratani and C. Alvarez-Lorenzo, *Biomaterials*, 2004, **25**, 1105–1113.
63. M. Oka, T. Noguchi, P. Kumar, K. Ikeuchi, T. Yamamuro, S. H. Hyon and Y. Ikada, *Clin. Mater.*, 1990, **6**, 361–382.
64. K. Smetana, *Biomaterials*, 1993, **14**, 1046–1050.
65. M. H. Alves, B. E. B. Jensen, A. A. A. Smith and A. N. Zelikin, *Macromol. Biosci.*, 2011, **11**, 1293–1313.
66. K. P. Andriano, Y. Tabata, Y. Ikada and J. Heller, *J. Biomed. Mater. Res.*, 1999, **48**, 602–612.
67. H. J. Chung, Y. H. Lee and T. G. Park, *J. Controlled Release*, 2008, **127**, 22–30.
68. S. J. de Jong, S. C. De Smedt, J. Demeester, C. F. van Nostrum, J. J. Kettenes-van den Bosch and W. E. Hennink, *J. Controlled Release*, 2001, **72**, 47–56.
69. R. Jin, C. Hiemstra, Z. Y. Zhong and J. Feijen, *Biomaterials*, 2007, **28**, 2791–2800.
70. T. Kissel, Y. X. Li and F. Unger, *Adv. Drug Delivery Rev.*, 2002, **54**, 99–134.
71. K. Nagahama, T. Ouchi and Y. Ohya, *Adv. Funct. Mater.*, 2008, **18**, 1220–1231.
72. J. Y. Choi, H. J. Lim, S. I. H. Abdi, H. Y. Chung and J. O. Lim, *J. Tissue Eng. Regener. Med.*, 2011, **8**, A16–A22.
73. Z. A. A. Hamid, A. Blencowe, B. Ozcelik, J. A. Palmer, G. W. Stevens, K. M. Abberton, W. A. Morrison, A. J. Penington and G. G. Qiao, *Biomaterials*, 2010, **31**, 6454–6467.
74. S. J. Im, Y. M. Choi, E. Subrarnanyarn, K. M. Huh and K. Park, *Macromol. Res.*, 2007, **15**, 363–369.
75. A. P. Zhu and M. B. Chan-Park, *J. Biomater. Sci., Polym. Ed.*, 2005, **16**, 301–316.
76. X. J. Loh, W. C. D. Cheong, J. Li and Y. Ito, *Soft Matter*, 2009, **5**, 2937–2946.

77. X. J. Loh, J. S. Gong, M. Sakuragi, T. Kitajima, M. Z. Liu, J. Li and Y. Ito, *Macromol. Biosci.*, 2009, **9**, 1069–1079.
78. X. Li, X. J. Loh, K. Wang, C. B. He and J. Li, *Biomacromolecules*, 2005, **6**, 2740–2747.
79. X. J. Loh, K. K. Tan, X. Li and J. Li, *Biomaterials*, 2006, **27**, 1841–1850.
80. X. J. Loh, X. Wang, H. Z. Li, X. Li and J. Li, *Mater. Sci. Eng., C*, 2007, **27**, 267–273.
81. J. Kisiday, M. Jin, B. Kurz, H. Hung, C. Semino, S. Zhang and A. J. Grodzinsky, *Proc. Natl. Acad. Sci. U. S. A.*, 2002, **99**, 9996–10001.
82. M. A. Bokhari, G. Akay, S. G. Zhang and M. A. Birch, *Biomaterials*, 2005, **26**, 5198–5208.
83. M. C. Branco and J. P. Schneider, *Acta Biomater.*, 2009, **5**, 817–831.
84. S. Koutsopoulos, L. D. Unsworth, Y. Nagaia and S. G. Zhang, *Proc. Natl. Acad. Sci. U. S. A.*, 2009, **106**, 4623–4628.
85. J. K. Kretsinger, L. A. Haines, B. Ozbas, D. J. Pochan and J. P. Schneider, *Biomaterials*, 2005, **26**, 5177–5186.
86. Y. Nagai, L. D. Unsworth, S. Koutsopoulos and S. G. Zhang, *J. Controlled Release*, 2006, **115**, 18–25.
87. D. Seliktar, A. H. Zisch, M. P. Lutolf, J. L. Wrana and J. A. Hubbell, *J. Biomed. Mater. Res., A*, 2004, **68**, 704–716.
88. H. G. Schild, *Prog. Polym. Sci.*, 1992, **17**, 163–249.
89. B. Jeong, S. W. Kim and Y. H. Bae, *Adv. Drug Delivery Rev.*, 2002, **54**, 37–51.
90. C. H. Chen, C. C. Tsai, W. S. Chen, F. L. Mo, H. F. Liang, S. C. Chen and H. W. Sung, *Biomacromolecules*, 2006, **7**, 736–743.
91. K. Haraguchi, T. Takehisa and M. Ebato, *Biomacromolecules*, 2006, **7**, 3267–3275.
92. Y. S. Kim, J. Y. Lim, H. J. Donahue and T. L. Lowe, *Tissue Eng.*, 2005, **11**, 30–40.
93. M. Nitschke, T. Gotze, S. Gramm and C. Werner, *Express Polym. Lett*, 2007, **1**, 660–666.
94. Z. L. Tang, Y. Akiyama, M. Yamato and T. Okano, *Biomaterials*, 2010, **31**, 7435–7443.
95. L. C. Gerstenfeld, D. M. Cullinane, G. L. Barnes, D. T. Graves and T. A. Einhorn, *J. Cell. Biochem.*, 2003, **88**, 873–884.
96. Y. Y. Yu, S. Lieu, C. Y. Lu, T. Miclau, R. S. Marcucio and C. Colnot, *Bone*, 2010, **46**, 841–851.
97. *Lumbar Spinal Stenosis, Your Orthopedic Connection*, orthoinfo.aaos.org/topic.cfm?topic=a00329>; accessed October 15, 2010.
98. J. M. Spivak, *J. Bone Joint Surg. Am.*, 1998, **80**, 1053–1066.
99. J. C. Fernyhough, J. J. Schimandle, M. C. Weigel, C. C. Edwards and A. M. Levine, *Spine (Philadelphia)*, 1992, **17**, 1474–1480.
100. J. C. Banwart, M. A. Asher and R. S. Hassanein, *Spine (Philadelphia)*, 1995, **20**, 1055–1060.
101. J. A. Goulet, L. E. Senunas, G. L. DeSilva and M. L. Greenfield, *Clin. Orthop. Relat. Res.*, 1997, 76–81.

102. P. Hardouin, K. Anselme, B. Flautre, F. Bianchi, G. Bascoulenguet and B. Bouxin, *Joint Bone Spine*, 2000, **67**, 419–424.
103. J. Fischer, A. Kolk, S. Wolfart, C. Pautke, P. H. Warnke, C. Plank and R. Smeets, *J Craniomaxill. Surg.*, 2011, **39**, 54–64.
104. R. E. Guldberg, *J. Bone Miner. Res.*, 2009, **24**, 1507–1511.
105. R. H. Li and J. M. Wozney, *Trends Biotechnol.*, 2001, **19**, 255–265.
106. J. O. Hollinger, J. M. Schmitt, D. C. Buck, R. Shannon, S. P. Joh, H. D. Zegzula and J. Wozney, *J. Biomed. Mater. Res.*, 1998, **43**, 356–364.
107. G. E. Friedlaender, C. R. Perry, J. D. Cole, S. D. Cook, G. Cierny, G. F. Muschler, G. A. Zych, J. H. Calhoun, A. J. LaForte and S. Yin, *J. Bone Jt. Surg., Am. Vol.*, 2001, **83A**, S151–S158.
108. T. A. Einhorn, *J. Bone Jt. Surg., Am. Vol.*, 2003, **85A**, 82–88.
109. O. P. Gautschi, S. P. Frey and R. Zellweger, *Aust. N. Z. J. Surg.*, 2007, **77**, 626–631.
110. P. V. Giannoudis and H. T. Dinopoulos, *J. Orthop. Trauma*, 2010, **24**, S9–S16.
111. E. J. Carragee, E. L. Hurwitz and B. K. Weiner, *Spine J.*, 2011, **11**, 471–491.
112. G. B. Bishop and T. A. Einhorn, *Int. Orthop.*, 2007, **31**, 721–727.
113. M. Simons and J. A. Ware, *Nat. Rev. Drug Discovery*, 2003, **2**, 863–871.
114. R. R. Chen and D. J. Mooney, *Pharm. Res.*, 2003, **20**, 1103–1112.
115. M. R. Lutolf, F. E. Weber, H. G. Schmoekel, J. C. Schense, T. Kohler, R. Muller and J. A. Hubbell, *Nat. Biotechnol.*, 2003, **21**, 513–518.
116. E. S. Place, N. D. Evans and M. M. Stevens, *Nat. Mater.*, 2009, **8**, 457–470.
117. K. C. Dee, T. T. Anderson and R. Bizios, *Biomaterials*, 1999, **20**, 221–227.
118. U. Hersel, C. Dahmen and H. Kessler, *Biomaterials*, 2003, **24**, 4385–4415.
119. C. Nathan and M. Sporn, *J. Cell Biol.*, 1991, **113**, 981–986.
120. U. Ripamonti and J. R. Tasker, *Curr. Pharm. Biotechnol.*, 2000, **1**, 47–55.
121. M. E. Byrne, K. Park and N. A. Peppas, *Adv. Drug Delivery Rev.*, 2002, **54**, 149–161.
122. A. K. A. Silva, C. Richard, M. Bessodes, D. Scherman and O. W. Merten, *Biomacromolecules*, 2009, **10**, 9–18.
123. K. Lee, E. A. Silva and D. J. Mooney, *J. R. Soc., Interface*, 2011, **8**, 153–170.
124. H. Bentz, J. A. Schroeder and T. D. Estridge, *J. Biomed. Mater. Res.*, 1998, **39**, 539–548.
125. S. Koch, C. Yao, G. Grieb, P. Prevel, E. M. Noah and G. C. M. Steffens, *J. Mater. Sci.: Mater. Med.*, 2006, **17**, 735–741.
126. P. Bailon, A. Palleroni, C. A. Schaffer, C. L. Spence, W. J. Fung, J. E. Porter, G. K. Ehrlich, W. Pan, Z. X. Xu, M. W. Modi, A. Farid and W. Berthold, *Bioconjugate Chem.*, 2001, **12**, 195–202.
127. F. M. Veronese, *Biomaterials*, 2001, **22**, 405–417.
128. A. H. Zisch, M. P. Lutolf, M. Ehrbar, G. P. Raeber, S. C. Rizzi, N. Davies, H. Schmokel, D. Bezuidenhout, V. Djonov, P. Zilla and J. A. Hubbell, *FASEB J.*, 2003, **17**, 2260–2262.

129. M. Ehrbar, A. Metters, P. Zammaretti, J. A. Hubbell and A. H. Zisch, *J. Controlled Release*, 2005, **101**, 93–109.
130. S. E. Sakiyama-Elbert, A. Panitch and J. A. Hubbell, *FASEB J.*, 2001, **15**, 1300–1302.
131. M. Ehrbar, V. G. Djonov, C. Schnell, S. A. Tschanz, G. Martiny-Baron, U. Schenk, J. Wood, P. H. Burri, J. A. Hubbell and A. H. Zisch, *Circ. Res.*, 2004, **94**, 1124–1132.
132. M. Ehrbar, S. M. Zeisberger, G. P. Raeber, J. A. Hubbell, C. Schnell and A. H. Zisch, *Biomaterials*, 2008, **29**, 1720–1729.
133. A. H. Zisch, U. Schenk, J. C. Schense, S. E. Sakiyama-Elbert and J. A. Hubbell, *J. Controlled Release*, 2001, **72**, 101–113.
134. C. C. Weber, H. Cai, M. Ehrbar, H. Kubota, G. Martiny-Baron, W. Weber, V. Djonov, E. Weber, A. S. Mallik, M. Fussenegger, K. Frei, J. A. Hubbell and A. H. Zisch, *J. Biol. Chem.*, 2005, **280**, 22445–22453.
135. A. H. Zisch, S. M. Zeisberger, M. Ehrbar, V. Djonov, C. C. Weber, A. Ziemiecki, E. B. Pasquale and J. A. Hubbell, *Biomaterials*, 2004, **25**, 3245–3257.
136. H. Hall and J. A. Hubbell, *Microvasc. Res.*, 2004, **68**, 169–178.
137. R. Pittier, F. Sauthier, J. A. Hubbell and H. Hall, *J. Neurobiol.*, 2005, **63**, 1–14.
138. T. Luhmann, P. Hanseler, B. Grant and H. Hall, *Biomaterials*, 2009, **30**, 4503–4512.
139. I. Arrighi, S. Mark, M. Alvisi, B. von Rechenberg, J. A. Hubbell and J. C. Schense, *Biomaterials*, 2009, **30**, 1763–1771.
140. K. M. Lorentz, L. R. Yang, P. Frey and J. A. Hubbell, *Biomaterials*, 2012, **33**, 494–503.
141. H. G. Schmoekel, F. E. Weber, J. C. Schense, K. W. Gratz, P. Schawalder and J. A. Hubbell, *Biotechnol. Bioeng.*, 2005, **89**, 253–262.
142. C. C. Lin and K. S. Anseth, *Adv. Funct. Mater.*, 2009, **19**, 2325–2331.
143. L. Uebersax, H. P. Merkle and L. Meinel, *Tissue Eng., Part B*, 2009, **15**, 263–289.
144. M. J. B. Wissink, R. Beernink, N. M. Scharenborg, A. A. Poot, G. H. M. Engbers, T. Beugeling, W. G. van Aken and J. Feijen, *J. Controlled Release*, 2000, **67**, 141–155.
145. G. Grieb, A. Groger, A. Piatkowski, M. Markowicz, G. C. M. Steffens and N. Pallua, *Cells Tissues Organs*, 2010, **191**, 96–104.
146. M. Markowicz, A. Heitland, G. C. M. Steffens and N. Pallua, *Int. J. Artif. Organs*, 2005, **28**, 1251–1258.
147. G. C. M. Steffens, C. Yao, P. Prevel, M. Markowicz, P. Schenck, E. M. Noah and N. Pallua, *Tissue Eng.*, 2004, **10**, 1502–1509.
148. P. B. van Wachem, J. A. Plantinga, M. J. B. Wissink, R. Beernink, A. A. Poot, G. H. M. Engbers, T. Beugeling, W. G. van Aken, J. Feijen and M. J. A. van Luyn, *J. Biomed. Mater. Res.*, 2001, **55**, 368–378.
149. M. J. B. Wissink, R. Beernink, J. S. Pieper, A. A. Poot, G. H. M. Engbers, T. Beugeling, W. G. van Aken and J. Feijen, *Biomaterials*, 2001, **22**, 2291–2299.

150. B. A. Bladergroen, B. Siebum, K. G. C. Siebers-Vermeulen, T. H. Van Kuppevelt, A. A. Poot, J. Feijen, C. G. Figdor and R. Torensma, *Tissue Eng., Part A*, 2009, **15**, 1591–1599.
151. U. Freudenberg, A. Hermann, P. B. Welzel, K. Stirl, S. C. Schwarz, M. Grimmer, A. Zieris, W. Panyanuwat, S. Zschoche, D. Meinhold, A. Storch and C. Werner, *Biomaterials*, 2009, **30**, 5049–5060.
152. T. Nie, A. Baldwin, N. Yamaguchi and K. L. Kiick, *J. Controlled Release*, 2007, **122**, 287–296.
153. N. Yamaguchi, L. Zhang, B. S. Chae, C. S. Palla, E. M. Furst and K. L. Kiick, *J. Am. Chem. Soc.*, 2007, **129**, 3040–3041.
154. J. P. G. Urban, A. Maroudas, M. T. Bayliss and J. Dillon, *Biorheology*, 1979, **16**, 447–459.
155. V. Rangaswamy and D. Jain, *Biotechnol. Lett.*, 2008, **30**, 493–496.
156. R. Bertagnoli, C. T. Sabatino, J. T. Edwards, G. A. Gontarz, A. Prewett and J. R. Parsons, *Spine J.*, 2005, **5**, 672–681.
157. C. Lauryssen, J. J. Yue, J. J. Jaramillo-de la Torre, A. Chen and A. Prewett, *Eur. Musculoskeletal Rev.*, 2010, **5**, 36–38.
158. J. Wang, Y. Zhou, Z. F. Zhang, C. Q. Li, W. J. Zheng and J. Liu, *Eur. Spine J.*, 2010, **20**, 623–628.
159. S. Genevay and S. J. Atlas, *Best Pract. Res., Clin. Rheumatol.*, 2010, **24**, 253–265.
160. C. M. Bono and A. R. Vaccaro, *J. Spinal Disord. Tech.*, 2007, **20**, 255–261.
161. D. H. Kim, M. Tantorski, J. Shaw, J. Martha, L. Li, N. Shanti, T. Rencu, S. Parazin and B. Kwon, *Spine (Philadelphia)*, 2010, **36**, E1080–1085.
162. H. R. Lin and Y. J. Yeh, *J. Biomed. Mater. Res., B*, 2004, **71**, 52–65.
163. R. Murugan and S. Ramakrishna, *Biomaterials*, 2004, **25**, 3829–3835.
164. S. C. Rizzi, D. T. Heath, A. G. A. Coombes, N. Bock, M. Textor and S. Downes, *J. Biomed. Mater. Res.*, 2001, **55**, 475–486.
165. D. Choi and P. N. Kumta, *J. Am. Ceram. Soc.*, 2006, **89**, 444–449.
166. B. Li, X. L. Wang, B. Guo, Y. M. Xiao, H. S. Fan and X. D. Zhang, in *Bioceramics 19*, Proceedings of the 19th International Symposium on Ceramics in Medicine, Chengdu, China, October 2006, Trans Tech Publications, Durnten-Zurich, Switzerland, 2007, pp. 235–238.
167. I. Mobasherpour, M. S. Heshajin, A. Kazemzadeh and M. Zakeri, *J. Alloys Compd.*, 2007, **430**, 330–333.
168. A. Sabokbar, R. Pandey, J. Diaz, J. M. W. Quinn, D. W. Murray and N. A. Athanasou, *J. Mater. Sci.: Mater. Med.*, 2001, **12**, 659–664.
169. S. Z. Fu, G. Gun, C. Y. Gong, S. Zeng, H. Liang, F. Luo, X. N. Zhang, X. Zhao, Y. Q. Wei and Z. Y. Qian, *J. Phys. Chem. B*, 2009, **113**, 16518–16525.
170. C. Trojani, F. Boukhechba, J. C. Scimeca, F. Vandenbos, J. F. Michiels, G. Daculsi, P. Boileau, P. Weiss, G. F. Carle and N. Rochet, *Biomaterials*, 2006, **27**, 3256–3264.
171. C. D. Mathers and D. Loncar, *PLoS Med.*, 2006, **3**, 2011–2030.
172. J. L. Drury and D. J. Mooney, *Biomaterials*, 2003, **24**, 4337–4351.

173. K. T. Nguyen and J. L. West, *Biomaterials*, 2002, **23**, 4307–4314.
174. T. Miyata, M. Jige, T. Nakaminami and T. Uragami, *Proc. Natl. Acad. Sci. U. S. A.*, 2006, **103**, 1190–1193.
175. S. V. Vinogradov, T. K. Bronich and A. V. Kabanov, *Adv. Drug Delivery Rev.*, 2002, **54**, 135–147.
176. M. Guvendiren and J. A. Burdick, *Biomaterials*, 2010, **31**, 6511–6518.
177. S. Ilkhanizadeh, A. I. Teixeira and O. Hermanson, *Biomaterials*, 2007, **28**, 3936–3943.
178. R. G. Wylie, S. Ahsan, Y. Aizawa, K. L. Maxwell, C. M. Morshead and M. S. Shoichet, *Nat. Mater.*, 2011, **10**, 799–806.
179. X.-J. Jiang, T. Wang, X.-Y. Li, D.-Q. Wu, Z.-B. Zheng, J.-F. Zhang, J.-L. Chen, B. Peng, H. Jiang, C. Huang and X.-Z. Zhang, *J. Biomed. Mater. Res., A*, 2009, **90**, 472–477.
180. W. D. Dai, L. E. Wold, J. S. Dow and R. A. Kloner, *J. Am. Coll. Cardiol.*, 2005, **46**, 714–719.
181. N. Landa, L. Miller, M. S. Feinberg, R. Holbova, M. Shachar, I. Freeman, S. Cohen and J. Leor, *Circulation*, 2008, **117**, 1388–1396.
182. J. Yu, Y. Gu, K. T. Du, S. Mihardja, R. E. Sievers and R. J. Lee, *Biomaterials*, 2009, **30**, 751–756.
183. R. Mukherjee, J. A. Zavadzkas, S. M. Saunders, J. E. McLean, L. B. Jeffords, C. Beck, R. E. Stroud, A. M. Leone, C. N. Koval, W. T. Rivers, S. Basu, A. Sheehy, G. Michal and F. G. Spinale, *Ann. Thoracic Surg.*, 2008, **86**, 1268–1277.
184. S. J. Yoon, Y. H. Fang, C. H. Lim, B. S. Kim, H. S. Son, Y. Park and K. Sun, *J. Biomed. Mater. Res., B*, 2009, **91**, 163–171.
185. M. E. Davis, J. P. M. Motion, D. A. Narmoneva, T. Takahashi, D. Hakuno, R. D. Kamm, S. G. Zhang and R. T. Lee, *Circulation*, 2005, **111**, 442–450.
186. K. L. Christman, H. H. Fok, R. E. Sievers, Q. H. Fang and R. J. Lee, *Tissue Eng.*, 2004, **10**, 403–409.
187. K. L. Christman, A. J. Vardanian, Q. Z. Fang, R. E. Sievers, H. H. Fok and R. J. Lee, *J. Am. Coll. Cardiol.*, 2004, **44**, 654–660.
188. J. J. Lopez, E. R. Edelman, A. Stamler, M. G. Hibberd, P. Prasad, R. P. Caputo, J. P. Carrozza, P. S. Douglas, F. W. Sellke and M. Simons, *J. Pharmacol. Exp. Ther.*, 1997, **282**, 385–390.
189. R. J. Laham, F. W. Sellke, E. R. Edelman, J. D. Pearlman, J. A. Ware, D. L. Brown, J. P. Gold and M. Simons, *Circulation*, 1999, **100**, 1865–1871.
190. A. Iwakura, M. Fujita, K. Kataoka, K. Tambara, Y. Sakakibara, M. Komeda and Y. Tabata, *Heart Vessels*, 2003, **18**, 93–99.
191. Y. Sakakibara, K. Tambara, G. Sakaguchi, F. L. Lu, M. Yamamoto, K. Nishimura, Y. Tabata and M. Komeda, *Eur. J. Cardio-Thorac. Surg.*, 2003, **24**, 105–111.
192. X. Hao, E. A. Silva, A. Mansson-Broberg, K.-H. Grinnemo, A. J. Siddiqui, G. Dellgren, E. Wardell, L. A. Brodin, D. J. Mooney and C. Sylven, *Cardiovasc. Res.*, 2007, **75**, 178–185.

193. K. L. Christman, Q. Z. Fang, M. S. Yee, K. R. Johnson, R. E. Sievers and R. J. Lee, *Biomaterials*, 2005, **26**, 1139–1144.
194. G. Zhang, Y. Nakamura, X. Wang, Q. Hu, L. J. Suggs and J. Zhang, *Tissue Eng.*, 2007, **13**, 2063–2071.
195. T. Wang, X. J. Jiang, T. Lin, S. Ren, X.-Y. Li, X.-Z. Zhang and Q.-Z. Tang, *Biomaterials*, 2009, **30**, 4161–4167.
196. T. Kofidis, J. L. de Bruin, G. Hoyt, Y. Ho, M. Tanaka, T. Yamane, D. R. Lebl, R. J. Swijnenburg, C. P. Chang, T. Quertermous and R. C. Robbins, *J. Heart Lung Transplant.*, 2005, **24**, 737–744.
197. J. B. Liu, Q. S. Hu, Z. L. Wang, C. S. Xu, X. H. Wang, G. R. Gong, A. Mansoor, J. Lee, M. X. Hou, L. P. Zeng, J. R. Zhang, M. Jerosch-Herold, T. Guo, R. J. Bache and J. Y. Zhang, *Am. J. Physiol.*, 2004, **287**, H501–H511.
198. S.-W. Cho, I.-K. Kim, S. H. Bhang, B. Joung, Y. J. Kim, K. J. Yoo, Y.-S. Yang, C. Y. Choi and B.-S. Kim, *Eur. J. Heart Failure*, 2007, **9**, 974–985.
199. J. Yu, K. T. Du, Q. Fang, Y. Gu, S. S. Mihardja, R. E. Sievers, J. C. Wu and R. J. Lee, *Biomaterials*, 2010, **31**, 7012–7020.
200. M. Cortes-Morichetti, G. Frati, O. Schussler, J.-P. Duong Van Huyen, E. Lauret, J. A. Genovese, A. F. Carpentier and J. C. Chachques, *Tissue Eng.*, 2007, **13**, 2681–2687.
201. D. Simpson, H. Liu, T.-H. M. Fan, R. Nerem and S. C. Dudley, Jr., *Stem Cells*, 2007, **25**, 2350–2357.
202. T. Kofidis, D. R. Lebl, E. C. Martinez, G. Hoyt, M. Tanaka and R. C. Robbins, *Circulation*, 2005, **112**, I173–I177.
203. K. Shapira, D. Dikovsky, M. Habib, L. Gepstein and D. Seliktar, *Biomed. Mat. Eng.*, 2008, **18**, 309–314.
204. K. Shapira-Schweitzer, M. Habib, L. Gepstein and D. Seliktar, *J. Mol. Cell. Cardiol.*, 2009, **46**, 213–224.
205. K. Shapira-Schweitzer and D. Seliktar, *Acta Biomater.*, 2007, **3**, 33–41.
206. G. Zhang, Q. S. Hu, E. A. Braunlin, L. J. Suggs and J. Y. Zhang, *Tissue Eng., Part A*, 2008, **14**, 1025–1036.
207. T. Wang, X. J. Jiang, Q. Z. Tang, X. Y. Li, T. Lin, D. Q. Wu, X. Z. Zhang and E. Okello, *Acta Biomater.*, 2009, **5**, 2939–2944.
208. G. Forte, F. Carotenuro, F. Pagliari, S. Pagliari, P. Cossa, R. Fiaccavento, A. Ahiuwalia, G. Vozzi, B. Vinci, A. Serafino, A. Rinaldi, E. Traversa, L. Carosella, M. Minieri and P. Di Nardo, *Stem Cells*, 2008, **26**, 2093–2103.
209. M. P. Lutolf and H. M. Blau, *Adv. Mater.*, 2009, **21**, 3255–3268.
210. K. C. Clause, J. P. Tinney, L. J. Liu, B. Gharaibeh, J. Huard, J. A. Kirk, S. G. Shroff, K. L. Fujimoto, W. R. Wagner, J. C. Ralphe, B. B. Keller and K. Tobita, *Tissue Eng., Part C*, **16**, 375–385.
211. T. P. Kraehenbuehl, P. Zammaretti, A. J. Van der Vlies, R. G. Schoenmakers, M. P. Lutolf, M. E. Jaconi and J. A. Hubbell, *Biomaterials*, 2008, **29**, 2757–2766.
212. L. E. Freed, G. C. Engelmayr, J. T. Borenstein, F. T. Moutos and F. Guilak, *Adv. Mater.*, 2009, **21**, 3410–3418.

213. L. D. Black, J. D. Meyers, J. S. Weinbaum, Y. A. Shvelidze and R. T. Tranquillo, *Tissue Eng., Part A*, 2009, **15**, 3099–3108.
214. D. A. Narmoneva, R. Vukmirovic, M. E. Davis, R. D. Kamm and R. T. Lee, *Circulation*, 2004, **110**, 962–968.
215. J. W. Nichol, G. C. Engelmayr, M. Y. Cheng and L. E. Freed, *Biochem. Biophys. Res. Commun.*, 2008, **373**, 360–365.
216. H. Aubin, J. W. Nichol, C. B. Hutson, H. Bae, A. L. Sieminski, D. M. Cropek, P. Akhyari and A. Khademhosseini, *Biomaterials*, 2010, **31**, 6941–6951.
217. A. Khademhosseini, G. Eng, J. Yeh, P. A. Kucharczyk, R. Langer, G. Vunjak-Novakovic and M. Radisic, *Biomed. Microdevices*, 2007, **9**, 149–157.
218. W. H. Zimmermann, K. Schneiderbanger, P. Schubert, M. Didie, F. Munzel, J. F. Heubach, S. Kostin, W. L. Neuhuber and T. Eschenhagen, *Circ. Res.*, 2002, **90**, 223–230.
219. W. H. Zimmermann, I. Melnychenko, G. Wasmeier, M. Didie, H. Naito, U. Nixdorff, A. Hess, L. Budinsky, K. Brune, B. Michaelis, S. Dhein, A. Schwoerer, H. Ehmke and T. Eschenhagen, *Nat. Med.*, 2006, **12**, 452–458.
220. T. Shimizu, M. Yamato, A. Kikuchi and T. Okano, *Biomaterials*, 2003, **24**, 2309–2316.
221. S. Masuda, T. Shimizu, M. Yamato and T. Okano, *Adv. Drug Delivery Rev.*, 2008, **60**, 277–285.
222. S. Blomdahl, B. M. Calissendorff, B. Tengroth and O. Wallin, *Acta Ophthalmol. Scand.*, 1997, **75**, 589–591.
223. A. Munier, T. Gunning, D. Kenny and M. O'Keefe, *Br. J. Ophthalmol.*, 1998, **82**, 630–633.
224. H. A. Quigley, *Br. J. Ophthalmol.*, 1996, **80**, 389–393.
225. C. J. White and M. E. Byrne, *Expert Opin. Drug Delivery*, 2010, **7**, 765–780.
226. M. Ruben and R. Watkins, *Br. J. Ophthalmol.*, 1975, **59**, 455–458.
227. J. S. Hillman, *Br. J. Ophthalmol.*, 1974, **58**, 674–679.
228. H. Hiratani and C. Alvarez-Lorenzo, *J. Controlled Release*, 2002, **83**, 223–230.
229. C. L. Schultz, T. R. Poling and J. O. Mint, *Clin. Exp. Optom.*, 2009, **92**, 343–348.
230. M. V. Natu, J. P. Sardinha, I. J. Correia and M. H. Gil, *Biomed. Mater.*, 2007, **2**, 241–249.
231. D. M. Garcia, J. L. Escobar, Y. Noa, N. Bada, E. Hernaez and I. Katime, *Eur. Polym. J.*, 2004, **40**, 1683–1690.
232. C. C. Peng, A. Ben-Shlomo, E. O. Mackay, C. E. Plummer and A. Chauhan, *Curr. Eye Res.*, **37**, 204–211.
233. S. Kojima, T. Sugiyama, S. Takai, D. Jin, M. Shibata, H. Oku, Y. Tabata and T. Ikeda, *Invest. Ophthalmol. Visual Sci.*, 2011, **52**, 7672–7680.
234. X. D. Xu, L. A. Liang, C. S. Chen, B. Lu, N. L. Wang, F. G. Jiang, X. Z. Zhang and R. X. Zhuo, *ACS Appl. Mater. Interfaces*, 2010, **2**, 2663–2671.

235. L. A. Liang, X. D. Xu, X. Z. Zhang, M. Feng, C. Peng and F. G. Jiang, *Biomed. Mat.*, 2010, **5**, 045008.
236. S. Gupta, M. K. Samanta and A. M. Raichur, *AAPS PharmSciTech*, 2010, **11**, 322–335.
237. *Spinal Cord Injury Explained*, http://www.streetsie.com/spinal-cord-injury/; accessed June 12, 2012.
238. M. Berry, W. L. Maxwell, A. Logan, A. Mathewson, P. McConnell, D. E. Ashhurst and G. H. Thomas, *Acta Neurochir. Suppl.*, 1983, **32**, 31–53.
239. J. W. Fawcett and R. A. Asher, *Brain Res. Bull.*, 1999, **49**, 377–391.
240. C. C. Stichel and H. W. Muller, *Cell Tissue Res.*, 1998, **294**, 1–9.
241. S. Woerly, G. Laroche, R. Marchand, J. Pato, V. Subr and K. Ulbrich, *J. Neural Transplant. Plast.*, 1995, **5**, 245–255.
242. S. Woerly, P. Petrov, E. Sykova, T. Roitbak, Z. Simonova and A. R. Harvey, *Tissue Eng.*, 1999, **5**, 467–488.
243. S. Woerly, G. W. Plant and A. R. Harvey, *Biomaterials*, 1996, **17**, 301–310.
244. S. Woerly, K. Ulbrich, V. Chytry, K. Smetana, P. Petrovicky, B. Rihova and D. J. Morassutti, *Cell Transplant.*, 1993, **2**, 229–239.
245. S. Woerly, E. Pinet, L. De Robertis, M. Bousmina, G. Laroche, T. Roitback, L. Vargova and E. Sykova, *J. Biomater. Sci., Polym. Ed.*, 1998, **9**, 681–711.
246. J. Peduzzi, V. D. Doan and S. Woerly, *J. Neurotrauma*, 1999, **16**, 1010.
247. J. M. Mazur and M. B. Menelaus, *Clin. Orthop. Relat. Res.*, 1991, 54–64.
248. E. Kunkelbagden, H. N. Dai and B. S. Bregman, *Exp. Neurol.*, 1992, **116**, 40–51.
249. E. Duconseille, A. Cressant, C. Kelche, S. Woerly, B. Will, B. Poucet and J. C. Cassel, *Restor. Neurol. Neurosci.*, 1999, **15**, 305–317.
250. E. Duconseille, S. Woerly, C. Kelche, B. Will and J. C. Cassel, *Restor. Neurol. Neurosci.*, 1998, **13**, 193–203.
251. S. Woerly, G. W. Plant and A. R. Harvey, *Neurosci. Lett.*, 1996, **205**, 197–201.
252. S. Woerly and D. J. Morassutti, *Neurosurg. Rev.*, 1993, **16**, 93–104.
253. M. K. Panni, J. Atkinson and R. D. Raymond, *Dev. Brain Res.*, 1994, **81**, 325–327.
254. G. W. Plant, T. V. Chirila and A. R. Harvey, *Cell Transplant.*, 1998, **7**, 381–391.
255. R. Bellamkonda, J. P. Ranieri, N. Bouche and P. Aebischer, *J. Biomed. Mater. Res.*, 1995, **29**, 663–671.
256. R. Cancedda, B. Dozin, P. Giannoni and R. Quarto, *Matrix Biol.*, 2003, **22**, 81–91.
257. J. A. Buckwalter and H. J. Mankin, in *Instructional Course Lectures*, The American Academy of Orthopaedic Surgeons, Rosemont, Illinois, 1998, vol. 47, pp. 487–504.
258. S. W. O'Driscoll, *J. Bone Jt. Surg., Am. Vol.*, 1998, **80A**, 1795–1812.
259. E. B. Hunziker, *Osteoarthr. Cartilage*, 1999, **7**, 15–28.
260. J. S. Temenoff and A. G. Mikos, *Biomaterials*, 2000, **21**, 431–440.

261. C. M. Vaz, S. van Tuijl, C. V. C. Bouten and F. P. T. Baaijens, *Acta Biomater.*, 2005, **1**, 575–582.
262. Z.-Q. Gu, J.-M. Xiao and X.-H. Zhang, *Biomed. Mater. Eng.*, 1998, **8**, 75–81.
263. K. S. Anseth, C. N. Bowman and L. BrannonPeppas, *Biomaterials*, 1996, **17**, 1647–1657.
264. P. H. Corkhill, A. S. Trevett and B. J. Tighe, *Proc. Inst. Mech. Eng., Part H*, 1990, **204**, 147–155.
265. J. V. Cauich-Rodriguez, S. Deb and R. Smith, *Biomaterials*, 1996, **17**, 2259–2264.
266. P. Lopour, Z. Plichta, Z. Volfova, P. Hron and P. Vondracek, *Biomaterials*, 1993, **14**, 1051–1055.
267. D. N. Ku, L. G. Braddon and D. M. Wootton, *U. S. Pat.* 5 981 826, 1999.
268. R. Jin, L. S. M. Teixeira, P. J. Dijkstra, M. Karperien, C. A. van Blitterswijk, Z. Y. Zhong and J. Feijen, *Biomaterials*, 2009, **30**, 2544–2551.
269. R. Jin, L. S. M. Teixeira, P. J. Dijkstra, C. A. van Blitterswijk, M. Karperien and J. Feijen, *Biomaterials*, 2010, **31**, 3103–3113.
270. R. Jin, L. S. M. Teixeira, A. Krouwels, P. J. Dijkstra, C. A. van Blitterswijk, M. Karperien and J. Feijen, *Acta Biomater.*, 2010, **6**, 1968–1977.
271. R. Jin, L. S. M. Teixeira, P. J. Dijkstra, Z. Y. Zhong, C. A. van Blitterswijk, M. Karperien and J. Feijen, *Tissue Eng., Part A*, 2010, **16**, 2429–2440.
272. R. Jin, L. S. M. Teixeira, P. J. Dijkstra, C. A. van Blitterswijk, M. Karperien and J. Feijen, *J. Controlled Release*, 2011, **152**, 186–195.
273. L. S. M. Teixeira, S. Bijl, V. V. Pully, C. Otto, J. Rong, J. Feijen, C. A. van Blitterswijk, P. J. Dijkstra and M. Karperien, *Biomaterials*, 2012, **33**, 3164–3174.
274. H. Betre, A. Chilkoti, L. A. Setton and Ieee, in *Second Joint EMBS-BMES Conference 2002*, IEEE, Washington, 2002, vol. 1, pp. 829–830.
275. H. Betre, L. A. Setton, D. E. Meyer and A. Chilkoti, *Biomacromolecules*, 2002, **3**, 910–916.
276. H. Betre, S. R. Ong, F. Guilak, A. Chilkoti, B. Fermor and L. A. Setton, *Biomaterials*, 2006, **27**, 91–99.
277. M. K. McHale, L. A. Setton and A. Chilkoti, *Tissue Eng.*, 2005, **11**, 1768–1779.
278. D. W. Lim, D. L. Nettles, L. A. Setton and A. Chilkoti, *Biomacromolecules*, 2007, **8**, 1463–1470.
279. D. W. Lim, D. L. Nettles, L. A. Setton and A. Chilkoti, *Biomacromolecules*, 2008, **9**, 222–230.
280. K. Trabbic-Carlson, L. A. Setton and A. Chilkoti, *Biomacromolecules*, 2003, **4**, 572–580.
281. D. L. Nettles, K. Kitaoka, N. A. Hanson, C. M. Flahiff, B. A. Mata, E. W. Hsu, A. Chilkoti and L. A. Setton, *Tissue Eng., Part A*, 2008, **14**, 1133–1140.
282. D. L. Nettles, M. A. Haider, A. Chilkoti and L. A. Setton, *Tissue Eng., Part A*, 2010, **16**, 11–20.

283. J. Cappello, J. Crissman, M. Dorman, M. Mikolajczak, G. Textor, M. Marquet and F. Ferrari, *Biotechnol. Prog.*, 1990, **6**, 198–202.

284. M. Haider, J. Cappello, H. Ghandehari and K. W. Leong, *Pharm. Res.*, 2008, **25**, 692–699.

285. C. Hrabchak, J. Rouleau, I. Moss, K. Woodhouse, M. Akens, C. Bellingham, F. Keeley, M. Dennis and A. Yee, *Acta Biomater.*, 2010, **6**, 2108–2115.

286. I. L. Moss, L. Gordon, K. A. Woodhouse, C. M. Whyne and A. J. M. Yee, *Spine*, 2011, **36**, 1022–1029.

287. K. K. Johe, T. G. Hazel, T. Muller, M. M. Dugich-Djordjevic and R. D. G. McKay, *Genes Dev.*, 1996, **10**, 3129–3140.

288. M. P. Lutolf and J. A. Hubbell, *Nat. Biotechnol.*, 2005, **23**, 47–55.

289. B. E. Reubinoff, M. F. Pera, C. Y. Fong, A. Trounson and A. Bongso, *Nat. Biotechnol.*, 2000, **18**, 399–404.

290. K. Willert, J. D. Brown, E. Danenberg, A. W. Duncan, I. L. Weissman, T. Reya, J. R. Yates and R. Nusse, *Nature*, 2003, **423**, 448–452.

291. K. Chatterjee, S. Lin-Gibson, W. E. Wallace, S. H. Parekh, Y. J. Lee, M. T. Cicerone, M. F. Young and C. G. Simon, *Biomaterials*, 2010, **31**, 5051–5062.

292. D. E. Discher, P. Janmey and Y. L. Wang, *Science*, 2005, **310**, 1139–1143.

293. D. E. Discher, D. J. Mooney and P. W. Zandstra, *Science*, 2009, **324**, 1673–1677.

294. A. J. Engler, S. Sen, H. L. Sweeney and D. E. Discher, *Cell*, 2006, **126**, 677–689.

295. B. Geiger, J. P. Spatz and A. D. Bershadsky, *Nat. Rev. Mol. Cell Biol.*, 2009, **10**, 21–33.

296. Y. S. Pek, A. C. A. Wan and J. Y. Ying, *Biomaterials*, 2010, **31**, 385–391.

297. F. Rehfeldt, A. E. X. Brown, M. Raab, S. S. Cai, A. L. Zajac, A. Zemel and D. E. Discher, *Integr. Biol.*, 2012, **4**, 422–430.

298. M. M. Stevens and J. H. George, *Science*, 2005, **310**, 1135–1138.

299. V. Vogel and M. Sheetz, *Nat. Rev. Mol. Cell Biol.*, 2006, **7**, 265–275.

CHAPTER 9

ReGel™ *Hydrogels for* **In Vivo** *Applications*

KIRK FOWERS

BTG International, Inc., Five Tower Bridge, 300 Barr Harbor Drive,
8th Floor, West Conshohocken, PA 19428-2998, USA
Email: kirk.fowers@btgplc.com

9.1 Introduction

Novel drug delivery systems optimize drug release profiles in a variety of
therapeutic areas, and are designed to improve safety, enhance efficacy, and
improve patient compliance. Polymer-based drug delivery systems include
microspheres, nanoparticles, solvent precipitating depots, and environmentally
sensitive systems.[1–6] Environmentally sensitive systems include pH-sensitive,
thermosensitive, and charge-sensitive (electrical or ionic) systems.

Thermosensitive polymers are a subset of smart or intelligent polymers that
exhibit a phase transition from a solution to a gel (sol–gel) when subjected to
alteration in temperature. Thermosensitive polymers are based on naturally
occurring and synthetic polymers, of which both non-degradable and biodegradable polymers are represented.[7] Initial polymers were non-biodegradable
Pluronics or Poloxamers (PEO-PPO-PEO), composed of poly(ethylene oxide)
and poly(propylene oxide). The sol↔gel transformation of Pluronics or
Poloxamers is exhibited by only a sub-set of this group of polymers within a
specific molecular weight range and hydrophilic/hydrophobic balance.[8–11] A
biodegradable thermosensitive polymer was felt to be beneficial, and building
blocks include chitosan, repeated protein segment "monomers", and synthetic

Monographs in Supramolecular Chemistry No. 11
Polymeric and Self Assembled Hydrogels: From Fundamental Understanding to Applications
Edited by Xian Jun Loh and Oren A. Scherman
© The Royal Society of Chemistry 2013
Published by the Royal Society of Chemistry, www.rsc.org

polymers. ReGel™ is an illustration of biodegradable block copolymers, where appropriate copolymer compositions were developed based on poly(ethylene glycol) and biodegradable poly(α-hydroxy acids) for site-specific delivery.[12–14]

Site-specific delivery achieves targeted delivery by concentrating the active pharmaceutical ingredient (API) at the injection site or providing controlled release for systemic delivery. OncoGel™, a combination of ReGel™ and paclitaxel, is the primary ReGel™ formulation studied. OncoGel™ targets injection accessible cancer indications, and capitalizes on physical targeting to achieve a local therapeutic end point.[15–19] A number of ReGel™ copolymers, in addition to the ReGel™ copolymer studied in OncoGel™, have been investigated, and will be further described.

ReGel™ building blocks were selected based on their widespread use and known degradation, excretion, and biocompatibility. Initial ReGel™ development focused on biodegradable block copolymers utilizing the biocompatible components poly(ethylene glycol) (PEG) and poly(D,L-lactide-*co*-glycolide) (PLG) with a defined molecular weight and hydrophilic/hydrophobic balance.

The first generation of ReGel™ polymers were triblock copolymers (ABA or BAB) prepared using PEG 1000 and various combinations of PLG blocks (ReGel™$_{\text{PEG1000}}$). The A-block consists of PLG (hydrophobic block) whereas the B-block is composed of PEG (hydrophilic block). The selected molecular weight and component ratios were selected based on predefined desirable properties, including: low molecular weight, water solubility, unique reversible sol $\leftrightarrow$ gel transition, and biocompatible biodegradation.[12–14] Working ranges for ReGel™$_{\text{PEG1000}}$ were identified for overall molecular weight, polydispersity ($M_{\text{w}}/M_{\text{n}}$), the PLG/PEG ratio, and the L/G ratio for both ABA and BAB ReGel™ copolymers. The noted parameters each affect the ability of ReGel™ to undergo a reversible sol $\leftrightarrow$ gel transition, and alterations in one property will alter the working range of other parameters. In addition to the ReGel™ copolymer composition, the catalyst concentration, reaction temperature, duration of reaction, and purification process are critical for the appropriate ReGel™ polymer characteristics.

The polymer constructs were specifically designed to attain a defined hydrophilic/hydrophobic balance, which enables ReGel™ to be water soluble (free-flowing, sol) at or below room temperature (RT) and allows it to gel (insoluble, bioerodible gel) when the temperature is raised to body temperature.[5] Clinical studies conducted with OncoGel™ were based on ReGel™$_{\text{PEG1000}}$. However, continued development of ReGel™ polymers included different PEG molecular weights, inclusion of an alternative monomer (caprolactone), combinations of multiple ReGel™ copolymers of different composition, and the ability to reconstitute ReGel™ at the point of use *versus* frozen storage (required for early generations of ReGel™). The properties of later generation ReGel™ polymers exhibited increased ease of handling, reconstitution at the point of use, higher drug solubility, and prolonged *in situ* duration.

Owing to the low molecular weight of ReGel™ copolymers, they can be reconstituted in water without the use of organic solvents. These polymers are free-flowing liquids below their gelation temperature, with a viscosity of < 1

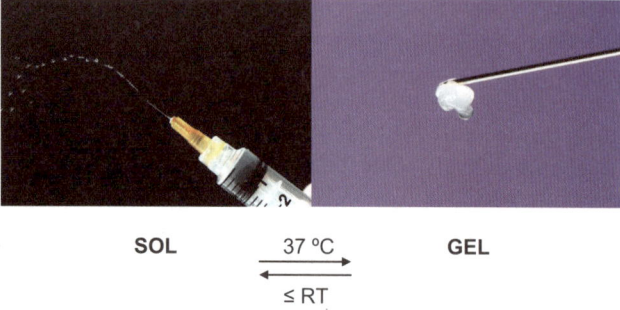

SOL   37 °C   GEL
                  ≤ RT

Figure 9.1 ReGel™ copolymers can be injected through small-gauge needles (23 gauge shown) and undergo a thermoreversible transition from a free-flowing solution (sol) to a viscous gel (gel).

poise. The sol–gel transition occurs without chemical modification of the tri-block copolymer or API as ReGel™ is a physically formed thermally reversible hydrogel (Figure 9.1). The gel state can be converted to the sol state simply by reducing the temperature below the gelation temperature of the particular ReGel™ copolymer. The transition time from a sol to a gel occurs over a few degrees Celsius and within seconds. The gel to sol transition back to a free-flowing solution is dependent on the particular ReGel™ polymer, and is closely associated with both the sol–gel transition temperature as well as the composition of the individual polymer.

The biodegradation mechanism for the hydrophobic A blocks occurs *via* hydrolysis (near zero order) into the constituent monomers that are eliminated through normal metabolic pathways. Once the ReGel™ copolymer has degraded sufficiently, the residual polymer becomes water soluble, diffuses from the site of injection, and is eliminated through the kidneys. The degradation rate is dependent on a number of factors, and can be tailored to occur from one week to approximately three months.

API release of compounds from the ReGel™ depot occurs through two primary mechanisms. The first mechanism is diffusion, followed by a combination of diffusion and erosion of the depot. For ReGel™$_{PEG1000}$, the predominant mechanism for highly water-soluble compounds is diffusion, while both mechanisms contribute significantly for moderate to nearly water-insoluble compounds. Later generations of ReGel™ were designed to increase the effect of depot erosion to control the release of highly water-soluble proteins. This was accomplished while maintaining a similar release profile for hydrophobic compounds, although in some cases the maximum drug loading capacity was enhanced in later generation ReGel™ polymers.

As a result of the amphiphilic nature of ReGel™ copolymers, they demonstrate desirable drug release characteristics for a broad range of APIs. The hydrophobic domains fulfill two distinct purposes related to drug release. First, the association of hydrophobic A-block cores facilitates drug solubilization, resulting in orders of magnitude increases in drug solubility when in the sol state. Hydrophobic drugs can be incorporated into ReGel™ copolymers simply by mixing the amorphous drug in a ReGel™ solution. Second, the hydrophobic core stabilizes hydrophobic drugs, and following the transition from sol to gel, maintains a hydrophobic association with the drug. The release rate is directly correlated with the drug's hydrophobicity, and the drug release can be modified by increasing the hydrophobicity of the API. The amphiphilic nature is also beneficial for controlling the release rate of peptides and proteins, as the hydrophobic nature of the peptide or protein will determine the degree of association with ReGel™. The environment is also mild for proteins, thus maintaining their native structure and activity. An additional benefit is the ability to increase the concentration of peptide or protein that can be successfully achieved without causing undesirable peptide–peptide or protein–protein interactions, resulting in aggregation due to self-association. Formulation work has demonstrated that ReGel™ may be used to deliver small hydrophobic molecules, peptides, and proteins.[20–23]

9.2 Polymer Development

ReGel™ copolymers were developed with distinctive properties to optimize release characteristics and match the desired formulation of the API. The *in situ* duration of ReGel™ and its hydrogel properties (*i.e.*, degradation rate, pore size, hydrophobicity) can be selected by the preparation of a specific ReGel™ copolymer. Modification of the formulation is also achieved by blending distinctive ReGel™ copolymers to obtain intermediate properties.

9.2.1 First Generation

First-generation ReGel™, designated ReGel™$_{PEG1000}$, accomplished a number of predetermined requirements: low molecular weight, water soluble, biodegradable by simple hydrolysis, ability to solubilize hydrophobic APIs, and biocompatibility. These properties were achieved using intelligent design based on the known characteristics of the constituents comprising ReGel™. A number of limitations were noted with ReGel™$_{PEG1000}$ and continued development improved ReGel™'s properties.

Limitations overcome by later ReGel™ copolymers include: (1) frozen storage, (2) requirement to thaw prior to use, (3) low sol–gel temperature, and (4) limited applicability to hydrophilic compounds. Modifications of ReGel™'s composition were conducted to broaden the formulation capability and improve handling properties.

9.2.2 Point-of-Use Reconstitution

ReGel™$_{PEG1000}$ is a viscous semi-solid material with a glass transition temperature (T_g) below room temperature, which prevents lyophilization of the material into a stable lyophilized powder or cake. The viscous semi-solid is readily soluble in water, but the required process involved mechanical stirring, temperature manipulation, and overnight mixing to obtain a solution. The process requires a time period longer than feasible for point-of-use reconstitution. As a result, frozen storage, thawing at 2–8 °C, and injection under a controlled temperature were required for ReGel™$_{PEG1000}$. Two strategies were evaluated to develop a reconstitutable ReGel™ copolymer.

9.2.3 Reconstitution Enhancer

A reconstitution enhancer was the first strategy to develop point-of-use reconstitution of ReGel™ copolymers. A lower molecular weight block copolymer composed of various low molecular weight PEG and PLG was added to the ReGel™ copolymer as a reconstitution enhancer. A second-generation ReGel™ copolymer was also developed, based on a higher molecular weight PEG (1000–2000 Daltons) to allow efficient reconstitution. An intermediate PEG molecular weight was facilitated by the synthesis of ReGel™ copolymers based on either PEG 1000 (ReGel™$_{PEG1000}$) or 2000 (ReGel™$_{PEG2000}$). The two ReGel™ copolymers were mixed at various ratios to obtain a specific average PEG molecular weight. The ReGel™ copolymer had a higher sol–gel transition temperature that allowed efficient reconstitution, and could be stored at room temperature owing to the absence of water. The sol–gel transition temperature was directly correlated with the composition of the ReGel™ copolymer (Figure 9.2). The resulting ReGel™ copolymer/reconstitution enhancer combination allowed rapid reconstitution without mechanical manipulation or alteration in temperature, thus improving limitations (1)–(3) noted above. However, there were additional limitations to the resulting ReGel™ copolymer. The strength of the resulting ReGel™ depot was diminished due to a reduction in the ReGel™ concentration and a lower PLG/PEG ratio that reduced the hydrophobic interactions between polymer chains. The reduced van der Waals interactions, as a result of reduced hydrophobicity, decreased the strength of the hydrogel, resulting in a shorter *in situ* duration and drug release. The polymer

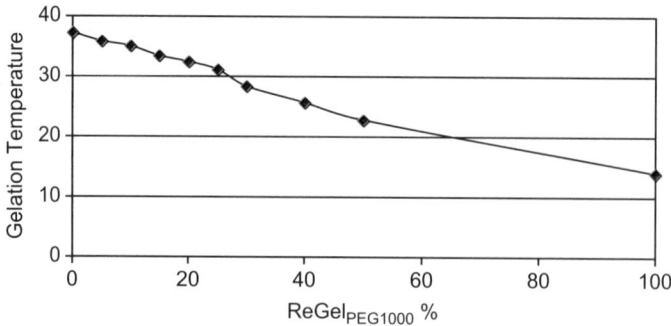

Figure 9.2 ReGel™ gelation temperature as a function of ReGel™$_{PEG1000}$ concentration. ReGel™$_{PEG1000}$ was mixed at various ratios (0–100%) with ReGel™$_{PEG2000}$.

Table 9.1 Paclitaxel, cyclosporin A, or a steroid analog were added to a 23% ReGel™ solution (w/w) at various concentrations. The equilibrium concentration was determined by HPLC.

Drug	Media	Solubility (at 5 °C)
Paclitaxel	Water	$<5\,\mu g\,mL^{-1}$
	ReGel™$_{PEG1000}$	$6\,mg\,mL^{-1}$
	ReGel™$_{CL}$	$12\,mg\,mL^{-1}$
Cyclosporin A	Water	$<5\,\mu g\,mL^{-1}$
	ReGel™$_{PEG1000}$	$>2\,mg\,mL^{-1}$
Steroid analog	Water	Practically insoluble
	ReGel™$_{PEG1000}$	$300\,\mu g\,mL^{-1}$
	ReGel™$_{CL}$	$6\,mg\,mL^{-1}$

development conducted in this area, however, highlighted the ability to blend discrete polymers to achieve average properties of the parent ReGel™ copolymers.

9.2.4 Decreasing ReGel™'s T_g

The second strategy to allow lyophilization of ReGel™ was to decrease ReGel™'s T_g by the use of an alternative monomer, caprolactone. Although caprolactone is not as widely used and studied as D,L-lactide or glycolide, it has demonstrated similar applicability and biocompatibility. Caprolactone was incorporated into a range of hydrophobic A blocks to determine the ideal composition in a ReGel™-containing copolymer (ReGel™$_{CL}$). The desired ReGel™$_{CL}$ properties were tested and compositions identified that facilitated room temperature storage, reconstitution, drug solubilization, and a sol–gel transition temperature above room temperature.

ReGel™$_{CL}$ overcame all four limitations, and has the ability to control the release of hydrophobic compounds due to a tighter pore structure following formation of the gel. The maximum solubility of hydrophobic active pharmaceutical agents that can be formulated with ReGel™$_{CL}$ is also enhanced. The combination of ReGel™$_{CL}$ and paclitaxel is one example where the maximum concentration was increased for a more hydrophobic API, from 6 to 12 mg mL^{-1} (Table 9.1).

In contrast to the ReGel™ copolymer based on a reconstitution enhancer, ReGel™$_{CL}$ was more hydrophobic, resulting in a stronger ReGel™ copolymer, higher drug solubilization potential, a longer *in situ* duration, and better controlled release properties for both hydrophobic and hydrophilic APIs.

9.3 ReGel™ Development for Protein Delivery

Earlier ReGel™ copolymers exhibited ideal characteristics for delivery of hydrophobic APIs, but lacked the ability to control the release of proteins. Alternative ReGel™ copolymers were developed to enable the controlled release of proteins by decreasing the pore size of the hydrogel formed following the sol–gel transition.

9.3.1 ReGel™ BAB Triblock Copolymer

ReGel™ BAB was developed to overcome limitations (3) and (4) specifically for protein delivery. The sol–gel transition temperature was above room temperature, and the API release curve for proteins was prolonged. The duration of protein release was directly correlated with the molecular weight of the protein, confirming a smaller pore structure (Figure 9.3). The ReGel™ BAB copolymer maintained the ability to control the release of hydrophobic compounds, but still exhibited limitations (1) and (2). The ReGel™ BAB outlines the ability to modify properties based on the hydrophobic/hydrophilic balance and structure of the copolymer, ABA or BAB.

9.3.2 Reconstitutable ReGel™CL

The primary goal of ReGel™CL was to obtain a point-of-use reconstitutable ReGel™ copolymer, although a secondary goal was to have a tighter pore structure than ReGel™PEG1000 with protein delivery in mind. ReGel™CL's tighter pore structure prolongs the release duration, which is correlated with the molecular weight of the protein.

9.3.3 ReGel™ Copolymer Blending

Studies with model hydrophilic compounds, fluorescein-labeled dextran of varied molecular weight and bovine serum albumin, demonstrated direct correlation with the molecular weight of the dextran from ReGel™ BAB. Similar short-release duration was demonstrated for dextran of differing molecular

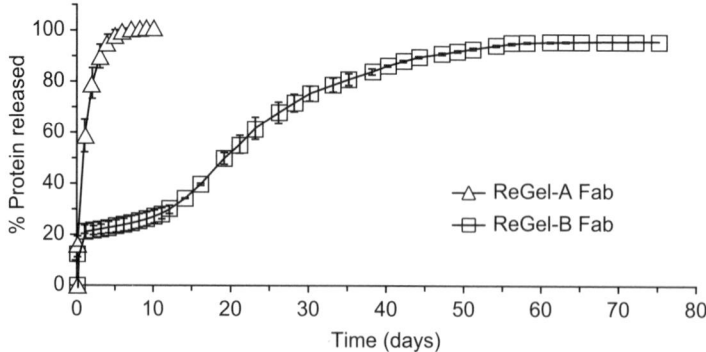

Figure 9.3   Comparison of a Fab protein release from ReGel™-A (ReGel™PEG1000, first generation ReGel™) and ReGel™-B (ReGel™-BAB), where ReGel™-B was designed to have a tighter hydrogel pore structure.

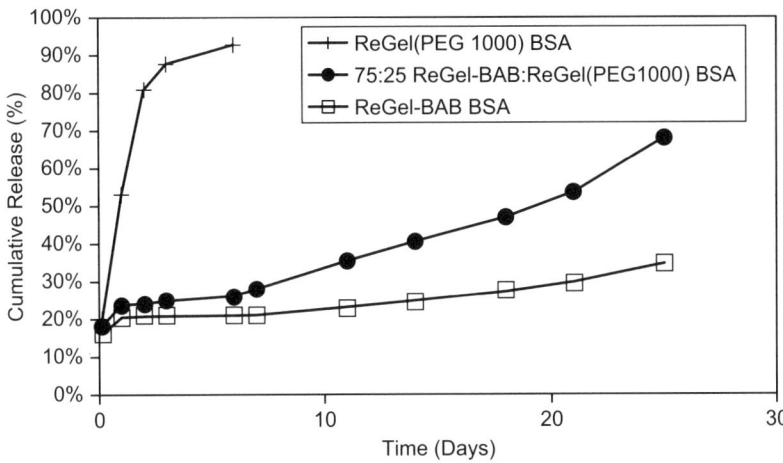

Figure 9.4 A blend of ReGel™$_{PEG1000}$ and ReGel™ BAB exhibits an intermediate release duration of bovine serum albumin (BSA) when compared to ReGel™$_{PEG1000}$ or ReGel™ BAB.

weight from ReGel™$_{PEG1000}$. When the two ReGel™ copolymers were blended, intermediate release profiles were obtained, as shown for bovine serum albumin (Figure 9.4).

9.4 Biocompatibility Studies

The biocompatibility of various ReGel™ copolymer formulations has been demonstrated in a wide array of biological sites. Initial nonclinical settings observed acute and chronic inflammation by histopathological examination of adjacent tissue.[20,22,24–28] Similar histopathological assessments to biodegradable sutures and microspheres based on PLGA noted in the literature were noted for ReGel™$_{PEG1000}$.[5,29] In addition, studies conducted in collaboration with pharmaceutical and biopharmaceutical companies and academic institutions have shown that ReGel™ was compatible with tissues at a range of anatomical sites following subcutaneous, intramuscular, intrarterial, intracranial, intramedullary, intraocular, and intralesional injection.[20–24,30–34]

9.5 Drug Solubilization and Stability Enhancement with ReGel™

A key ReGel™ property makes it ideally suited to deliver hydrophobic APIs due to the ability to increase the solubility by orders of magnitude. ReGel™ is a non-ionic surfactant that spontaneously forms polymeric micelles containing a hydrophobic core when dissolved in an aqueous solution. Formulation with hydrophobic drugs occurs as the drug migrates from the aqueous environment

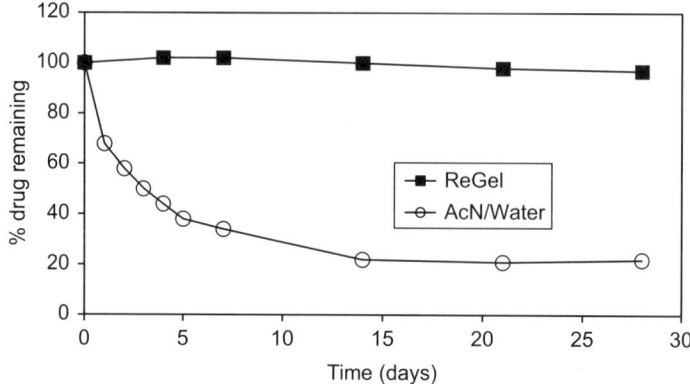

Figure 9.5 The stability of paclitaxel was measured at specific time points following incubation in either ReGel™$_{PEG1000}$ [23% solution (w/w)] or a 75% acetonitrile (AcN) solution at 37 °C.

into ReGel™. As ReGel™ undergoes a sol–gel transition, hydrophobic drugs associate with the hydrophobic constituents of ReGel™ to form a controlled-release depot.[5]

ReGel™ significantly increases the solubility (>2000-fold) of hydrophobic drugs, such as paclitaxel and cyclosporin A (Table 9.1), and significantly improves the stability of paclitaxel in ReGel™ (Figure 9.5).[5] ReGel™ maintained >95% of paclitaxel in its parent form over a 28-day stability study, while paclitaxel rapidly degraded in an aqueous environment to approximately 20% during the same time period.

9.6   ReGel™ Drug Delivery

ReGel™ has demonstrated broad formulation capability.[5,14,21,35,36] ReGel™ formulations have successfully been developed for hydrophobic APIs, peptides, and proteins. The lead ReGel™ formulation incorporated a hydrophobic small molecule, OncoGel™ (ReGel™/paclitaxel), for local tumor management which has completed a number of clinical studies. ReGel™ has also been evaluated *in vitro* and *in vivo* for a variety of indications with peptides and proteins.

9.6.1   Small-Molecule Hydrophobic Drugs

Nonclinical studies using MDA-231 (human breast tumor xenografts) tumors in mice showed similar or superior efficacy with a 10-fold lower dose compared to the maximum tolerated systemic dose. This improved efficacy was accompanied by minimal side effects and no toxic deaths.[5] The diffusion of paclitaxel from OncoGel™ was monitored in a [^{14}C]paclitaxel biodistribution study in tumor bearing mice,[5] and following injection into normal pig pancreata.[20] The biodistribution study demonstrated a half-life within the tumor of approximately three weeks, while paclitaxel levels in plasma, urine, and distal tissues were <5% of the total dose. Distribution within the pancreas exhibited therapeutic paclitaxel levels at distances up to 3–5 cm from the depot site at 14 days post-injection. The residual OncoGel™ depot and tissue contained paclitaxel in its parent form, confirming *in vitro* data demonstrating the ability of ReGel™ to stabilize hydrolytically unstable compounds.[20] Studies of OncoGel™ in the MDA-MB-231, 9L gliosarcoma and CRL-1666 breast cancer adenocarcinoma xenograft models were well tolerated and efficacious in these models.[32,37]

Nonclinical studies of OncoGel™ were conducted in support of clinical studies in a variety of cancer indications: a phase I study in superficial tumors (various tumor types), a phase I/II study in recurrent gioblastome multiforme, a phase IIa neoadjuvant breast cancer study prior to surgery, a phase IIa adjuvant study in esophageal cancer, and a phase IIb study in esophageal cancer comparing chemoradiotherapy to chemoradiotherapy plus OncoGel™.

Various hydrophobic small molecules were evaluated with ReGel™ copolymers and exhibited favorable solubility and release duration. The release duration for hydrophobic small molecules is dependent primarily on the hydrophobicity of the API. Even for ReGel™ BAB and ReGel™$_{CL}$ copolymers the primary mode of release is affected by the hydrophobicity of the API. The pore size for these two ReGel™ copolymers is small enough to provide controlled release for proteins, but the hydrophobic domains control the release of hydrophobic small molecules. Evaluation of hydrophilic small molecules have demonstrated they are not candidates for controlled release formulations in ReGel™. The pore size of ReGel™ hydrogels is insufficient to control the release of hydrophilic small molecules, and owing to their hydrophilic nature the polymer/API interaction is limited. Diffusion of hydrophilic small-molecule APIs occurs over a few hours.

9.6.2   Peptide Drugs

In addition to polymer development, another strategy was identified to decrease the drug release of hydrophilic small-molecule APIs. Work focused on increasing the hydrophobicity of the API to enhance interactions with the hydrophobic domains of ReGel™, which increased the duration of release and coupled release closer to the biodegradation rate of ReGel™ *versus* diffusion.

9.6.2.1   Hydrophobic Peptide Salts

Increasing the hydrophobicity of the API works well for compounds which can form hydrophobic salts. Peptides are often complexed with hydrophilic salts to increase their solubility in a formulation, while for controlled release formulations the desire is to increase the hydrophobicity of the complex. Significant increases in the duration of the drug release for select APIs were obtained using this principle (Figure 9.6). The solubility of this example of hydrophilic peptide/hydrophobic salt was reduced by a factor of 10 by

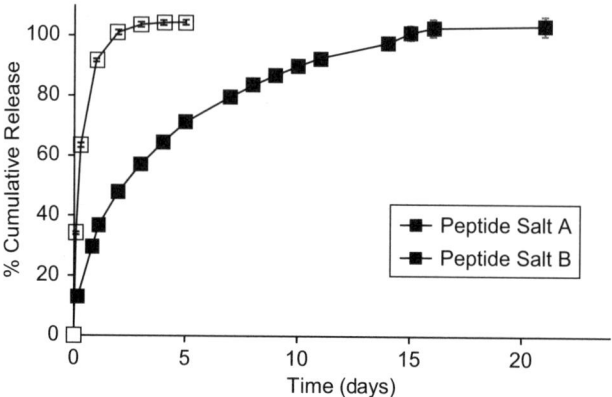

Figure 9.6 Increased hydrophobicity of a peptide by exchanging the salt increases the drug release duration from ReGel™$_{PEG1000}$.

complexation with a hydrophobic *versus* hydrophilic salt. The corresponding *in vitro* release duration was increased by a factor of 10. Alteration in the release profile can also be accomplished by altering the ReGel™ formulation. Utilizing one or both techniques, altering the hydrophobicity of the API and utilizing different ReGel™ copolymers, may allow the release duration to be tailored to a specific therapeutic target. A variety of both hydrophilic and hydrophobic salts are available and can be formulated with compatible peptides to prepare a complex with the desired solubility.

9.6.2.2 Glucagon-Like Peptide-1

A further example of the ability of ReGel™$_{PEG1000}$ delivery systems to enhance peptide delivery was demonstrated using glucagon-like peptide-1 (GLP-1). GLP-1 is known to be highly unstable in aqueous solutions in the concentration range required for efficient administration, and commonly undergoes aggregation/fibrillation. While non-complexed synthetically prepared GLP-1 is released from ReGel™ in ~3 days (data not shown), a suspension of metal-complexed GLP-1 in ReGel™$_{PEG1000}$ provides for complete release of the drug over a greater than 2-week period in its bioactive form (Figure 9.7). *In vivo* studies in ZDF rats demonstrated the ability of this formulation to significantly lower blood glucose levels over a period of 10–12 days (Figure 9.8). A variety of APIs possess similar metal cation binding domains, and often the binding results in an increased hydrophobicity and/or formation of multi-peptide complexes. The hydrophobic domains of ReGel™ copolymers function to stabilize the hydrophobic complexes formed between peptides and metal cations, resulting in a prolonged release profile. The principle is readily tested *in vitro*, although *in vivo* studies are required to confirm that the *in vivo* release duration is controlled by this mechanism. In some cases, the equilibrium of the metal

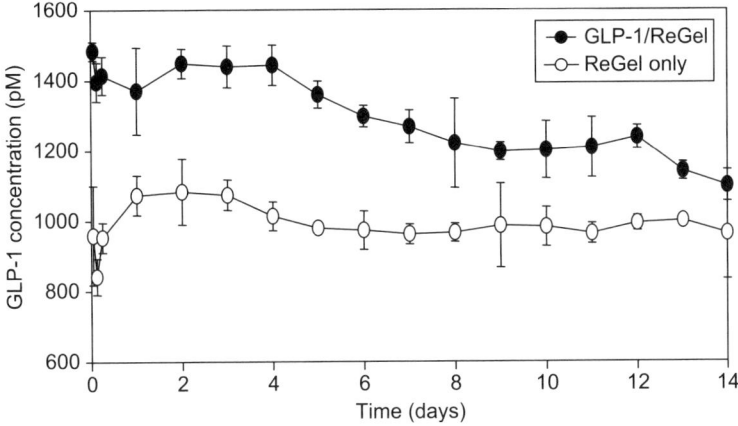

Figure 9.7 GLP-1 concentrations following injection of GLP-1/ReGel™$_{PEG1000}$ in ZDF diabetic rats.

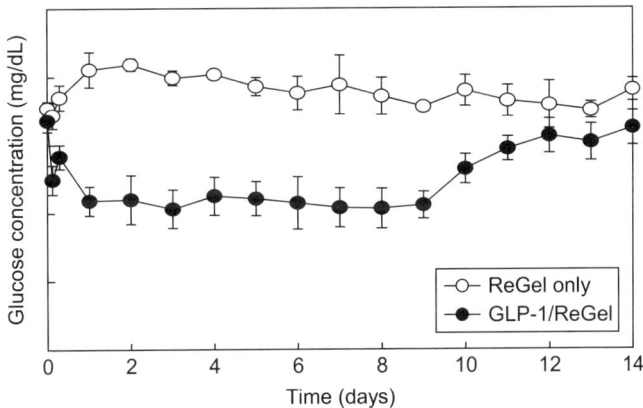

Figure 9.8 Glucose concentrations following injection of GLP-1/ReGel™$_{PEG1000}$ in ZDF diabetic rats.

cation/peptide complex will shift based on the *in vivo* conditions, resulting in reduced drug release duration. The effect on the metal cation/peptide complex formulated in ReGel™ varies by complex and is dependent on the strength of the formed complex.

9.6.3 Proteins

ReGel™ has been formulated with a wide range of peptides and proteins.[5,21,35,36] ReGel™ is a totally aqueous-based formulation (no organic solvents) that creates a mild environment conducive to maintaining native protein structure and function. The examples illustrated are for various ReGel™ copolymers as noted.

9.6.3.1   Metal Cation–Protein Complexes

Both porcine growth hormone (pGH) and human growth hormone (hGH) were formulated in ReGel™ copolymer formulations. pGH interacts to a greater degree with ReGel™ than hGH, resulting in the latter exhibiting reduced-term drug release from ReGel™ copolymer formulations. In an attempt to modify their release profiles, both pGH and hGH were complexed with metal cations to improve the drug release duration. pGH exhibited a two-week *in vitro* release duration independent of the inclusion of metal cations, which was mirrored *in vivo* in a bioefficacy study using hypophysectomized rats.[5] Identical total pGH doses were administered either as a single ReGel™$_{EG1000}$/pGH injection or divided into 14 equal daily injections that exhibited similar efficacy (*i.e.*, growth curves). In contrast, both the *in vitro* and *in vivo* release durations for hGH were prolonged when metal cations were complexed with hGH. The *in vivo* effect was not as dramatic as the *in vitro* effect, as hGH exhibited a shorter drug release in animal models. The decreased release duration is a function of the ability of the inherent *in vivo* equilibrium of metal cation/protein complexes being driven to free hGH. In the case of hGH, the formulation was modified to enhance the stability of metal cation/hGH complexes. As well as metal cation inclusion in the metal cation/hGH complex, additional insoluble metal cation salts were formulated with the hGH/ReGel™ copolymer. As a result, the drug release duration was prolonged by providing a metal cation reservoir within the ReGel™ controlled release depot.

Insulin is another example of a protein that forms metal complexes, resulting in increased hydrophobicity. When formulated with ReGel™, a prolonged release duration was demonstrated. Initial studies of ReGel™/insulin formulations extended the release duration and enhanced the stability of metal cation–insulin from the ReGel™$_{PEG1000}$ depot over a period of two weeks. Similar to hGH, further improvement in sustaining the *in vivo* pharmacokinetic release profile of insulin was achieved using excess metal cation as a reservoir, and a concurrent drop in blood glucose levels was demonstrated in diabetic ZDF rat studies.[35]

9.6.3.2   Interleukin-2

Interleukin-2 (IL-2) is approved for systemic administration to treat renal carcinoma and melanoma, but is associated with significant toxicities including vascular leak syndrome.[38] An immunomodulatory localized peritumoral delivery system was investigated based on a combination of IL-2 in ReGel™$_{PEG1000}$/ReGel™$_{PEG2000}$. *In vitro* release experiments demonstrated that IL-2 is released over a 3- to 4-day period in fully bioactive form as measured by a cell proliferative assay, and induces lymphocyte cytotoxicity against target cells in an identical manner to reconstituted Proleukin®. Nonclinical studies in a number of tumor cell lines demonstrated enhanced efficacy (4-fold) and reduced toxicity over conventional methods of IL-2 delivery.[21]

Four-fold increases in efficacy were demonstrated in four murine IL-2-resistant tumor types: MethA fibrosarcoma (ascites in mouse peritoneum) and solid subcutaneous tumors (RD 995 squamous carcinoma, B16 melanoma, and RENCA renal carcinoma). Local tumor growth was arrested when ReGel™/IL-2 was administered as weekly peritumoral injections in the solid subcutaneous tumors. Both tumor stasis and overall survival were statistically significant compared to control or Proleukin.[21] Pharmacokinetic data demonstrated IL-2 retention in the peritumoral region for protracted periods, significantly lower kidney excretion, and a prolonged half-life in blood as compared to peritumoral administered soluble IL-2.

9.6.3.3 *ReGel™ BAB and ReGel™_{CL} Protein Delivery*

ReGel™ BAB and ReGel™_{CL} copolymers were evaluated with product feasibility studies using proteins as well as peptides and hydrophobic small molecules. ReGel™ BAB and ReGel™_{CL} have a tighter pore structure than ReGel™_{PEG1000} or blends of ReGel™_{PEG1000}/ReGel™_{PEG2000} and were able to alter the release profile for proteins from 1 to 3 days to periods of weeks to months, depending on the formulation. The release profiles are different for the two ReGel™ copolymers, which make them valuable tools in identifying the optimal ReGel™ protein formulation.

9.7 Nonclinical Safety and Efficacy Evaluation of OncoGel™

Four goals were outlined for nonclinical studies with OncoGel™. First, safety and improved tolerability *versus* systemic administration of paclitaxel; second, localization within and around the tumor site; third, safety and efficacy as a stand-alone treatment; and fourth, efficacy and tolerability in combination with other therapies.

9.7.1 Safety Studies

OncoGel™ safety studies were conducted in normal tissue in three species: rat, dog, and pig. Studies included administration into a variety of tissues: skin (subcutaneous injection),[28] central nervous system (CNS), both intracranial[33] and spinal cord,[32,33,37] and the pancreas.[20,24] Initial studies in rats and dogs were conducted to determine the no observable adverse event level (NOAEL) and the maximum tolerated dose (MTD) in normal tissues to support an investigational new drug application and select the starting dose for clinical dose escalation studies of OncoGel™.[19]

Rapidly dividing cells (primarily cancer cells) were the target for local delivery of OncoGel™, but it is important to establish local tolerability of the system in normal tissue. Nonclinical safety studies were conducted and showed that dose-limiting toxicities (DLTs) were local in nature; therefore, systemic toxicities did not need to be considered in the selection of the dose for clinical studies. Local tolerability results for OncoGel™ were consistent with safety results conducted with other locally delivered anticancer therapies.[30,33,39–41]

9.7.2 Biodistribution, Tissue Distribution, and Pharmacokinetics

Biodistribution, tissue distribution, and pharmacokinetics were conducted for OncoGel™ in a variety of *in vivo* animal models.[19] The following studies were conducted with a single OncoGel™ formulation: 6.3 mg of paclitaxel per mL of ReGel™. An ADME (adsorption, distribution, metabolism, and excretion) study of paclitaxel was performed following OncoGel™ administration to the MDA-MB-231 breast tumor xenograft in mice.[5] The study was conducted over 42 days, two half-lives, and the results confirmed that paclitaxel was localized primarily to the tumor site and would function as a local regional therapy with sustained release of paclitaxel.

A subsequent study was conducted in a porcine pancreas model to monitor the OncoGel™ depot and paclitaxel's distribution distance within the tissue.[20,24] The OncoGel™ depot and tissue levels were examined at 7 and 14 days, and cytotoxic paclitaxel concentrations were noted at 30–50 mm from the Onco-Gel™ depot.[20] A corresponding increase in tissue concentration was noted with increasing OncoGel™ injection volumes. Paclitaxel recovered from the surrounding tissue was predominantly ($>90\%$) in its parent form. The study demonstrated that ReGel™ stabilizes paclitaxel *in vivo* prior to release and that, when released into the adjacent tissue, paclitaxel was stable up to two weeks in surrounding tissue, confirming the *in vitro* results.[20]

Pharmacokinetics were conducted in a breast tumor xenograft (MDA-MB-231) following repeat administration, every 2 or 3 weeks, and demonstrated low systemic levels ($<1\%$ of the dose administered). Serum levels correlated with the OncoGel™ injection volume and demonstrated sustained release of paclitaxel [data on file].

9.7.3 Efficacy Studies

OncoGel™ was investigated as a monotherapy in three nonclinical studies, subcutaneous breast tumor xenograft in mice, and two CNS tumor models (spinal cord metastases and intracranial in rats).[5,26,30,33] Enhanced survival and improvement in normal function were noted as a result of OncoGel™ administration. Dose response was found in each tumor model, local tolerability was dependent on the site of administration, and the spinal tumor model exhibited the highest sensitivity of normal tissue to OncoGel™.[26,27] OncoGel™ monotherapy demonstrated acceptable local tolerability and no systemic toxicity in nonclinical efficacy studies.

Paclitaxel's efficacy is well established and cancer agents are often combined to optimize synergistic activity. Toxicities noted for each chemotherapy agent or radiation treatment may also be better managed through combination therapies. Dose alteration is utilized to achieve targeted therapeutic endpoints, while simultaneously reducing toxicity. OncoGel™'s low systemic levels make it an ideal combination therapy candidate, as the potential synergistic activity is not hampered due to attendant systemic toxicity.

OncoGel™ combinations were evaluated in an adjuvant setting in various animal models (surgery, radiation therapy, an additional local chemotherapy delivery system, and systemic chemotherapy).[27,30,33,37] Studies allowed further demonstration of increased efficacy for synergistic combination therapies, without increasing systemic toxicity.

OncoGel™ as an adjuvant to surgery in a spinal column metastasis model (CRL-1666 breast adenocarcimona cell line) statistically enhanced the surgical effect ($p < 0.001$) *versus* surgery alone. In the intracranial model (9L gliosarcoma cell line), the addition of OncoGel™ to radiation therapy (RT) statistically increased survival *versus* OncoGel™ or RT alone ($p < 0.001$).[33,37]

Based on the success of these studies, further combination therapies were evaluated which involved up to three treatment combinations. Options included OncoGel™, RT, surgery, oral chemotherapy, and an additional controlled release chemotherapy drug delivery system. Various combinations were evaluated to determine the optimal synergy between the treatment therapies.[27,30,33,37] The OncoGel™ dose established in both the spinal cord metastases and intracranial models as a stand-alone treatment was equally tolerated in dual or tri-modality therapies, further supporting the tolerability of local OncoGel™ administration in these models. The nonclinical program conducted has been instrumental in designing clinical studies, as well as identifying potential combination therapies and indications for clinical investigation.

9.8 Clinical Development of OncoGel™

OncoGel™ (ReGel™/paclitaxel) was investigated for local treatment of solid tumors in four clinical trials in the US and Europe in three indications (esophageal, neoadjuvant-breast cancer, and recurrent glioblastome multiforme).[19,25,28] Two trials were conducted exclusively in either the US or Europe and two were multi-national trials. The selection of the indications was based on identifying those that would allow local drug delivery *via* injection, imaging with standard techniques, and a clear clinical benefit associated with a local regional therapy. The latter was due to the local paclitaxel diffusion distance (< 5 cm) from OncoGel™, and the absence of a systemically relevant paclitaxel concentration.[19,25,28] As possible cancer indications were considered, an evaluation was conducted of whether the metastatic potential is negligible or if systemic therapies are expected to predominantly target metastatic tumor cells.[20,24,26,27,30,33] Those indications where systemic therapies were required then determined the appropriate systemic chemotherapy, the potential synergy with local regional paclitaxel, and the ability to differentiate the effects of OncoGel™ *versus* the standard of care.

There are considerations when utilizing local regional therapy that must be taken into account. Traditional dosing regimens are not followed, and as a result there will be additional inherent patient variability *versus* systemic therapy. The dosing algorithm for OncoGel™ does not follow a typical che-motherapy regimen (dose m^{-2} body surface area), as the dosing is based on the tumor volume (dose cm^{-3} tumor volume). The dose is directly correlated with patient tumor volume, such that each patient receives the same dose based on tumor volume, but the absolute amount of paclitaxel administered may vary widely from patient to patient. OncoGel™ dosing ranged from less than 1 mL to 68 mL, which equates to a range between 6 and 428 mg, across the clinical studies using the highest OncoGel™ concentration, $6.3\,mg\,mL^{-1}$.

9.8.1 Phase I Studies

OncoGel™ phase I studies were conducted in two indications. The identified endpoint for the multi-center US phase I dose escalation study of OncoGel™ was safety in superficial palpable tumors, and was well tolerated for eight different tumor types.[28] Intratumoral concentrations of paclitaxel following OncoGel™ injection were 0.06, 0.25, 0.63, and 2.0 mg paclitaxel per cm^3 of tumor. All 16 patients were evaluable for toxicity. Twelve adverse events reported were considered related to OncoGel™, and all were reflective of local responses to OncoGel™ administration. Overall, OncoGel™ was well tolerated, and paclitaxel release was localized in nature, as demonstrated by low or undetectable paclitaxel levels systemically. Fourteen of 16 patients were eva-luable for efficacy, and based on modified World Health Organization, six patients had stable disease and eight patients had progressive disease.

Phase I/II and phase II studies included safety as a component, but also evaluated the antitumor efficacy. A multi-center US phase I/II study in recur-rent glioblastome multiforme patients evaluated OncoGel™ placement in the tumor cavity following surgical resection of the tumor. While there were indications of efficacy, OncoGel™ administered in this setting resulted in brain edema late in the course of paclitaxel release/ReGel™ biodegradation (4–6 weeks), which limited the application of OncoGel™ for these patients [data on file]. As the definitive reason for edema was not able to be determined within the clinical study, clinical evaluation in the intracranial indication was dis-continued for OncoGel™.

9.8.2 Phase II Studies

OncoGel™ phase II studies were conducted in two indications, neoadjuvant breast cancer and esophageal cancer. The multi-center, multi-national European phase IIa study of OncoGel™ as a neoadjuvant therapy in breast cancer patients prior to surgery was also well tolerated, but further development in this indication was not pursued following evaluation of the current clinical protocols for these patients [data on file]. Standard neoadjuvant therapy is often used to determine which chemotherapy agent would be administered following surgery for breast cancer patients, and it was felt OncoGel™ would mask the effects of systemic chemotherapy in a neoadjuvant setting. While the study was positive, it highlights the complexities of utilizing local regional therapy in cancer where systemic therapy is utilized.

The multi-center, multi-national US phase IIa dose escalation study of OncoGel™ in combination with RT conducted in inoperable esophageal cancer patients demonstrated OncoGel™ was well tolerated in combination with external beam RT (50.4 Gy).[25] In order to isolate OncoGel™'s safety and efficacy, without confounding effects of systemic therapies, patients who were not candidates to receive chemotherapy were enrolled. No treatment-limiting toxicities were observed, and OncoGel™ did not appear to add an increased risk to RT. In addition, RT did not alter OncoGel™'s expected pharmacokinetic profile. Results were interpreted with caution as an RT-only treatment arm was not included in the study.[25] Based on the results of this study, future clinical studies included a comparative evaluation of standard of care *versus* a standard of care plus OncoGel™. This study formed the basis for conducting a phase IIb study in combination with chemoradiotherapy in potentially curable patients.

Pharmacokinetics in the esophageal phase IIa dose escalation study confirmed sustained release of paclitaxel with low systemic paclitaxel concentrations, as noted in nonclinical animal models.[25,28] The pharmacokinetic data further confirmed OncoGel™ as a local regional therapy, and its ability to be added to concurrent chemotherapy regimens without contributing to associated systemic toxicities. In the phase IIa study, intratumoral concentrations of paclitaxel following OncoGel™ injection were 0.48, 1.0, and 2.0 mg paclitaxel per cm^3 of tumor. Paclitaxel was detectable in all 11 patients at day 1, in 10 patients through day 3, and in 6 patients through week 3. Peak plasma concentrations, C_{max}, ranged from 0.53 to 2.73 ng mL^{-1} and were directly related to the absolute amount of paclitaxel administered, where the OncoGel™ administered volume is based on the tumor volume of individual patients.[25] The T_{max} occurred from 3 to 24 h post-OncoGel™ injection, with a sustained release thereafter.

Patients' efficacy assessment included improvement in dysphagia score (ability to swallow liquids, semi-solids, or solid food) and tumor response. Of 11 patients, nine (82%) had improvement in dysphagia, while six patients (55%) had a two-point improvement in a five-point scale (secondary endpoint) and three patients had a one-point improvement.[25] Tumor response was assessed by endoscopic ultrasound (EUS) and modified response evaluation criteria in solid tumors (mRECIST). Five patients were classified as having a partial response (PR) and five as having stable disease (SD), with one patient not evaluable due to the inability to pass the endoscope for EUS assessment.[25] An example of an exophytic tumor prior to and following combination therapy of OncoGel™ intratumoral injection and RT is shown in Figure 9.9. The patient was classified as having a PR, and had a two-point improvement in dysphagia

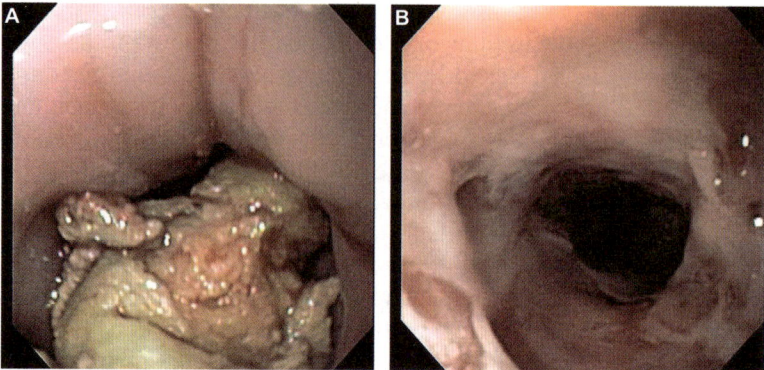

Figure 9.9   Image of an esophageal squamous cell carcinoma prior to (A) and following (B) administration of OncoGel™ and external beam radiation
therapy (50.4 Gy).

score. Biopsies were collected from 10 of 11 patients, and four patient biopsies
were negative for carcinoma. Two of the four patients were downstaged to the
extent they were considered resectable, and one patient underwent esopha-
gectomy. The second patient had a second primary tumor (rectal cancer) and
underwent resection for this tumor.[25]

The multi-center, multi-national phase IIb study consisted of a control
group receiving the standard of care, namely 5-FU, cisplatin, and RT
(chemoradiotherapy), and the treatment arm of OncoGel™ in combination with
chemoradiotherapy. Patients were surgical candidates and OncoGel™ plus or
minus chemoradiotherapy constituted neoadjuvant therapy. The study was a
multi-center, multi-national study and was targeted for patients who would
benefit from local regional therapy. The primary endpoint was tumor response,
with secondary endpoints of safety, survival, and pathological complete
response (pCR) of the tumor. The latter is documented in the literature as a
surrogate for survival.[42,43]

The combination of OncoGel™ plus chemoradiotherapy did increase the
number of adverse events in a subset of reported events, but overall the therapy
was well-tolerated. The primary endpoint was tumor response, overall response
(OR), and the OncoGel™ arm and standard of care arm had an OR of 12.5%
and 20%, respectively. Evaluation of a local endpoint, pCR, for these patients
was selected as a potential surrogate endpoint, and it allowed a clear differentiation between the control arm and the treatment arm. pCR has demonstrated a close correlation to overall survival in this patient population.

Unfortunately, the treatment arm demonstrated a decreased number of
patients experiencing a pCR, 12.5%, *versus* the control arm, 27.7%. Despite the
correlation of pCR and survival noted in the literature, it was not reflected in
the 12-month survival for these patients, as overall survival was similar between
the two groups. The effect on overall survival is somewhat confounded by
occult metastases that are known to drastically shorten overall survival in this

patient population. The control arm exhibited increased overall survival, although it did not reach significance [data on file].

A possible explanation for the decreased percentage of patients experiencing pCR in patients receiving OncoGel™ in combination with chemoradiotherapy (treatment arm) was not readily noted during evaluation of the clinical data. While antagonistic activity has been noted previously in combination therapies in cancer, it is noted only in a minority of studies and the mechanism is varied and often unknown. The altered delivery mechanism of paclitaxel from OncoGel™, sustained release for >6 weeks, may have changed the effect on tumor cells *versus* systemic administration of paclitaxel in combination with chemoradiotherapy. Chemoradiotherapy with paclitaxel as one of the chemotherapy agents has increased the effect of chemoradiotherapy, which was the rationale for including OncoGel™ in combination with chemoradiotherapy.[42,43] As a result of a lack of clinical efficacy, OncoGel™ development for esophageal cancer was discontinued.

9.9 Final Comments

The variety of ReGel™ copolymers developed outline the broad capability and flexibility of the ReGel™ family of copolymers. Understanding the underlying principles of the thermosensitive sol–gel transition and the contribution of the hydrophobic/hydrophilic balance, monomer composition, polydispersity, molecular weight, and copolymer block structure allowed for intelligent design of ReGel™ copolymers with tailored properties.

Controlled release formulations have an accepted niche of indications where they provide clinical benefit. In addition to the unique nature of the areas where they are utilized, there are specific considerations that need to be taken into account as their utility is evaluated. The clinical development of OncoGel™ highlights the advantages and considerations required in evaluating a local regional therapy in cancer.

References

1. L. Klouda and A. G. Mikos, *Eur. J. Pharm. Biopharm.*, 2008, **68**, 34.
2. J. H. Park, M. Ye and K. Park, *Molecules*, 2005, **10**, 146.
3. F. Mohamed and C. F. Van der Walle, *J. Pharm. Sci.*, 2008, **97**, 71.
4. L. S. Nair and C. Laurencin, *Adv. Biochem. Eng. Biotechnol.*, 2006, **102**, 47.
5. G. M. Zentner, R. Rathi and C. Shih, *J. Controlled Release*, 2001, **72**, 203.
6. B. Jeong, S. W. Kim and Y. H. Bae, *Adv. Drug Delivery Rev.*, 2002, **54**, 37.
7. M. K. Nguyen and D. S. Lee, *Macromol. Biosci.*, 2010, **10**, 563.
8. R. Bhardwaj and J. Blanchard, *J. Pharm. Sci.*, 1996, **85**, 915.
9. K. A. Fults and T. P. Johnston, *J. Parenter. Sci. Technol.*, 1990, **44**, 58.
10. T. P. Johnston, M. A. Punjabi and C. J. Froelich, *Pharm. Res.*, 1992, **9**, 425.

11. S. Miyazaki, Y. Ohkawa, M. Takada and D. Atwood, *Chem. Pharm. Bull.*, 1992, **40**, 2224.
12. R. C. Rathi and G. M. Zentner, *U.S. Pat.* 6 004 573, 1999.
13. R. C. Rathi, G. M. Zentner and B. Jeong, *U.S. Pat.* 6 117 949, 2000.
14. R. C. Rathi, G. M. Zentner and B. Jeong, *U.S. Pat.* 6 201 072, 2001.
15. L. R. Rogers, J. P. Rock, A. K. Sills, M. A. Vogelbaum, J. H. Suh, T. L. Ellis, V. W. Stieber, A. L. Asher, R. W. Fraser, J. S. Billingsley, P. Lewis, D. Schellingerhout, E. G. Shaw and Brain Metastasis Study Group, *J. Neurosurg.*, 2006, **105**, 375.
16. F. J. Attenello, D. Mukherjee, G. Datoo, M. J. McGirt, E. Bohan, J. D. Weingart, A. Olivi, A. Quinones-Hinojosa and H. Brem, *Ann. Surg. Oncol.*, 2008, **15**, 2887.
17. H. C. Lawson, P. Sampath, E. Bohan, M. C. Park, N. Hussain, A. Olivi, J. Weingart, B. A. Kell and H. Brem, *J. Neurooncol.*, 2007, **83**, 61.
18. H. Malhotra and G. L. Plosker, *Drugs Aging*, 2001, **18**, 787.
19. N. L. Elstad and K. D. Fowers, *Adv. Drug Delivery Rev.*, 2009, **10**, 785.
20. K. Matthes, M. Mino-Kenudson, D. V. Sahani, N. Holalkere, K. D. Fowers, R. Rathi and W. R. Brugge, *Gastrointest. Endosc.*, 2007, **65**, 448.
21. W. E. Samlowski, J. R. McGregor, M. Jurek, M. Baudys, G. M. Zenter and K. D. Fowers, *J. Immunother.*, 2006, **29**, 524.
22. W. Zhu, T. T. Masaki, Y. H. Bae, R. Rathi, A. K. Cheung and S. E. Kern, *J. Biomed. Mater. Res., B*, 2006, **77**, 135.
23. S. Choi, M. Baudys and S. W. Kim, *Pharm. Res.*, 2004, **21**, 827.
24. E. Linghu, K. Matthes, M. Mino-Kenudson and W. R. Brugge, *Endoscopy*, 2005, **37**, 1140.
25. G. A. DuVall, D. Tarabar, R. H. Seidel, N. L. Elstad and K. D. Fowers, *Anticancer Drugs*, 2009, **20**, 89.
26. C. A. Bagley, M. J. Bookland, J. A. Pindrik, T. Ozmen, Z. L. Gokaslan and T. F. Witham, *J. Neurosurg. Spine*, 2007, **7**, 194.
27. B. Gok, M. McGirt, D. M. Sciubba, G. Garces-Ambrossi, C. Nelson, J. Noggle, A. Bydon, T. F. Witham, J. P. Wolinsky and Z. L. Gokaslan, *Neurosurgery*, 2009, **65**, 193.
28. S. J. Vukelja, S. P. Anthony, J. C. Arseneau, B. S. Berman, C. C. Cunningham, J. J. Nemunaitis, W. E. Samlowski and K. D. Fowers, *Anticancer Drugs*, 2007, **18**, 283.
29. M. S. Shive and J. M. Anderson, *Adv. Drug Delivery Rev.*, 1997, **28**, 5.
30. A. K. Vellimana, V. R. Recinos, L. Hwang, K. D. Fowers, K. W. Li, Y. Zhang, S. Okonma, C. Eberhant, H. Brem and B. Tyler, Combination paclitaxel thermal gel depot with Temozolomide and radiotherapy significantly prolongs survival in an experimental rodent glioma model, *J. Neurosurgery*, submitted.
31. C. M. Terry, L. Li, H. Li, I. Zhuplatov, D. K. Blumenthal, S. E. Kim, S. C. Owen, E. G. Kholmovski, K. D. Fowers, R. Rathi and A. K. Cheung, *J. Control. Rel.*, 2012 Jun 28; **160**(3), 459–467. Epub 2012 Mar 17.
32. B. M. Tyler, A. Hdeib, J. Caplan, F. G. Legnani, K. D. Fowers, H. Brem, G. Jallo and G. Pradilla, *J. Neurosurg. Spine*, 2012, **16**, 93.

33. B. Tyler, K. D. Fowers, K. W. Li, V. R. Recinos, J. M. Caplan, A. Hdeib, R. Grossman, L. Basaldella, K. Bekelis, G. Pradilla, R. Legnani and H. Brem, *J. Neurosurg.*, 2010, **113**, 210.
34. A. K. Cheung, C. Terry and L. Li, *J. Ren. Nutr.*, 2008, **18**, 140.
35. S. Choi and S. W. Kim, *Pharm. Res.*, 2003, **20**, 2008.
36. T. Masaki, R. Rathi, G. Zentner, J. K. Leypoldt, S. F. Mohammad, G. L. Burns, L. Li, S. Zhuplatov, T. Chirananthavat, S. J. Kim, S. Kern, J. Holman, S. W. Kim and A. K. Cheung, *Kidney Int.*, 2004, **66**, 2061.
37. B. Gok, M. J. McGirt, D. M. Sciubba, G. Garces-Ambrossi, C. Nelson, J. Noggle, A. Bydon, T. F. Witham, J. P. Wolinsky and Z. L. Gokaslan, *Neurosurgery*, 2009, **65**, 193.
38. F. P. Ognibene, S. A. Rosenberg, M. Lotze, J. Skibber, M. M. Parker, J. H. Shelhamer and J. E. Parrillo, *Chest*, 1988, **94**, 750.
39. N. I. Marupudi, J. E. Han, K. W. Li, V. M. Renard, B. M. Tyler and H. Brem, *Expert Opin. Drug Saf.*, 2007, **6**, 609.
40. E. Ruel-Gariépy, M. Shive, A. Bichara, M. Berrada, D. Le Garrec, A. Chenite and J. C. Leroux, *Eur. J. Pharm. Biopharm.*, 2004, **57**, 53.
41. E. Harper, W. Dang, R. G. Lapidus and R. I. Garver, Jr., *Clin. Cancer Res.*, 1999, **5**, 4242.
42. M. Koshy, B. D. Greenwald, P. Hausner, M. J. Krasna, N. Horiba, R. J. Battafarano, W. Burrows and M. Suntharalingam, *Am. J. Clin. Oncol.*, 2011, **34**, 259.
43. D. E. Gannett, R. F. Wolf, G. W. Takahashi, L. Jeannie, R. C. Wagner, F. S. Ey, M. M. Owens, W. E. Johnson, D. W. Cook and R. E. Alberty, *Gastrointest. Cancer Res.*, 2007, **1**, 132.

CHAPTER 10

Biomedical Applications of Hydrogels: Poly(vinyl alcohol)-Based Hydrogels for Embolotherapy and Drug Delivery

ANDY LEWIS* AND CLARE HEAYSMAN

Biocompatibles UK Ltd (A BTG Ltd Group Company), Farnham Business Park, Weydon Lane, Farnham, Surrey GU9 8QL, UK
*Email: andrew.lewis@biocompatibles.com

10.1 Introduction

Poly(vinyl alcohol) (PVA; Figure 10.1) is a synthetic water-soluble polymer made by the free radical polymerization of the monomer vinyl acetate, followed by a partial or complete saponification (hydrolysis) of the acetate groups to yield a multi-pendent hydroxylated polymer backbone. Direct polymerization of vinyl alcohol is not possible owing to the instability of the monomer, which tautomerizes to the acetaldehyde form. The use of vinyl acetate as the precursor material means there is normally always a component of residual acetate in the final PVA, which is important in disrupting the hydrogen bonding between the hydroxyl groups and maintaining water solubility. This polymer has widespread application due to its excellent adhesive, film-forming and emulsifying

Monographs in Supramolecular Chemistry No. 11
Polymeric and Self Assembled Hydrogels: From Fundamental Understanding to Applications
Edited by Xian Jun Loh and Oren A. Scherman

Published by the Royal Society of Chemistry, www.rsc.org

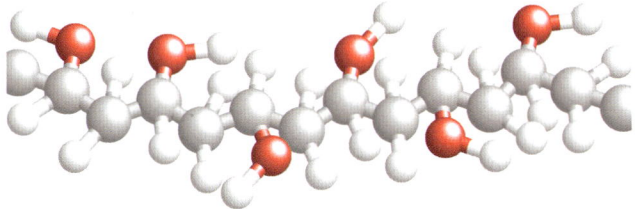

Figure 10.1   Ball-and-stick representation of a section of a PVA chain.

properties. Its inherent hydrophilicity renders the material useful as a basis for hydrogel formation, for which it must be cross-linked by some means. This has been achieved by a wide variety of approaches, including chemical cross-linking with agents, such as formaldehyde, glutaraldehyde or epichlorohydrin, by the use of interlayered cations such as calcium or anions such as borate, by UV, e-beam or γ-irradiation induced cross-linking, or by freeze–thawing the material to induce physical cross-links by crystallite formation.[1] The process of cross-linking may be selected on the basis of the toxicity of the residuals or species formed during the reaction, particularly if the end product is for use in a pharmaceutical or biomedical application. The properties of the ensuing hydrogel will be governed largely by the method and extent of the cross-linking involved, but fundamental is a sufficient degree of polymer chain inter-connection such that the material will swell but not completely dissolve in water or other biological fluids. It is the high water content and corresponding physicomechanical properties imparted to the hydrogel network that confer on PVA its excellent biocompatibility and usefulness for the fabrication of products that come into contact with body tissues.

A brief Pubmed search for PVA reveals over 4000 references to the use of this polymer in medical-oriented research. It has been described in the literature for use in a myriad of applications, and as far back as 1949 as a biocompatible packing material post-pulmonectomy.[2] Over 700 of these references are connected to the use of PVA in embolization therapy; this is a minimally invasive non-surgical technique performed mostly by interventional radiologists (IRs), for the selective occlusion of particular blood vessels.[3] A range of different embolization devices are commonly used, depending upon the vessel type, size, blood flow and purpose for the occlusion. These include metallic coils, occlusion balloons, autologous blood clots, cyanoacrylate glue and particulate or microspherical embolization agents. A cross-linked PVA foam known as Ivalon® was the first synthetic particulate embolization agent described for clinical use in 1975.[4] These PVA particles were fabricated by polymerization in the presence of foaming agents to create a porous sponge-like structure which was cross-linked into shape by use of formaldehyde. The compressed PVA sponge blocks were then rasped into irregular shaped particles that were separated into a series of size ranges useful to the IR by passing through a sieve stack. A range of post-surgical absorbent products based on PVA for packing various cavities, such as the ear, nose and sinus, are available today under the same brand name, although the particulate embolization particles are not. The irregular nature of the foam particles, although efficient at blocking blood vessels, can also induce undesirable blockage of the microcatheters used for delivery of the device if handled incorrectly. A microspherical PVA sponge-like product was developed in an attempt to overcome these deficiencies (Contour SE®, Boston Scientific), but it was found to be overly compressible in clinical use, leading to a more distal, less predictable level of vessel occlusion.[5,6] Sponge-like particles have therefore been largely superseded by hydrogel-based microspherical embolization agents, again available in a range of sizes but with a lower propensity for microcatheter blockage due to their compressible nature, yet a more predictable level of occlusion within a vessel due to their elastic recovery properties.[7] Bead Block® and DC Bead® embolization devices are two such microspherical products, which are based on a modified PVA hydrogel. This chapter is devoted to the description of the chemistry of these products, structure–property relationships and how the ensuing characteristics dictate clinical performance.

10.2   Chemical Structure and Property Relationships

10.2.1   The Chemistry of PVA Hydrogels Based Upon Nelficon

A novel class of PVA hydrogels, based upon chemistry first developed by Ciba-Geigy in 1995, was described in a patent relating to methods for the preparation of photocross-linkable prepolymers for use in the manufacture of contact lenses. The prepolymers (also known as macromers) were based on PVA of at least 2000 molecular weight that comprised a 1,3-diol basic structure in which a certain percentage of the 1,3-diol units had been modified to a 1,3-dioxane having in the 2-position a radical that was polymerizable but not polymerized. The polymerizable radical was especially preferred to be an aminoalkyl radical having a polymerizable group bonded to the nitrogen atom. This was essentially achieved by the synthesis of *N*-acryloylaminoacetaldehyde dimethyl acetal (NAAADA), a species that could be used in a subsequent acid-catalysed reaction with the 1,3-diol units on the PVA backbone to form a stable cyclic acetal structure from which a pendent reactive acrylamide group is suspended (Figure 10.2). This allowed for bulk photopolymerization of the prepolymer and a suitable UV initiator in a mould to form a PVA-based hydrogel contact lens, the generic material known as Nelfilcon® and the basis of Ciba Vision's Focus® Dailies® lens.

10.2.2   Preparation of PVA-Based Hydrogel Beads

The Nelfilcon chemistry was adapted to enable the polymerization to occur in the form of a water-in-oil suspension in order to give rise to hydrogel beads of different sizes. The PVA macromer was dissolved in the aqueous phase, to which other acrylic monomers could optionally be added in order to form a graft copolymer network. Selection of the co-monomer(s) is crucial in determining some of the eventual properties of the hydrogel beads, the limitation

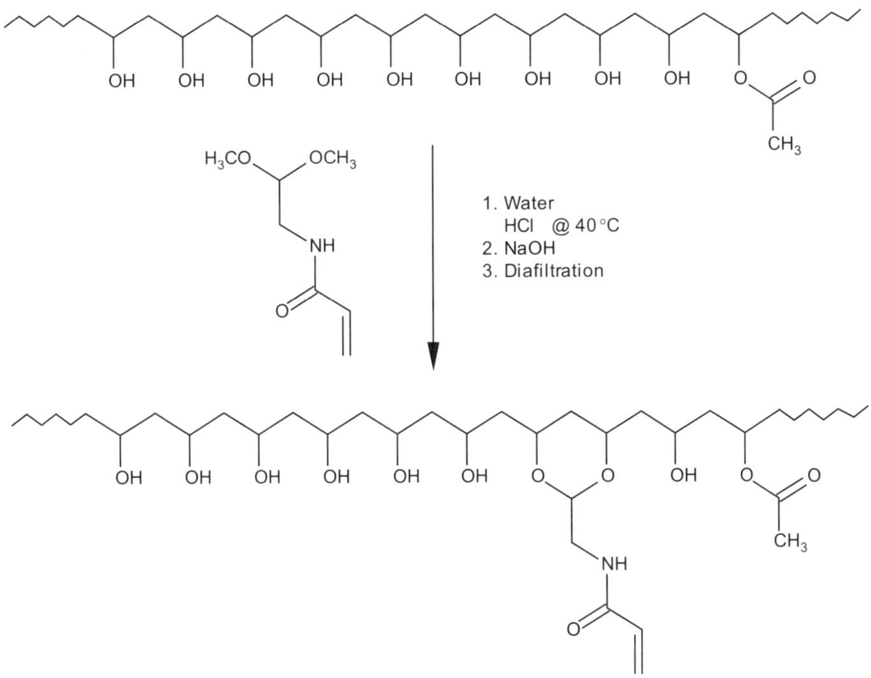

Figure 10.2 Reaction scheme illustrating the formation of macromer by functionalization of PVA.

being that the monomer must be soluble in the aqueous macromer solution and have a similarly matched reactivity of the double bond to allow for efficient copolymerization. The suspension polymerization is sensitive to a number of variables: choice of the oil phase, type and amount of suspension stabilizer, reaction vessel and stirrer paddle geometry, stirring speed and reaction temperature. The reaction can be initiated by choice of a suitable water-soluble thermal initiator, or more commonly by a suitable redox initiator system. Unlike many other classic approaches to the formation of PVA hydrogels, these microspheres are not conventionally cross-linked by inclusion of a reactive bifunctional species, but rather the acrylic polymer chains are grafted to the PVA chains by virtue of the pendent NAAADA functionalities (Figure 10.3).

10.2.3 Synthesis and Properties of Poly[(vinyl alcohol)-*graft*-(2-acrylamido-2-methylpropanesulfonate sodium salt)] (PVA-*g*-AMPS) Beads

The PVA macromer strategy allows great flexibility in the fabrication of a wide variety of hydrogel structures. It is tolerant of a variation in different polymerization methods in order to yield different morphologies and can accommodate a multitude of co-monomer constituents to impart a range of

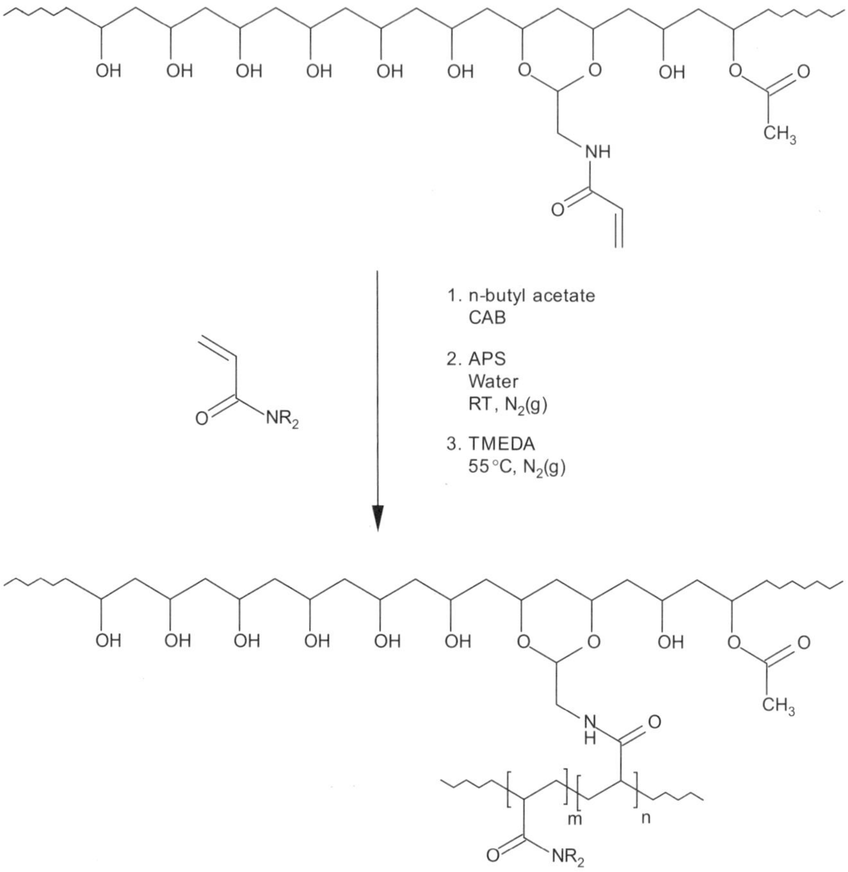

Figure 10.3 Proposed reaction scheme for the synthesis of PVA beads with a generic acrylamide as co-monomer. The end product is a simplified representation of the polymer structure where *n* is fixed in the preparation of the macromer and *m* is varied in each formulation.

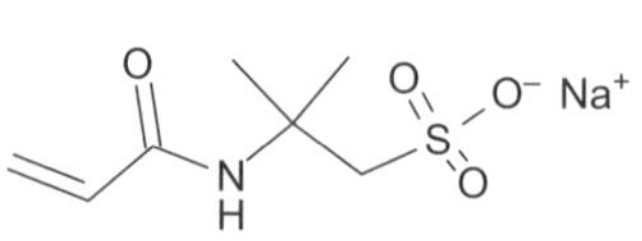

Figure 10.4 Chemical structure of 2-acrylamido-2-methylpropanesulfonate sodium salt (AMPS).

physiochemical properties. The Bead Block and DC Bead embolization systems are microspherical medical hydrogels based on 2-acrylamido-2-methylpropanesulfonate sodium salt (AMPS) (Figure 10.4). The high water-absorbing and swelling capacity of AMPS-based hydrogels are key to their medical applications. AMPS-based hydrogels have been shown to possess uniform conductivity, low electrical impedance, cohesive strength, appropriate skin adhesion, biocompatibility and are capable of repeated use. They have been used, for instance, in electrocardiograph (ECG) electrodes, defibrillation electrodes, electrosurgical grounding pads and iontophoretic drug delivery electrodes. In addition, the absorbing nature of polymers derived from AMPS is used for exudate absorption as a tackifier component of some wound dressings. The sulfonate group in the AMPS monomer provides for a high degree of polarity, hydrophilicity and anionic character across a wide range of pH. In addition, the AMPS absorbs water readily, imparting enhanced water content and transport characteristics to the hydrogel polymers. The geminal dimethyl group and the sulfomethyl group combine to sterically hinder the amide functionality and provide both hydrolytic and thermal stabilities to AMPS-containing polymers. When AMPS is incorporated into hydrogels, high compression strength, elasticity and elongation are achieved at high water content. The high water content, tissue-like modulus and possession of heparin-like sulfonate residues on the polymer backbone mean that AMPS-containing hydrogels are highly biocompatible in nature; a crucial attribute for a permanent implant material.

10.2.3.1   *Copolymerization of PVA-g-AMPS Hydrogel Beads*

Beads can be synthesized using a wide range of AMPS content in the formulations, in good yield and with elemental analyses commensurate with their theoretical compositions, suggesting efficient copolymerization. A useful method for obtaining information regarding structure of such copolymer systems is simple optical microscopy under illumination with light. If the light is absorbed by the sample it may be possible to observe an image. Samples which contain greater than 95% water (as is the case for these systems as shown in Section 10.2.3.2 absorb very low amounts of light and often only faint images can be seen. To capture suitable images of PVA-based microspheres with varying AMPS content, phase contrast light microscopy has been used (Figure 10.5). As the AMPS content is increased, the copolymer microspheres appear more opaque. This is due to incompatibility between the PVA and AMPS polymers, as the acrylamide chains become longer and may preferentially self-

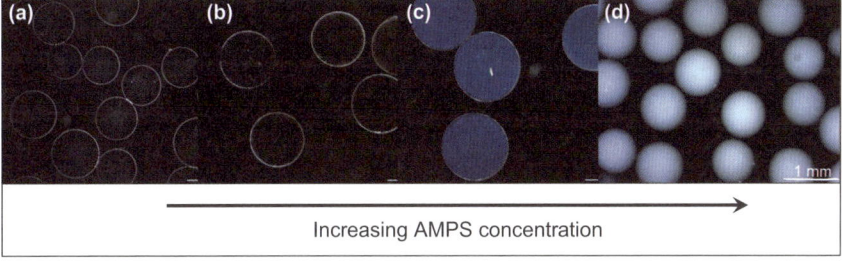

Figure 10.5   Phase contrast microscopy images of beads with increasing AMPS content.

associate, subsequently leading to phase separation between these components.[8] Incompatibility was observed by Adoor *et al.*[9] in systems of PVA and
poly(methyl methacrylate) blends. In their systems they observed incompatibility between phases when there was less than 60% PVA in the blend.

Demchenko *et al.*[10] have studied the compatibility of components within
PVA-PAA graft copolymers, where PAA is poly(acrylamide). Their work has
shown that the copolymer structure is stabilized by different types of hydrogen
bonds, typically between amide groups of grafted PAA chains and hydroxyl
groups of the main PVA chain. Increasing the number of grafts above a critical
number resulted in a decrease in hydrogen bonding between grafted chains and
the main chain. In addition, this resulted in an increase in hydrogen bonding
between grafted chains, which formed PAA domains. An increase in graft chain
length was also shown to alter the hydrogen bonding within the systems and a
reduction in hydrogen bonding between the graft chains and the main chain
resulted in significant swelling in water.[11] These findings can be applied to the
bead formulations prepared in this study, as the phase separation between PVA
and AMPS may be driven by a similar change in hydrogen bonding. In
microspheres with low AMPS content, hydrogen bonding may exist between
amide groups of the acrylamides, AMPS and NAAADA. There may also be
interaction between the amide groups and the hydroxyl groups of the PVA
backbone. As the AMPS content in the formulation increases, long chains of
AMPS will form. There may then be an increase in hydrogen bonding between
amide groups amongst the chains, *i.e.* self-association, creating AMPS-rich
domains.

For the lower AMPS formulations there is no phase separation and the high
water content of the hydrogel means the beads have a similar refractive index to
water and renders them virtually invisible when suspended in aqueous media.
Whilst this is a prerequisite if used in contact lens fabrication, it is a significant
disadvantage for embolization beads, as the IR will need to be able to visualize
the product during administration to ensure an even suspension for a smooth
and consistent delivery. These PVA hydrogel beads can therefore be optionally
tinted by use of a reactive dye such as Reactive Blue 4 (typically used to impart
tint to contact lenses to allow visualization whilst handling). This dye possesses
a chlorotriazine functionality which can covalently react with the hydroxyl
groups on the PVA backbone for permanent colouration. Both Bead Block and
DC Bead are tinted using this approach to provide a blue handling tint for
improved visualization. The beads are sieved into defined size fractions by use
of a vibrating sieve stack, measured into either vials or syringes and finally
sterilized using steam (Figure 10.6).

10.2.3.2   *Effects of AMPS Content on the Physicomechanical Properties of Hydrogel Bead Formulations*

Increasing the content of the hydrophilic AMPS monomer in the copolymer
formulation leads to higher equilibrium water content in the resultant hydrogel

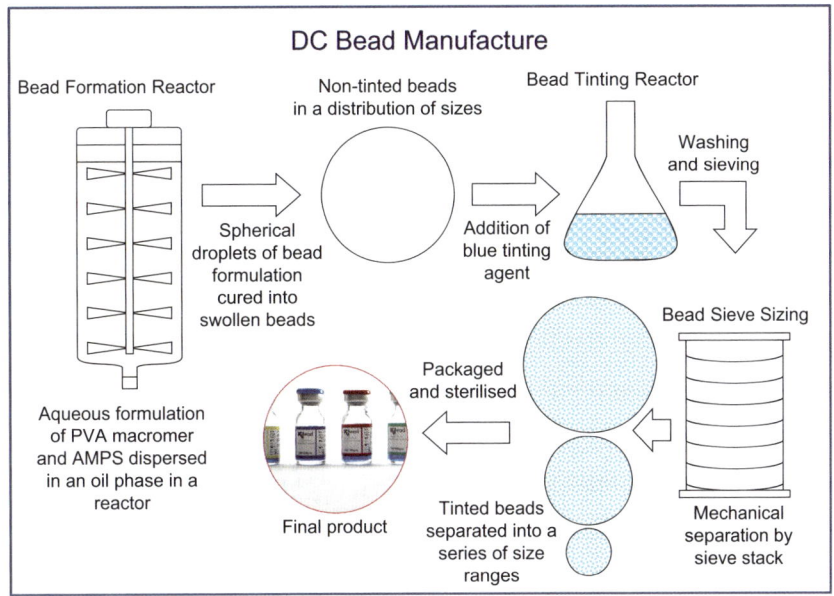

Figure 10.6   Outline of the manufacturing process for DC Bead.

beads. Interestingly, there is also a corresponding increase in the resistance of the beads to compressive deformation. A swelling pressure develops within the microspheres as they absorb water, which is a balancing force between the tendency to expand and the rigidity of the elastic matrix.[12,13] The increase in resistance to compression may be correlated with the increase in the AMPS phase within the polymer fraction, as the monomer acts as a spacer, essentially increasing the length between graft points. This contributes to the expansion and when swollen creates a rigid structure that is more resistant to compression. A similar increase in swelling ratio has been described with the use of PEG spacers in cellulose-based hydrogels.[14]

The mechanical properties of the beads are of fundamental importance in determining their usefulness and performance as embolization devices. Beads that are easily compressed can deform on passage through the tiny lumens of the microcatheters used for embolotherapy and allow for effortless, controlled delivery during procedures without blockage of the catheter. Bead Block has been developed for the embolization of arteriovenous malformations and hypervascular tumours and is often used in the treatment of uterine fibroids. The uterine arteries that require occlusion during this procedure are relatively large and typically beads in the size ranges 700–900 μm and 900–1200 μm would be selected. These are large in comparison with the inner lumen diameters of the typical microcatheters used in the procedure, and hence Bead Block is synthesized from a formulation containing a low AMPS component in order to increase the compressibility/deformability of the beads and aid in their catheter deliverability. It is important to note that these properties will also have a major impact on the *in vivo* performance of the product, as the more deformable the beads, the more they will be able to conform and squeeze down a vessel and hence the greater the distal penetration of the device. As important as the requirement for larger beads to deform during microcatheter administration is their ability to elastically recover their size and shape post-delivery. A recent study comparing elasticity and viscoelasticity of different commercially available embolization microspheres concluded that whilst Bead Block had relatively low rigidity under compression, it was more elastic than the other products with a significantly higher relaxation half-time.[7] The Contour SE spherical PVA sponge-like embolization product mentioned in Section 10.1 failed in the clinic, largely due to an inability to recover elastically. It is therefore crucial for the IRs to understand the behaviour of particulate embolization devices *in vivo* in order for them to predict and control the level of desired occlusion.

In contrast to Bead Block, DC Bead has a higher AMPS component, leading to a product with higher compressive resistance.[15] DC Bead has been developed primarily for the treatment of hypervascular malignancies of the liver and clinical data support the use of smaller bead size ranges for such procedures. The maximum recommended size would therefore be 500–700 µm (although the majority of cases now use 100–300 µm or 300–500 µm), which is relatively easily delivered down a microcatheter. As we will see in the following section, the higher level of AMPS is desired, more for the ion-exchange capacity it brings to the beads, allowing for the loading of cationically charged drugs into the bead structure. Drug loading results in a further reduction of the average bead diameter and hence the viscoelastic properties of DC Bead are not as crucial to product performance as they are for Bead Block.

10.2.3.3   *Permeability of PVA-g-AMPS Hydrogel Beads*

Given the very high water content of the PVA-g-AMPS hydrogel beads, there was an expectation that the water-swollen network structure may be amenable as a depot and for the transport of various species, a property particularly attractive from a drug delivery perspective. The ability of the beads to deliver small molecule species is discussed in detail in Section 10.3; the initial question to be answered was: what size restriction was placed on a particular species of interest? The diffusion of fluorescein isothiocyanate dextrans (FITC-Ds) of various molecular weights (4–250 kDa) into beads of various AMPS content has been studied to determine their molecular weight cut-off (MWCO) point, above which diffusion of molecules into the beads cannot occur. Confocal laser scanning microscopy (CLSM) has been used in this analysis as a non-destructive method of optical sectioning of samples in their hydrated state, as described by Vandenbossche *et al.*,[16] to determine the location of these model hydrophilic macromolecules within each microsphere type. Applied to these relatively thick samples, thin optical sections were produced by removing noise that often obscures images in conventional microscopy. In Figure 10.7 the confocal regions of interest of centralized sections of beads with increasing

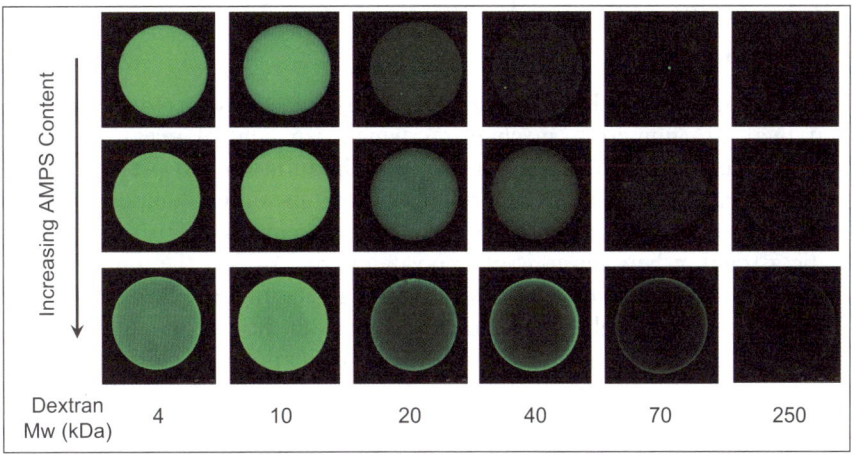

Figure 10.7   Internal CLSM sections showing the effect of AMPS content on FITC-D diffusion.

AMPS content are depicted. From these images it is possible to see that diffusion of the FITC-D with a molecular weight of 20 kDa into the centre of all bead formulations is possible. Higher than the MWCO range, no FITC-D could be observed within the optical section of the microsphere, although fluorescence was detected in the surrounding solution. Macromolecules with molecular weights above this cut-off point should not be able to pass into the microspheres by diffusion alone. Moreover, the MWCO point increases with progressively higher AMPS content.

FITC-Ds are readily available water-soluble macromolecules and are used as models for drugs, such as immunostimulants, and enzymes.[17–19] Vandenbossche *et al.*[16] have described FITC-Ds as "rod-like" structures and with increasing molecular weight they expand into coils. Therapeutic compounds are available in a variety of different shapes and sizes and although all beads were shown to be permeable to FITC-Ds of at least 20 kDa, the applicability to more globular structures was unclear. Therefore, to provide an indication to the permeability of other shapes of macromolecules, lysozyme was employed. The results of a fluorescence spectrophotometry study demonstrated that the beads were capable of actively loading lysozyme at pH 7. Lysozyme has a molecular weight of 14.7 kDa and is a globular protein. The beads were shown to load at least 25 mg of lysozyme, with greater than 99% efficiency, from an initial 5 mg mL^{-1} solution. High loading efficiency is characteristic of an ion-exchange system and it is likely that lysozyme, which is a cationic protein below pH 11.2, is able to interact with the negatively charged sulfonate groups of the copolymer.[20] The result reveals that the beads are capable of actively loading large therapeutic compounds and that the hydrogel matrix is accessible to globular structures. DC Bead exploits these efficient ion-exchange properties in its application as a drug delivery embolization system, which will now be discussed in further detail in the following section.

10.3 Drug Delivery Using PVA-*g*-AMPS Hydrogel Beads

Polymer microspheres have been commonly used as platforms for drug delivery and there are numerous mechanisms which can control release from such systems, including response to external stimuli, diffusion and swelling.[21,22] The rate of release can be altered by many factors, including composition of the polymer or any interaction of the drug. Ion-exchange systems utilize electrostatic attraction between ions of opposite charge to bind the ionic drug to a solid. It is an equilibrium process that allows the gradual release of the drug from its carrier through the exchange of ions from the surrounding solution.[23] This method of immobilization of the drug to the polymer platform does not generally involve permanent change to the drug's structure and allows it to be released in its initial form.[24] Loading with high efficiency can be achieved easily by immersion of the pre-formed ion-exchange microspheres in the drug solution. In simple ion-exchange systems the amount of drug loaded can be controlled by altering the concentration of the loading solutions. In situations where maximum loading is desired, excess drug may be used to saturate all of the loading sites. In all drug delivery systems, factors such as drug solubility, molecular weight or drug–polymer interaction can also influence the total amount of drug loaded and contribute to changing a drug's release profile.[25] The impact of osmolarity, pH, volume or ionic strength of the media should be considered during release, especially when creating an *in vitro* model intended to mimic an *in vivo* situation.

10.3.1 Interaction of PVA-*g*-AMPS Hydrogel Beads with Doxorubicin Hydrochloride

Doxorubicin hydrochloride (Dox) is an anthracycline chemotherapeutic agent used to treat a variety of solid tumours, including primary liver cancer. It exists as a hydrochloride salt by virtue of protonation of the amine group on the glucosamine ring, which increases the water solubility but also renders it cationically charged at neutral pH. This enables the drug to be loaded into the PVA-*g*-AMPS hydrogel beads by an active ion-exchange process, the rate of which can be controlled by the concentration of the drug solution. After placing the beads into an aqueous Dox solution, the active loading process causes the beads to turn an intense red colour due to the presence of the highly coloured drug. This is accompanied by a corresponding depletion in colour of the Dox loading solution, as first reported by Lewis *et al.*[26] Increasing the ionic strength of the loading solution is commonly reported to reduce the amount of loading in ion-exchange systems. Loading should be performed using deionized solutions where possible, as competitive binding reduces the loading ability of the systems.[27] Loading studies have shown that where there is no AMPS content in the hydrogel beads, the drug passes by diffusion into the bead structure until equilibrated in concentration with the loading solution and then

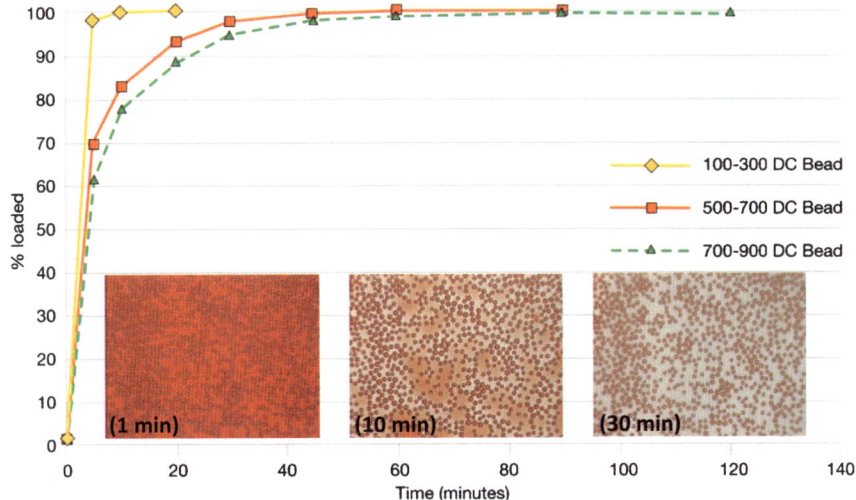

Figure 10.8   Loading profile of DC Bead with Dox over time (*inset*: optical micrographs of beads immersed in Dox solution).

subsequently releases the drug rapidly as there is no interaction between drug and polymer. In contrast, AMPS-containing beads were shown to actively uptake drug from solution, with a gradual depletion in colour of the loading liquid as the positively charged drug interacts with the negatively charged sulfonate groups from the AMPS (Figure 10.8). The rate of Dox uptake increases with decreasing size of the beads, as there is a greater surface area per unit volume for the diffusion and ion exchange to occur.

10.3.2   Drug Distribution within PVA-*g*-AMPS Hydrogel Beads

The inherent fluorescent properties of Dox enables the use of confocal scanning laser microscopy (CSLM) to optically section the bead internal structure in order to visualize the distribution of the drug. In all sections of all the bead formulations a ring can be visualized, indicating that the drug is more concentrated at the periphery, although present throughout the entire bead structure (Figure 10.9). This effect has been previously observed by Lewis *et al.*[26] using CLSM and Raman analysis of Dox-loaded DC Beads. These authors have suggested that the existence of a more concentrated outer ring of drug is related to a higher density of AMPS moieties within the polymer structure which binds the drug. During polymerization the two components of redox initiation, the persulfate and TMEDA, are initially separated in the two phases of the system. They meet at the interface of the droplets and polymerization starts here and progresses inwards. A polymer gradient is then observed, with a higher concentration of AMPS in the outer regions.

FTIR imaging analysis was also performed on 10 µm sections of beads with and without AMPS (Figure 10.9). The spectra of both bead types were

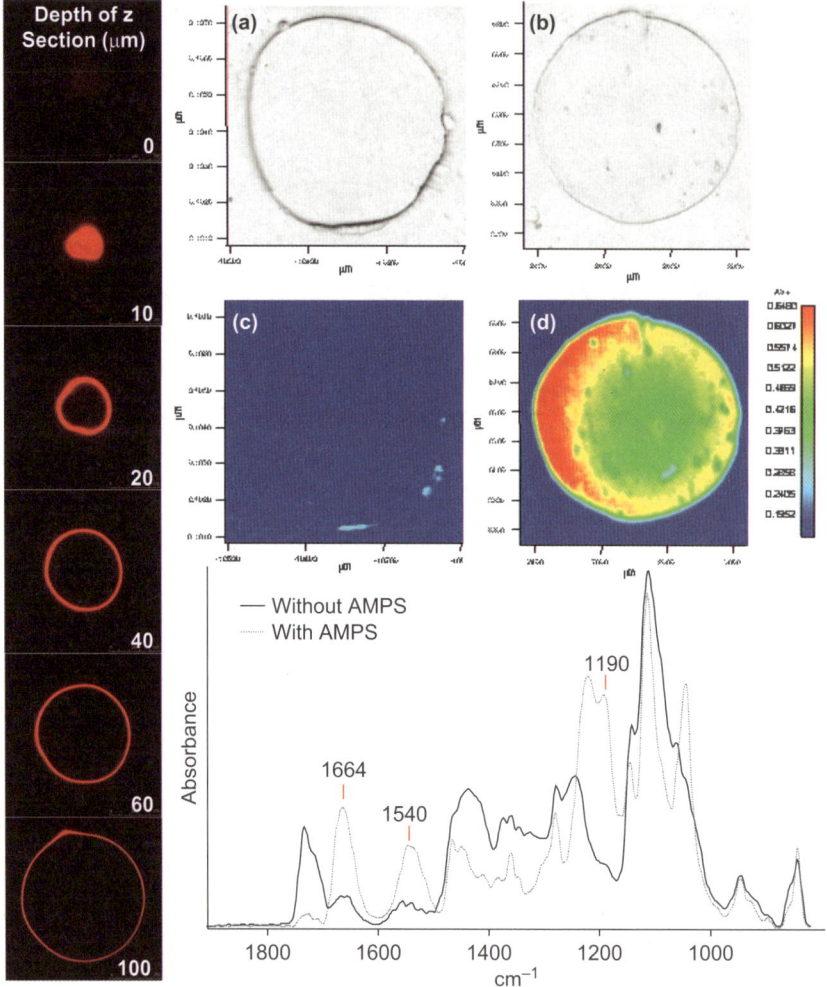

Figure 10.9   *Left*: confocal sections of a Dox-loaded bead. *Right*: optical (a, b), FTIR image (c, d) and FTIR spectrum of a bead without AMPS and an AMPS-containing bead.

compared to identify the wavenumber of characteristic peaks. Key peaks present due to AMPS were identified at 1664, 1540 and 1190 cm^{-1}, characteristic of amide I stretching, amide II stretching and symmetrical stretching of S=O, respectively.[28,29] The characteristic amide I and S=O peaks were mapped across the bead sections. The FTIR imaging in Figure 10.9 shows an example map of the 1190 cm^{-1} peak (S=O) for beads without AMPS. There is no absorbance represented in these images for beads without AMPS. The sulfonate peak mapping for AMPS-containing beads, however, demonstrates a higher absorbance at the periphery of the section and confirms that AMPS is indeed more concentrated in the outer areas of the bead.

10.3.3 Effect of Drug Loading on PVA-*g*-AMPS Hydrogel Bead Mechanical Properties

Lewis *et al.*[26] have previously described a reduction in microsphere diameter when loading Dox into DC Bead. This effect is as a result of a reduction in EWC, as when Dox molecules are bound to the sulfonate groups of the copolymer there is displacement of water molecules. Dox is a planar, aromatic molecule that is well documented to self-associate in solution to form aggregates.[30–32] In different solutions, Dox association has been shown to be dependent on concentration, ionic strength, temperature and pH.[31–35] Gonzalez Fajardo[36] has shown that when Dox is bound by the sulfonate moiety of AMPS, neutralization of the positive charge results in hydrophobic interactions between the aromatic rings of the Dox within the beads. This interaction has also been described in sulfopropyl dextran microspheres and is reportedly stronger than the ionic interaction between the polymer and drug.[37] The stacking of Dox molecules alters the internal structure of the beads by acting to "crosslink" the polymer physically, dispelling water and making the structure more fixed and rigid and reducing the overall diameter.

10.3.4 Doxorubicin Elution from PVA-*g*-AMPS Hydrogel Beads

The loading and release of Dox from the anionic microspheres studied in this chapter can be described using eqn (10.1). Ion exchange is a reversible process and during elution the bound drug is released when a suitable counter ion is present in the elution medium, *e.g.* Na$^+$. After exchange the drug molecule is able to diffuse through the microsphere to the surrounding solution:

$$\text{Dox}^+\text{SO}_3^- + \text{Na}^+ \rightleftharpoons \text{Dox}^+ + \text{Na}^+\text{SO}_3^- \qquad (10.1)$$

Bead size has an impact on the release rate of the drug, as the higher surface area per unit volume offered by smaller bead sizes leads to more rapid elution kinetics (Figure 10.10). Drug release *in vitro* is also dependent upon factors such as ionic strength and type of elution medium, temperature of the medium and the type of elution model employed.

It is recognized that in a closed release system the rate of elution is altered by the volume and ionic composition of the elution medium. It is not necessarily a suitable model to evaluate the release kinetics or predict *in vivo* behaviour of the beads, but is a useful and simple method for formulation comparison. Many elution models use large volumes of highly ionic solutions and vigorous stirring conditions to minimize the effect of ion diffusion through the film adherent to the particles. In these models the diffusion of the counter ions within the particles instead becomes the limiting factor of release.[38] It is difficult to accurately model the release of a drug from microspheres administered subcutaneously, intramuscularly or during embolization as they are often closely packed and

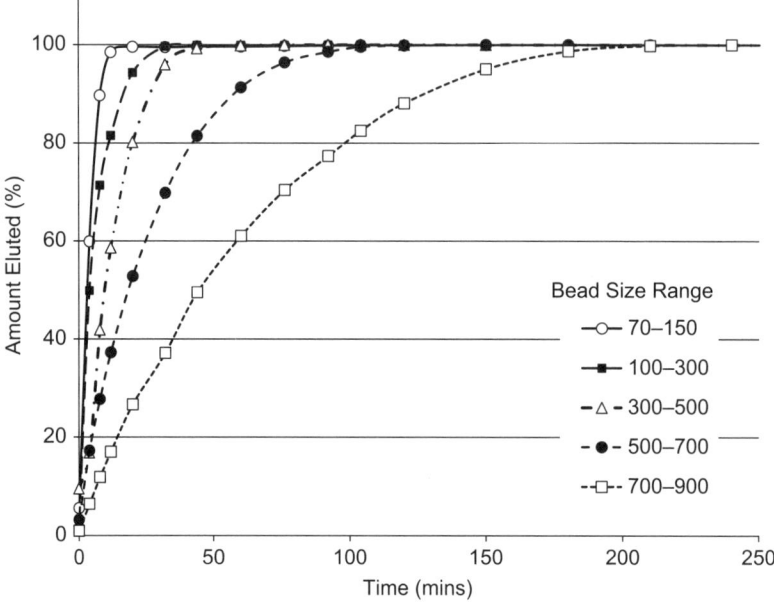

Figure 10.10   Effect of bead size on *in vitro* Dox release.

not mixed in solution. Accumulation of the drug and interaction with surrounding microspheres may reduce the release rate.[39] Amyot and colleagues
have previously described the use of a T-shaped apparatus that was designed to
mimic the *in vivo* conditions of embolization.[40] The apparatus has a small well
at the bottom where the beads are placed and the drug is able to diffuse into the
surrounding static environment. Above the cavity an area of convection is
present before the main flow junction. Gonzalez *et al.*[41] have recently used a
T-apparatus to study the release of Dox from DC Bead. The release data were
best represented using a slow release model and it was demonstrated that
the half-life of elution was over 1500 hours. A linear correlation between the
in vitro release data and the *in vivo* pharmacokinetic data from 15 patients
treated with the loaded microspheres was observed.

10.4   Application of PVA-*g*-AMPS Hydrogel Beads for Embolotherapy and Drug Delivery

10.4.1   Bead Block

We have seen from Section 10.2.3.2 that it is possible to alter the hydrogel bead
formulation in order to arrive at a desired set of performance characteristics
commensurate with the intended use of the product. Bead Block is provided in
colour-coded syringes for easy immediate selection of the desired size range

(Figure 10.11A). The chemical stability of the beads are such that the product can be sterilized directly in its hydrated form by use of steam, with no detectable signs of hydrolytic degradation. The beads are supplied in 5 mL of phosphate buffered saline and can be simply mixed with a similar volume of a suitable contrast agent to yield a diluted isobouyant suspension. Bead Block is coloured blue so that the user can see that an even suspension of beads has been attained prior to delivery through a microcatheter (Figure 10.11B).

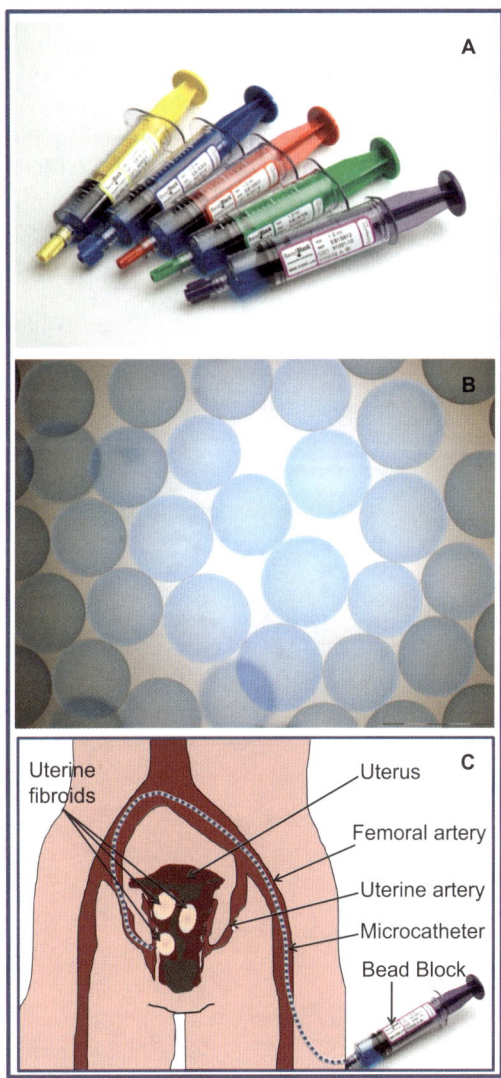

Figure 10.11 (A) The Bead Block range of products is supplied in color-coded syringes according to the size of the product; (B) Photomicrograph of Bead Block PVA-based microspheres; (C) Representation of a uterine artery embolization.

This product is formed from low AMPS formulations, which we have shown to be softer and able to deform more easily, a feature useful for the delivery of larger sized beads through narrow lumen microcatheters such as required for uterine artery embolization (Figure 10.11C). Bead Block has therefore been developed especially with the proposed application in mind and provides the IR with a useful balance of properties that translates into clinically successful outcomes.[42,43]

10.4.2 DC Bead

Section 10.3 has revealed a key property of higher AMPS-containing formulations: the ability to actively and rapidly sequester, and subsequently release, cationically charged species from aqueous solution by an ion-exchange mechanism. DC Bead (known as LC Bead in the USA) has been developed to exploit the properties of the higher AMPS formulation and furnish the IR with an ideal product for a one-step transarterial chemoembolization (TACE). The product is provided in a vial as 2 mL of hydrated beads in 5 mL total phosphate buffered saline volume; they are tinted blue as for Bead Block for ease of visualization (Figure 10.12A). The upper size range available is 500–700 μm,

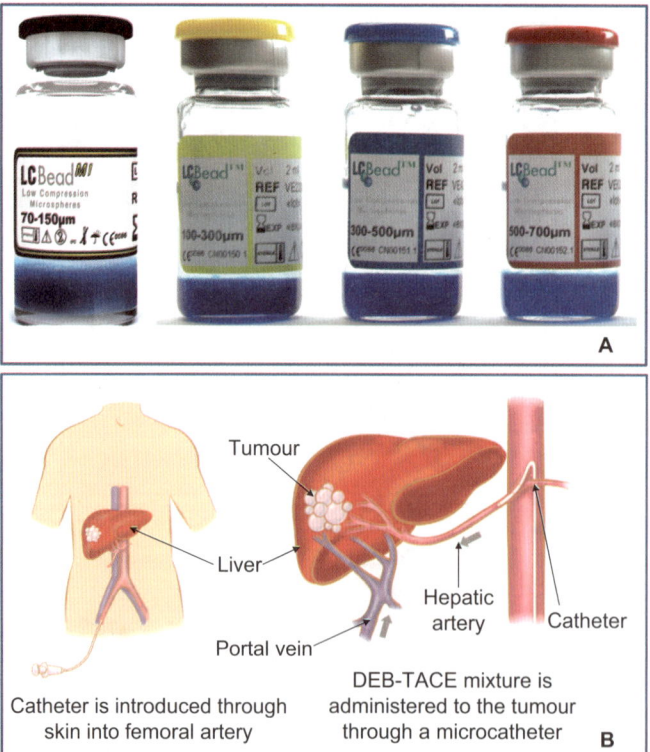

Figure 10.12   (A) The LC Bead range of products is supplied in vials with color-coded caps and labelling; (B) Representation of a DEB-TACE procedure.

with a new offering in the lower size range of 70–150 μm, known as DC Bead[M1]. TACE has been practised for many years but generally with an enormous variability in choice of embolic materials, drugs and procedural techniques. This has made comparison between clinical studies and identification of the most efficacious treatment regimens practically impossible. The procedure usually involves intra-arterial delivery of an unstable emulsion of a chemotherapeutic agent and oily contrast agent into the feeding arteries of a tumour, followed by a second step to deliver an embolization agent into the same vessels to occlude the arteries, blocking flow of nutrients and oxygen to the tumour whilst minimizing drug wash-out.[44,45] The use of DC Bead allows for the loading of the drug directly into the embolization agent itself, which can then be targeted into the tumour vessels where the beads will lodge and elute the drug over a period of many weeks to months (often referred to as drug-eluting bead TACE or DEB-TACE; Figure 10.12B).[46–49] Numerous pre-clinical and clinical studies have shown this one-step procedure delivers a more consistent dose, reduces systemic exposure to the toxic drugs, increases local drug concentrations in the tumour tissue and leads to improved tumour response.

The use of DC Bead with Dox (sometimes referred to as DEBDOX) has been predominantly in the treatment of primary liver cancer, or hepatocellular carcinoma.[50–57] Success in this indication has seen its use extend to some metastatic cancers of the liver such as neuroendocrine,[58] uveal[59] and breast metastases. Colorectal metastases to the liver is the most prevalent of these metastatic tumours and for the treatment of this condition the use of irinotecan hydrochloride and DC Bead (often called DEBIRI) is preferred over doxorubicin, as the drug has greater activity in this cancer type.[60–65] This product is also being evaluated in the clinic outside the area of embolotherapy, but rather as a drug delivery device for direct injection into the resection margin after surgical removal of glioblastoma multiforme (primary brain tumour).[66–68]

10.5   Conclusions

This chapter has described the effect of PVA-*g*-AMPS composition on hydrogel bead properties, utilizing novel chemistry adapted from contact lens technology. By varying the balance between the components in the composition, hydrogel beads with a range of different physicomechanical properties can be attained. This has allowed for the development of a family of medical devices with product performance characteristics specifically tailored to their intended application. By appropriate selection of co-monomer, ion-exchange capabilities can be introduced which render the resulting beads particularly useful for drug delivery purposes. These novel products have required the parallel development of new techniques and models in order to properly investigate their characteristics, which is particularly challenging given their small size. This has included a range of different microscopy and imaging methods, tests for determining physical parameters of the hydrogels, methods for measuring drug uptake and elution and animal models for safety and efficacy assessments.

Figure 10.13 The Prince Philip Award for Plastics in the Service of Mankind endorsed by the Institute of Materials Minerals and Mining.

At the time of writing, these products are used in over 50 different countries worldwide, have been used to treat over 100 000 patients and DC Bead has recently received the Prince Philip Award for Plastics in the Service of Mankind (Figure 10.13).

Acknowledgements

Some of the studies presented in the chapter were performed under a SEEDA-supported EPSRC CASE award (C.H.) with the University of Brighton. The authors would therefore like to thank Professor Andrew Lloyd and Dr Gary Phillips for their valuable input into this work.

References

1. *Hydrogels in Medicine and Pharmacy*, ed. N. A. Peppas, CRC Press, Boca Raton, FL, 1987, vol. 2.
2. J. H. Grindlay and O. T. Clagett, *Mayo Clin. Proc.*, 1949, **24**, 538.
3. P. Landwehr, *et al.*, *Radiologe*, 2008, **48**, 73–97.
4. S. M. Tadavarthy, J. H. Moller and K. Amplatz, *Am. J. Roentgenol. Radium Ther. Nucl. Med.*, 1975, **125**, 609–616.
5. A. Laurent, *et al.*, *Cardiovasc. Intervent. Radiol.*, 2010, **33**, 995–1000.

6. A. Laurent, *et al.*, *Invest. Radiol.*, 2006, **41**, 8–14.
7. K. Hidaka, *et al.*, *J. Mech. Behav. Biomed. Mater.*, 2011, **4**, 2161–2167.
8. A. Arun and B. S. R. Reddy, *Biomaterials*, 2005, **26**, 1185–1193.
9. S. G. Adoor, *et al.*, *J. Appl. Polym. Sci.*, 2006, **100**, 2415–2421.
10. O. G. Demchenko, *et al.*, *Macromol. Symp.*, 2001, **166**, 117–122.
11. T. Zheltonozhskaya, *et al.*, *Macromol. Symp.*, 2003, **203**, 173–182.
12. F. Helfferich, *Ion Exchange*, Dover, New York, 1995.
13. R. Paterson, *An Introduction to Ion Exchange*, Heyden, London, 1970.
14. A. Sannino and L. Nicolais, *J. Appl. Polym. Sci.*, 2003, **90**, 168–174.
15. A. L. Lewis, *et al.*, *J. Mater. Sci.: Mater. Med.*, 2007, **18**, 1691–1699.
16. G. M. R. Vandenbossche, P. Van Oostveldt and J. P. Remon, *J. Pharm. Pharmacol.*, 1991, **43**, 275–277.
17. G. Xu and M. J. Groves, *J. Pharm. Pharmacol.*, 2001, **53**, 49–56.
18. P. D. Thornton, G. McConnell and R. V. Ulijn, *Chem. Commun.*, 2005, 5913–5915.
19. J. Kress, *et al.*, *Chem.–Eur. J.*, 2002, **8**, 3769–3772.
20. M. Y. Arica and G. Bayramoglu, *Process Biochem.*, 2005, **40**, 1433.
21. R. S. Langer and N. A. Peppas, *Biomaterials*, 1981, **2**, 201–214.
22. J. Heller and A. S. Hoffman, in *Biomaterials Science: An Introduction to Materials in Medicine*, ed. B. D. Ratner, *et al.*, Elsevier Academic Press, London, 2004.
23. S. Borodkin, in *Polymers for Controlled Drug Delivery*, ed. P. J. Tarcha, CRC Press, Boca Raton, FL, 1991, p. 215.
24. V. Anand, R. Kandarapu and S. Garg, *Drug Discovery Today*, 2001, **6**, 905–914.
25. S. Freiberg and X. X. Zhu, *Int. J. Pharm.*, 2004, **282**, 1–18.
26. A. L. Lewis, *et al.*, *J. Mater. Sci.: Mater. Med.*, 2007, **18**, 1691–1699.
27. T. Tarvainen, *et al.*, *Biomaterials*, 1999, **20**, 2177–2183.
28. W. Kemp, *Organic Spectroscopy*, Macmillan, London, 1984.
29. D. A. Skoog, *Principles of Instrumental Analysis*, Saunders, New York, 1985.
30. M. Dalmark and H. H. Storm, *J. Gen. Physiol.*, 1981, **78**, 349–364.
31. M. Menozzi, *et al.*, *J. Pharm. Sci.*, 1984, **73**, 766–770.
32. S. Eksborg, *J. Pharm. Sci.*, 1978, **67**, 782–785.
33. M. Dalmark and P. Johansen, *Mol. Pharmacol.*, 1982, **22**, 158–165.
34. E. Hayakawa, *et al.*, *Chem. Pharm. Bull.*, 1991, **39**, 1009–1012.
35. E. Hayakawa, *et al.*, *Chem. Pharm. Bull.*, 1991, **39**, 1282–1286.
36. M. V. Gonzalez Fajardo, PhD thesis, University of Brighton, 2006.
37. Z. Liu, *et al.*, *J. Controlled Release*, 2001, **77**, 213–224.
38. A. Sawaya, *et al.*, *J. Microencapsulation*, 1988, **5**, 255–267.
39. X. Y. Wu, G. Eshun and Y. Zhou, *J. Pharm. Sci.*, 1998, **87**, 586–593.
40. F. Amyot, *et al.*, *ITBM-RBM*, 2002, **23**, 285–289.
41. M. V. Gonzalez, *et al.*, *J. Mater. Sci.: Mater. Med.*, 2008, **19**, 767–775.
42. T. J. Kroencke, *et al.*, *J. Vasc. Intervent. Radiol.*, 2008, **19**, 47–57.
43. R. L. Worthington-Kirsch, *et al.*, *Cardiovasc. Intervent. Radiol.*, 2011, **34**, 493–501.

44. J. M. Llovet, *et al.*, *Lancet*, 2002, **359**, 1734–1739.
45. C. M. Lo, *et al.*, *Hepatology*, 2002, **35**, 1164–1171.
46. J. Kettenbach, *et al.*, *Cardiovasc. Intervent. Radiol.*, 2008, **31**, 468–476.
47. A. L. Lewis and M. R. Dreher, *J. Controlled Release*, 2012, **161**, 338–350.
48. E. Liapi and J. F. Geschwind, *Cardiovasc. Intervent. Radiol.*, 2011, **34**, 37–49.
49. K. Malagari, *Expert Rev. Anticancer Ther.*, 2008, **8**, 1643–1650.
50. J. Lammer, *et al.*, *Cardiovasc. Intervent. Radiol.*, 2010, **33**, 41–52.
51. K. Malagari, *et al.*, *Abdom. Imaging*, 2008, **33**, 512–519.
52. K. Malagari, *et al.*, *Cardiovasc. Intervent. Radiol.*, 2010, **33**, 541–551.
53. R. C. Martin, *et al.*, *Korean J. Hepatol.*, 2011, **17**, 51–60.
54. R. C. Martin, *et al.*, *J. Am. Coll. Surg.*, 2011, **213**, 493–500.
55. R. T. P. Poon, *et al.*, *Clin. Gastroenterol. Hepatol.*, 2007, **5**, 1100–1108.
56. D. K. Reyes, *et al.*, *Cancer J.*, 2009, **15**, 526–532.
57. M. Varela, *et al.*, *J. Hepatol.*, 2007, **46**, 474–481.
58. T. de Baere, *et al.*, *J. Vasc. Intervent. Radiol.*, 2008, **19**, 855–861.
59. G. Fiorentini, *et al.*, *In Vivo*, 2009, **23**, 131–137.
60. C. Aliberti, *et al.*, *Anticancer Res.*, 2006, **26**, 3793–3795.
61. G. Fiorentini, *et al.*, presented at ASCO 2009 Gastrointestinal Cancers Symposium, San Francisco, January 2009.
62. G. Fiorentini, *et al.*, *In Vivo*, 2007, **21**, 1085–1092.
63. R. C. Martin, *et al.*, *Ann. Surg. Oncol.*, 2011, **18**, 192–198.
64. R. C. Martin, *et al.*, *J. Oncol.*, 2009, 539–795.
65. R. C. Martin, *et al.*, *World J. Surg. Oncol.*, 2009, **7**, 80.
66. S. Baltes, *et al.*, *J. Mater. Sci.: Mater. Med.*, 2010, **21**, 1393–1402.
67. S. Vinchon-Petit, *et al.*, *Int. J. Pharm.*, 2010, **402**, 184–189.
68. S. Glage, *et al.*, *Clin. Transl. Oncol.*, 2012, **14**, 50–59.

CHAPTER 11
Outlook

XIAN JUN LOH* AND OREN A. SCHERMAN

Melville Laboratory for Polymer Synthesis, Department of Chemistry,
University of Cambridge, Lensfield Road, Cambridge CB2 1EW, UK
*Email: xianjun_loh@scholars.a-star.edu.sg

> *Day after day, day after day,*
> *We stuck, nor breath nor motion;*
> *As idle as a painted ship*
> *Upon a painted ocean.*
>
> *Water, water, every where,*
> *And all the boards did shrink;*
> *Water, water, every where,*
> *Nor any drop to drink.*

When Samuel T. Coleridge penned the *Rime of the Ancient Mariner* in 1798, little did he realise that the woes and troubles that befell the Ancient Mariner at sea would so poetically reflect the trials and tribulations that hydrogel research scientists face today in their quest for perfection. In spite of their huge promise, hydrogels have not yet come through the storm as, most publications on hydrogels to date, focus only on the potential biological applications in surgery and drug delivery. They have rarely been considered alongside polymers and composites as performance materials and substantial challenges remain in this field to try to address this issue.

Hydrogels also have several limitations. The low tensile strength of many hydrogels limits their use in load-bearing applications and the stability of the

Monographs in Supramolecular Chemistry No. 11
Polymeric and Self Assembled Hydrogels: From Fundamental Understanding to Applications
Edited by Xian Jun Loh and Oren A. Scherman
© The Royal Society of Chemistry 2013
Published by the Royal Society of Chemistry, www.rsc.org

hydrogel remains an issue in many applications. In order to circumvent this, doubly networked hydrogels have been synthesized and characterized. Doubly networked hydrogels have remarkable strength and have been highlighted as a possible solution to solve this problem.[1] Double network hydrogels are able to withstand stresses of up to 10–20 MPa. Some of these gels are also resistant to being sliced by a sharp cutter.

"Smart" hydrogels have been gaining increasing attention due to their potential to be controlled by the application of various stimuli. However, the use of the word "smart" appears to be an overstatement to describe a material that can only swell and shrink in response to a single or sometimes dual stimuli. Clearly having a hydrogel that can swell and shrink with a change in temperature is at best as intelligent as a pump. Furthermore, the response times of these hydrogels are usually slow and the hysteresis associated with the "on" and "off" states is a further drawback of these systems. Although the response time can be improved by having thinner and smaller hydrogels, it places a size limitation on the potential applications of these hydrogels.

One of the major challenges relates to the improvement of the safety profile of hydrogels for clinical usage. In order to form a biocompatible hydrogel, the components that are required for their assembly must also be non-toxic. In spite of their wide usage, the toxicity of the cross-linking agents such as formaldehyde and glutaraldehyde remain a concern.[2–4] An epoxy-based ethylene glycol diglycidyl ether was also found to be unsuitable for cell proliferation.[4] Triethylene glycol dimethacrylate (TEGDMA), a common cross-linking agent, was identified as a mutagenic compound in mammalian cells.[5,6] Gene mutations were induced by this compound, possibly because of the covalent binding to DNA *via* Michael addition. Monomers such as methyl methacrylate (MMA) and glycidyl methacrylate (GMA) were found to be cytotoxic and genotoxic.[7,8] In particular, Yoshii has published the toxicity of 39 different acrylates and methacrylates.[8] The results do indicate that serious attention has to be paid to the post-fabrication process of the hydrogels to ensure that no residual monomers remain. Not only are the monomers and the cross-linkers toxic, but the photoinitiators can also be cytotoxic. It has been recently shown that Irgacure 651 or 2,2-dimethoxy-2-phenylacetophenone (DMAP) at concentrations of 0.01% (w/w) are very cytotoxic to chondrocytes and 3T3 fibroblast cultures.[9] However, Darocur 2959 [2-hydroxy-4′-(2-hydroxyethoxy)-2-methylpropiophenone] has been suggested as a cyto-compatible photoinitiator in the same study. Nevertheless, it highlights the fact that it is important to reduce the residual levels of photoinitiator in the gels to as low a level as possible. However, 1-ethyl-3-(3-dimethylaminopropyl)carbodiimide hydrochloride and genipin appear not to be as toxic to the cells and have been suggested as alternative cross-linking agents.[10–12] Enzymatic cross-linking reactions have also recently gained increasing interest as bio-friendly gelation agents.[13] It would be ideal to make hydrogels biocompatible such that a hydrogel implantation in the body does not set off immunogenic reactions in the body. For this reason, creating hydrogels from the extracellular matrix (ECM) or ECM-mimetic materials could allow the hydrogels to evade attack from the immune system and integrate seamlessly with the host tissue.

In spite of all these challenges, the work and enthusiasm of researchers in this field remain undaunted. They, like the Ancient Mariner, who despite all the hardships faced, still persevered to return from a withering voyage fraught with disappointments, dangers and disasters. Part of their journey has been captured in this book which we, as the editors, hope could serve as a guide to other voyagers as it offers a tantalizing glimpse of exotic lands just beyond the horizon. It has charted the steps already taken so that others can build upon them to surmount current challenges and seek new and novel vistas in their search towards tomorrow. Above all, it seeks to inspire hope that break-throughs and exciting cutting edge science is well within grasp.

References

1. J. P. Gong, Y. Katsuyama, T. Kurokawa and Y. Osada, *Adv. Mater.*, 2003, **15**, 1155–1158.
2. P. Calero, E. Jorge-Herrero, J. Turnay, N. Olmo, I. L. de Silanes, M. A. Lizarbe, M. M. Maestro, B. Arenaz and J. L. Castillo-Olivares, *Biomaterials*, 2002, **23**, 3473–3478.
3. J. E. Gough, C. A. Scotchford and S. Downes, *J. Biomed. Mater. Res.*, 2002, **61**, 121–130.
4. H. W. Sung, D. M. Huang, W. H. Chang, R. N. Huang and J. C. Hsu, *J. Biomed. Mater. Res.*, 1999, **46**, 520–530.
5. H. Schweikl and G. Schmalz, *Mutat. Res., Genet. Toxicol. Environ. Mutagen.*, 1999, **438**, 71–78.
6. H. Schweikl, G. Schmalz and K. Rackebrandt, *Mutat. Res., Genet. Toxicol. Environ. Mutagen.*, 1998, **415**, 119–130.
7. H. W. Yang, L. S. S. Chou, M. Y. Chou and Y. C. Chang, *Biomaterials*, 2003, **24**, 2909–2914.
8. E. Yoshii, *J. Biomed. Mater. Res.*, 1997, **37**, 517–524.
9. S. J. Bryant, C. R. Nuttelman and K. S. Anseth, *J. Biomater. Sci., Polym. Ed.*, 2000, **11**, 439–457.
10. C. L. Tsai, S. H. Hsu and W. L. Cheng, *Artif. Organs*, 2002, **26**, 18–26.
11. L. L. H. Huang, H. W. Sung, C. C. Tsai and D. M. Huang, *J. Biomed. Mater. Res.*, 1998, **42**, 568–576.
12. F. L. Mi, Y. C. Tan, H. F. Liang and H. W. Sung, *Biomaterials*, 2002, **23**, 181–191.
13. L. S. M. Teixeira, J. Feijen, C. A. van Blitterswijk, P. J. Dijkstra and M. Karperien, *Biomaterials*, 2012, **33**, 1281–1290.

Subject Index

References to tables are given in **bold** type. References to figures are given in *italic* type.